Wolfgang Linert (ed.)

Highlights in Solute-Solvent Interactions

Springer-Verlag Wien GmbH

Prof. Dr. Wolfgang Linert
Institute of Inorganic Chemistry, Technical University of Vienna,
Vienna, Austria

Printed on acid-free and chlorine-free bleached paper

SPIN: 10850855

With 87 Figures

CIP data applied for

Special Edition of
Monatshefte für Chemie/Chemical Monthly Vol. 132, No. 11, 2001

ISBN 978-3-211-83731-3 ISBN 978-3-7091-6151-7 (eBook)
DOI 10.1007/978-3-7091-6151-7

Editorial

Some time ago, at an international conference on coordination chemistry, I was invited to select a winning poster in the field of solution chemistry. More than 200 posters were presented in the section "Solution Chemistry", but remarkably few were concerned with what is generally understood by this term. Yes, many workers described chemistry involving solutions in that they studied reactions or preparations carried out in liquid media, but we could hardly automatically describe work carried out in glass vessels as "Silica chemistry"!

Did this paucity of papers imply that interest in solution chemistry has abated? By no means! It reflects, rather, that the interest in solute-solvent interactions is no longer the province of chemistry alone. In biology, for instance, the role of the solvent, *i.e.* water, is of enormous importance, particularly with respect to the interplay between hydrophobic and hydrophilic phenomena. This has led to renewed interest in the structure of water itself and therefore provoked many theory-based studies, both thermodynamical and quantum mechanical. This aspect is clearly reflected in some of the work presented here. Furthermore, it means that much of the work in this field presently being carried out by chemists is now increasingly presented in journals and at conferences devoted to biology, physics, and medicine. Finally it has to be noted, with a little sadness, that solution chemistry research is increasingly being taken over by those working in the field of theoretical physics.

Modern technology often requires reactions to be carried out in mixed solvents, and so considerable interest has arisen in the way in which these interact with solutes. Thus, papers are included in this issue that deal with preferential solution in solvent mixtures and with thermodynamic aspects of such solvation phenomena.

Recent investigations of thermochromism and solvatochromism in solution have also been considered – a rapidly expanding field of study, both of intrinsic interest and suited for practical applications, *e.g.* for heat sensitive coatings. Furthermore, coming to the boundaries of solution chemistry, a paper in which even insoluble material such as powdered silica can exhibit solvent-like properties when suspended in organic solvents has been included.

Of course, this special issue covers but a small range of the vast variety of work in progress on the structure of liquids and their behaviour towards solutes. Nevertheless it is hoped that it will wet the appetite of the readers and encourage them to take interest in this fundamental aspect of modern physics and physical chemistry.

Wolfgang Linert

Preface

Most organic molecules retain their integrity when dissolved, and even though in such cases the effects exerted by solvents are, in the language of the coordination chemist, of the "outer sphere" kind, the choice of solvent can be critical to the successful outcome of an operation or preparation. Solubilities of reactants and products must be taken into account, and even if the organic principals in the reactions retain their integrity, many of the reagents are electrolytes, and their state of aggregation will affect their reactivity. In testifying to the importance of understanding solute-solvent interactions I draw attention to a large class of inorganic species for which the involvement in the chemical and physical properties by the solvent is even more deeply seated. It is comprised by the large body of metal atoms in low oxidation states for which solvent molecules intervene as reagents. At the same time, because the ions carry charges, the effects arising from outer sphere interactions are usually greater than they are for neutral molecules. To cite an example: when $FeCl_3(s)$ is dissolved in water to form a dilute – say $0.010\,M$ – solution there is a complete reorganization of the coordination sphere of the cation. Whereas in the solid each cation is surrounded by six chloride ions, in the solution the dominant form is $[Fe(H_2O)_6]^{3+}$ followed by $[Fe(H_2O)_5Cl]^{2+}$, $[Fe(H_2O)_4Cl_2]^{+}$, *etc.* in rapidly decreasing abundance. I infer from the titles of the contributions by *Ohtaki* and *Marcus* to the volume that some of the consequences of this kind of solvent intervention in the behaviour of the solutions are dealt with in their papers. Perhaps the considerations which follow complement what appears in these articles.

To understand the properties of such solutions, a description of the molecules present is essential. In addressing the systems by experiment, a critical item of information is just how rapidly the adjustments in the composition of the coordination sphere of the ions take place. By contrast to the behaviour of $FeCl_3(s)$, though, the driving force for the reorganization of $CrCl_3(s)$ to the hexaquo ion and $3Cl^-(aq)$ is somewhat greater than it is for $FeCl_3(s)$, when $CrCl_3(s)$ is introduced to water. Even wetting of the solid is a slow process, let alone its dissolution. This example illustrates the disparity between thermodynamic and kinetic stability which prevails in the chemistry of the metal species being considered. As exemplified by the exchange of water between the inner coordination sphere of the tripositive ions, the rates span a range of about 10^{19}. A change of the oxidation state by one unit can also affect the lability profoundly: for example, the $1e^-$ reduction of $Cr(H_2O)_6^{3+}$ increases the rate of exchange of the bound water by a factor of about 10^{14}. Whereas solutions of cations which undergo substitution slowly can be investigated using traditional experimental approaches, those which undergo substitution very rapidly require more sophisticated methods.

As defined in my ready-to-hand dictionary the facets which comprise the science of chemistry are "the structure, composition, and properties of substances and of their transformations". The central position of chemistry among the other physical sciences, resulting from the extensive overlap of its subject matter with them, is often referred to, especially by chemists. Of the four facets mentioned, the last stands apart as being most particularly the province of the chemist. In carrying out reactions, whether for the purpose of changing compositions or for harnessing the energy associated with them, the most important arena is the liquid state, especially as a liquid acting as a solvent provides the required mobility for the components of the reaction. Understanding solvent interactions, affecting as they do both the equilibrium state and the rates of reaction, is vital to developing the kind of understanding which is needed to optimize the desired outcome of the chemical change.

Near the end of the 4th decade of the last century, progress began to be made toward understanding the reactions of inorganic species in solution, beginning with those involving substitution. Data on the stability of inorganic complexes began to appear in the literature, and somewhat later reports devoted to understanding rates and mechanisms of substitution reactions. Whereas such rates are easy to measure when they are slow – unless of course they are very slow – those posed by the substitution of labile complexes were a serious challenge which was met by *Eigen* and coworkers. Despite this important advance and other early rapid progress, our understanding of basic issues, such as the extent of bond making as against bond breaking in the activated complexes for diverse systems, or of the electron redistribution factors which affect them, is far from satisfactory. While this state of affairs might be expected to result in continued research activity, interest in the field relative to other preoccupations has seriously declined. This is true also of the other major reaction class, that of redox reactions, taken as an important subject in its own right. Great advances were made early in the 5th decade by virtue of theoretical advances by *Hush*, *Marcus*, *Sutin*, and others for the simplest of such reactions, those involving $1e^-$ changes during which the coordination spheres of the reactants remain intact. Thanks to their success in dealing with the inner and outer sphere barriers to electron transfer imposed by the requirements of the *Franck-Condon* restriction, a *semi*-quantitative understanding of such reactions has been arrived at. By contrast, understanding redox reactions in the course of which primary bonds are ruptured, among them the important class involving oxygen transfer in the oxidation of organic molecules, has languished. The realization by the reader that understanding solvent-solute interactions is basic to progress in understanding the reactions of the substitution labile cations in solution will contribute to the survival of research activity in the subject.

Henry Taube, Stanford, July 2001
Nobel Laureate in Chemistry 1983

Contents

Invited Review

Ionic Solvation in Aqueous and Nonaqueous Solutions

Hitoshi Ohtaki*

Department of Applied Chemistry, Faculty of Science and Engineering, Ritsumeikan University, Kusatsu 525-8577, Japan

Summary. The history of studies on ionic solvation is briefly reviewed, and structural and dynamic properties of solvated ions in aqueous and nonaqueous solutions are discussed. An emphasis is placed on ionic solvation in nonaqueous mixed solvents in which preferential solvation of ions takes place. A new parameter for expressing the degree of preferential solvation of an ion is proposed.

Keywords. Ionic solvation; Nonaqueous solutions; Mixed solvents; Preferential solvation; Preferential solvation quotient.

Introduction

All ions and molecules in solution are more or less solvated, and thus solvated ions are the most fundamental species for ionic reactions in solution. A number of monographs have been published in recent years on this subject [1–8], and two important reviews have appeared recently [9, 10]. Two review books have been published in Japanese [11, 12].

Although many electrolytes dissolve in water, they are usually hardly soluble in nonaqueous solvents. Some organic solvents with relatively high dielectric constants or large donor and acceptor properties can be good solvents for electrolytes. In mixed solvents, ions are preferentially solvated by one solvent compared to the other. Such preferential solvation has been phenomenologically observed in many experiments such as conductivity, enthalpy, viscosity measurements, *etc*. However, no quantitative discussion has been undertaken for describing the degree of preferential solvation, because the solvent composition in the solvation sphere of ions can hardly be explored by experiments. Development of X-ray diffraction techniques for solution allows to see individual solvent molecules in the solvation shell, and accumulation of data for preferential solvation of ions in mixed solvents has allowed to discuss to some extent causes of preference of solvents recently. In this article, ionic hydration from the past to the present is briefly reviewed from static as well as dynamic points of view, and the results of ionic solvation in neat and mixed nonaqueous solvents are

* E-mail: ohtaki@pop.se.ritsumei.ac.jp

summarized. Experimental results so far reported are rather limited for solvation of ions in nonaqueous solvents and very limited in mixed solvents. Nevertheless, it may be very interesting and useful to discuss preferential solvation phenomena in a quantitative way. A special emphasis is placed on the introduction of a parameter representing the degree of preferential solvation which relates to the *Gibbs* energies of solvation of ions in two solvents. Exceptions from the representation of preferential solvation in terms of solvation energies of an ion in individual solvents in a mixture are discussed by taking into account specific ion-solvent and solvent-solvent interactions and steric effects of bulkiness of solvent molecules.

Solution Chemistry in the Time of *Arrhenius*

More than one century has passed since the publication of the Dissociation Theory of Electrolytes by *Arrhenius* [13]. Before *Arrhenius*, it was believed that electrolytes did not dissociate into ions in water until current was passed. An electric field produced between two electrodes could separate cations and anions in the electrolytes according to the considerations of *Clausius*. *Faraday*'s study on electrolysis in 1833 and the law of independent migration of ions proposed by *Kohlrausch* in 1855 were explained on the basis of this concept. The theory of ionic dissociation proposed by *Arrhenius* in 1887, in which he stated that electrolytes spontaneously separate into positive and negative ions in water against the electrostatic attraction between the ions, was not well accepted by the majority of scientists, and even *Arrhenius* himself could not explain the reason why positive and negative ions could separate against the *Coulomb* attraction between them. The source of the energy for separation of ions was unknown at that time.

Structural studies of water on the molecular level have been examined at the beginning of 1930 [14–16] which gave experimental evidence to the theoretical study of *Bernal* and *Fowler* [17]. However, elucidation of ion-water interactions in solution at the molecular level came much later.

Development of Solution Chemistry until 1957

At the beginning of the 20th century, many chemists studied solution chemistry (of course the term solution chemistry was not yet created, and their studies were recognized as those of physical chemistry), and *van't Hoff, Kohlrausch, Arrhenius*, and *Ostwald* are but some examples. Most of them employed methods of electrochemistry, and their discussions mostly concerned the static point of view. Some investigations based on dynamic properties of ions such as elecrtroconductivities and viscosities of solution had also been carried out in the period of 1900–1920. Electrochemistry was very useful for studies on solution chemistry until 1930–1940.

An important study was published by *Debye* and *Hückel* in 1923 [18]. Their theory of ionic activities in solution played a key role for explaining non-ideality of ion-ion interactions. The size of ions introduced by them as 'the distance of closest approach', which was later changed to 'ion size parameter', suggested that the sizes of ions in solution must differ from that of the naked ones due to ion-solvent interactions. The *Debye-Hückel* theory contributed to the elucidation of not only static properties of ionic activities but also dynamic ones of ionic mobilities and diffusion coefficients. The theory of ion-pair formation by *N. Bjerrum* [19] was induced by the *Debye-Hückel* theory. *Born*'s theory [20] for ionic solvation was welcom by many physical chemists due to its good fitting to experimental results in water and in water-like alcohols and its simple form, although the defect of the *Born* theory was pointed out when enthalpies and entropies of ionic solvation were measured and other nonaqueous solvents than alcohols having iso-dielectric constants were used.

However, in all theories solvents were assumed to be a dielectric continuum with specific properties, and no concept on a molecular level had been introduced. The contradiction between ionic sizes reduced with respect to the ion size parameters predicted by the *Debye-Hückel* theory or *Stokes* radii of ions obtained from the mobility of ions in solution and the radii of naked ions for some alkali and halide ions could not be solved until *Frank-Wen* [21] and *Samoilov* [22] proposed their theories for structure-making and structure breaking of ions and negative hydration, respectively, at the *Faraday Discussion* in 1957. Then, ionic solvation was discussed by taking into consideration the structure of solvents, especially the water structure for ionic hydration.

Ionic Hydration

The time from 1930 to 1940 may be called the first golden period of solution chemistry. In this period, various classical methodologies had been applied in studies of ionic hydration. Since electrolyte solutions contain, at least, two ionic components, *i.e.* cations and anions, and thermodynamic properties of both ions are not separately measured by experiments, some assumptions have to be introduced when measured properties are to be separated into cationic and anionic ones; otherwise, hydration numbers of individual ions are not determinable.

The nuclear magnetic resonance (NMR) method is very useful for probing the hydration number of ions, and many authors have reported the hydration numbers of various ions determined this way since 1960. However, without complicated corrections, this method will only determine the number of water molecules strongly bound to ions, *i.e.* those with substitution reaction rate constants below $10^9\ s^{-1}$. Most halide ions have much higher exchange rate constants, and thus their hydration number was assumed to be practically zero. Although this assumption does not introduce any serious errors for the determination of hydration numbers of cations, it seemed strange to see that most anions had no hydrated water molecules in solution. Moreover, this assumption is contradictory to the theories of *Frank-Wen* and *Samoilov*.

Solution X-ray (XD) and neutron diffraction (ND) methods and the extended X-ray absorption fine structure (EXAFS) method can overcome this difficulty and provide hydration numbers of individual ions without any assumption. The solution X-ray diffraction method, which is much easier to handle than the neutron diffraction method, has been used quite widely for the determination of hydration numbers of various ions. The method can also determine the lengths of ion-oxygen bonds in the hydrated species. By ND we can obtain the same structural parameters as well as the tilt angle of water molecules around the ions by knowing ion-oxygen and ion-proton distances in the solvation shell. In these methods, different from others previously employed for the determination of hydration numbers, the number of water molecules around an ion is found from the analysis of radial distribution functions, independent of the strength of the ion-water bonds. An ion with a large ionic size is determined to have a large hydration number, and the reverse is true for small ions. Therefore, the number of water molecules in the first solvation shell of a small ion such as Li^+ or Be^{2+} is low (four in these cases), but a large ion such as caesium, which is a typical structure breaker and weakly combined with water molecules, is found to have a large number of water molecules around the ion (about eight). It should be noted that the hydration numbers determined by the diffraction methods represent the time- and space-averaged number of water molecules around ions, as are those found by any other method, and distribution of hydration numbers of ions in the same solution and fluctuation of the number with time are not experimentally determinable at present. Computer simulations such as molecular dynamics (MD) simulations have to be used to observe molecular arrangements and movements of hydrated ions in more detail.

The methods employed for the determination of solvation numbers of ions are summarized in Table 1, with comments upon what they are measuring. The solvation numbers determined by most thermodynamic methods are rather ambiguously defined. By these methods, not only solvent molecules in the first solvation shell but also those in the second or even in further shells are frequently detected.

Some examples for the way of estimation of hydration numbers of ions are illustrated below:

1) The mean activity coefficients obtained from the *Debey-Hückel* theory may be divided into separate activity coefficients of cations and anions. Both include ion-size parameters ($\mathring{a}$) which reflect the sizes of their hydrated ions in aqueous solution, and thus the thickness of the hydration sphere can be estimated from the values of the ion-size parameter and the ionic radius of the naked ions. However, it is obvious that from these data the number of water molecules in the hydration shell cannot be determined because the volume of one water molecule in the hydration sphere cannot be evaluated. The ion-size parameter is a parameter evaluated under the assumption of a continuum of the bulk water, which is far from reality.
2) The density of aqueous electrolyte solutions may be explained in terms of bulk water of normal density and high-density water in the solvation sphere. However, the density of the solvation sphere may gradually change with the distance from the center of an ion; the first solvation shell has the highest density, and then the density should decrease in the next shells. The change in the densities in these spheres and the mass of the electrolyte cannot be evaluated.
3) The hydration numbers of some ions have been evaluated from the hydration entropies of the ions under the assumption that the hydration entropy per molecule equals that of freezing. Of course, this assumption is too naive.
4) The compressibility of solutions has been divided into two terms: compressible bulk water and incompressible hydrated water. The two-state assumption is too simple, and water phases of various compressibilities may exist from the first hydration shell to the bulk.

Table 1. Methods employed for studies on ionic hydration

Method	Region
Diffraction (X-ray, neutron)	Mainly the 1st shell; when the isomorphous substitution method is applied, the 2nd shall can be studied
X-ray absorption (EXAFS, XANES)	1st shell
Electron diffraction	1st shell of ions in quickly frozen solution (solid)
Computer simulations (Molecular dynamics, Monte Carlo)	1st and 2nd shells
Spectroscopy	
Visible, ultraviolet, infrared	1st shell
Raman	1st shell
NMR (chemical shift, line broadening)	1st shell
Mössbauer	1st shell
Thermodynamics	
Enthalpy and entropy of hydration	Not clear
Compressibility	Mainly the 1st shell
Activity	Mainly the 1st shell, but not clear
Vapor pressure, boiling and melting points	Mainly the 1st shell, but not clear
Density	Not clear
Solubility, salting out	Mainly the 1st shell, but not clear
Mobilities	
Transport number	Not clear
Mobility, diffusion coefficient	Not clear
Reaction kinetics	
Isotope dilution	1st shell
NMR	1st shell

These are just examples, and one can easily consider how simple and brave assumptions have been employed in studies of hydration numbers of ions in the past.

NMR measurements introduced in the sixties of the last century have also been employed in the study of ionic hydration, the method being first used for the determination of the hydration structure (static structure) and then for investigation of the dynamic behavior of hydrated water molecules. The NMR method is much advantageous compared with thermodynamic methods, because more direct information is obtainable for the number of water molecules which are practically fixed to ions, and the method allows to use ^{2}H and ^{17}O instead of ^{1}H and ^{16}O in water. Since the frequency of the magnetic field applied to a magnet in the NMR apparatus is in the order of 10^8 Hz, the NMR method can be used without special correction for solutions of Cr^{3+} and Al^{3+} whose rates of the solvent exchange reaction are very slow. The hydration numbers of ions in solution, the exchange rate constant of which is in the order of 10^3–10^7 s^{-1}, may be determined rather safely with suitable corrections. However, for ions with much faster solvent exchange rates, the NMR method has difficulties in correctly determining the hydration number of ions. Thus, the hydration number of anions is usually assumed to be one to zero, which is quite ridiculous. If the hydration number of anions were zero, an electrolyte were not soluble in water.

The most reasonable values for hydration numbers of ions have been determined by XD. Early works by XD are sometimes rather doubtful, because investigators did not well understand chemical equilibria in solution, and sometimes they neglected ion-pair and complex formation reactions between cations and anions and hydrolysis of cations in a neutral solution. Most of reliable results for the hydration numbers of ions are reported from the beginning of seventies.

In solution XD measurements, the hydration structure of cations and anions can be obtained independently and without any assumption, because most cations and anions have different ionic sizes which produce peaks at different positions in the radial distribution curve of a solution. In XD measurements, the hydration number of an ion is defined to be the number of water (solvent) molecules around an ion, independent of the bond strength between ion and hydrated water molecules. Therefore, the number of hydrated water molecules determined by the XD method, as well as by ND and EXAFS, is different from that evaluated by thermodynamic and other methods previously employed. If the attention is focused on the number of water molecules in the first hydration shell, the hydration number of a caesium ion is larger than that of a lithium ion due to the large surface area of the former ion with respect to the latter. The bond strength of hydrated water molecules to the central ions is given in terms of the temperature factor of the ion-water pair.

The orientation of water molecules around an ion cannot be determined by XD; only ND is capable to evaluate the structure parameters on the orientation of water molecules in the hydration shell, although sometimes experimentally obtained structural parameters do not coincide well with the results found by molecular dynamics simulations, which usually provide a reasonable agreement with results from XD for other structural properties of hydrated ions.

The ND method has a great advantage compared with XD and EXAFS measurements, because the former can use isotopes with different neutron scattering cross sections. The isotopic substitution method in ND measurements can provide very detailed structural information of hydrated ions when experiments are carried out very precisely and accurately. The reliability of ND data is sometimes lower than that of XD, because the scatter of measured neutron intensities at each point is much larger than that of XD. The *Jahn-Teller* distorted structure of the hexahydrated copper(II) ion in water was revealed by XD [23], but by ND the elongated $Cu–OH_2$ bonds at the axial position were not well seen [24]. The neutron scattering experiments are more enthusiastically employed for studies of dynamic properties of hydrogen atoms and other light atoms in molecules by using the quasi-elastic neutron scattering (QENS) technique.

The Extended X-ray Absorption Fine Structure (EXAFS) method is now often used for the determination of structures of hydrated ions and complex species in solution. The method is widely used for complicated systems which can hardly be studies by XD and ND because of its high selectivity of atoms and high sensitivity. Many biological samples have been investigated by this method. However, EXAFS can detect almost only the first neighbor atoms, and thus the structure of the second coordination shell and other long range interatomic interactions can hardly be discussed by this method. The X-ray Absorption Near Edge Structure (XANES) method is also widely employed to study the structure of complex species and the electronic state of the X-ray absorbing atom.

Computer Simulations Applied to Ionic Solvation

In the last decades, computer simulations have been rapidly developed, inspired by the progress of high-speed electronic computers. Computer simulations applied to solution chemistry may be classified into three groups: 1) Molecular Mechanics (MM), 2) Monte Carlo (MC), and 3) Molecular Dynamics (MD) simulations.

The first one is usually used for an estimation of the most stable configuration of a given molecule, and solute-solvent interactions are not well taken into consideration. However, for molecules containing many atoms and having a complicated form (mostly organic molecules), application of MM may be useful for estimating the most stable molecular form of an isolated molecule, which may not be changed in solution. Since MM mostly focuses its attention on the atomic configuration in a molecule and on the search for the lowest atom-atom interaction energy of a molecule, it is not convenient to discuss the structure of liquids and solutions containing a large number of atoms and molecules. Moreover, MM does not give clear information for the dynamic behavior of liquids and solutions.

Monte Carlo simulations have widely been used for discussing atomic arrangements in solution of lowest energy. However, MC calculations intend to reproduce molecular arrangements of the lowest energy in a system by changing atomic positions step by step with the help randomly generated numbers. Moreover, since the change in the molecular arrangement is time-independent, it is not possible to discuss dynamic properties of liquids and solutions.

Compared with the above two methods, MD simulations are much more useful for elucidating both static and dynamic properties of liquids and solutions. Combination of Molecular Orbital (MO) calculations for relatively small ensembles of the central ions and solvent molecules around the ion with MD calculations for the whole molecular system in a unit cell may provide a new way for structural studies on solvated ions and molecules and for the investigation of dynamic properties of the solvated species in solution.

The reliability of results obtained by computer simulations must be checked by comparing them with experimental results. For static properties, *i.e.* the structure of solutions, results from XD and ND methods are usually employed. For dynamic properties, *i.e.* the time of residence of solvent molecules in the solvation shell of an ion and rates of solvent exchange reactions, mostly NMR data are quoted. However, the time scale of MD simulations is too short (ps to ns time scale) compared with the time scale of the residential time of solvent molecules around an ion and the rate of solvent exchange reaction of most ions (usually 10^{-10} s to μs time scale, or even ms), and thus, it is not easy to certify the results obtained by MD simulation on the basis of experimental results.

Structure of Hydrated Ions

A sufficient accumulation of data for the structure and dynamic properties of hydrated ions has been seen in the past decades since 1960. The hydration structure of most single atomic cations and anions in water has been determined by MD, ND, and EXAFS methods. Orientation of water molecules in the solvation shells is not well elucidated, however. Although ND can determine the distances between the central ion and hydrogen atoms and between the ion and the oxygen atom of a solvated water molecule from which the tilt angle of the water molecule can be estimated, only a limited number of data are available, and they are not always in good agreement with the results found from MD simulations. On the other hand, dynamic properties of hydrated water molecules have been compiled with insufficient accuracy due to experimental difficulties. Rates of exchange of hydrated water molecules in solution have been experimentally determined after the works by *Eigen et al.* [25]. The development of the NMR method in measurements of dynamic properties of solvated ions accelerated the accumulation of experimental results. The data have been summarized in many books [1–8, 11, 12] and reviews [9, 10], but some of the typical results are illustrated in the following sections.

The hydration numbers and distances between central ions and oxygen atoms in the hydrated water molecules have been determined by NMR, XD, ND, and EXAFS methods. The tilt angles of hydrated water molecules around some metal ions which have suitable isotopes have been found by ND, although discrepancies of the experimental results by ND and MD have been observed in some cases, probably due to large scattering of experimental data in ND measurements, and thus the reliability of the results for the orientation of hydrated water molecules may be lower than that for the hydration numbers and bond distances of hydrated ions.

Table 2 summarizes selected results among various experimental results reported [10]. Cases with more than one entry show that it cannot be concluded which result is more reliable from the literature (*e.g.* Li^+ and NH_4^+) or that different results have been found from different methods (for example, by XD, ND, and EXAFS for H^+, Ca^{2+}, Fe^{2+}, Ni^{2+}, Cr^{3+}, *etc.*).

Generally speaking, results by NMR agree with those by the diffraction methods (XD, ND, EXAFS). However, if we define the hydration number to be the number of water molecules in the nearest neighbor of an ion, the results obtained by the diffraction methods are more reliable than those by NMR, because the NMR method cannot well measure the number of water molecules moving fast within less than *ca.* 10^{-9} s. The hydration numbers of most anions are given to be one to zero by NMR. The results found by XD, ND, and EXAFS usually agree with those drawn on the basis of coordination chemistry. However, the hydration numbers found for alkali and alkali earth metal ions are not always in good agreement with the results derived by *Frank-Wen*'s and *Samoilov*'s considerations, because the hydrated water molecules determined by the diffraction methods do not always imply water molecules strongly combined with the central ions. On the other hand, NMR measurements detect only water molecules strongly combined with ions and moving slowly with respect to the NMR time scale.

Non-integer hydration numbers are obtained for some ions. Although no clear explanation has been given for such non-integer hydration numbers found by NMR, non-integer hydration numbers obtained by the diffraction methods, including EXAFS, have been well illustrated in terms of (*i*) experimental errors, (*ii*) the use of symmetrical *Gaussian* functions at the analysis of radial distribution functions, (*iii*) an ion having, at least, two different hydration numbers, and (*iv*) an ion having a

Table 2. Selected values of hydration numbers and bond angles of ion-water interactions in the first hydration shell of cations and anions determined by NMR, X-ray, neutron diffraction, and EXAFS[a]

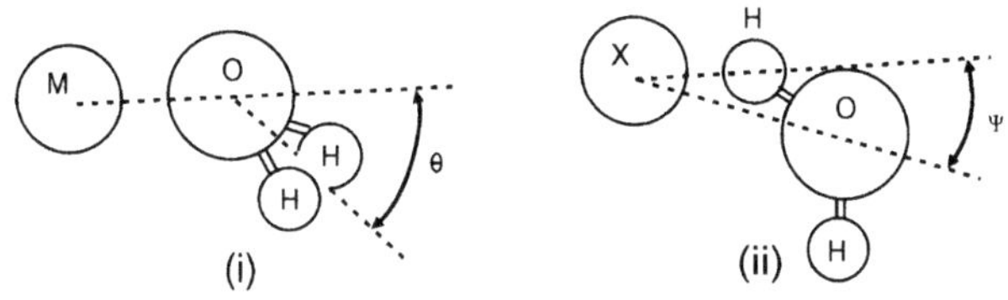

Orientation of water molecule: (*i*) M = cation, (*ii*) X = anion

	NMR					X-Ray, neutron diffraction, EXAFS				
Ion	Hydration number n	Nucleus	T/°C	Method	Ref.	Hydration number n	Bond length ($r_{M\text{–}O}$, $r_{X\text{–}O}$)/pm	Bond angle (θ, Ψ)/°	Method	Ref.
H^+	2.5	^1H	broad T	cs	[1]	{3	{244		X	[13]
						1	290			
						4	288	$\theta = 0$	N	[14]
Li^+	3.0	^1H	broad T	cs	[1]	3.3	195	$\theta = 52$	N	[15]
						5.5	195	$\theta = 40$	N	[15]
Na^+	3.5	^1H	broad T	cs	[1]	6	244–248		X	[16]
K^+	3.0	^1H	broad T	cs	[1]	6	280		X	[17]
Rb^+	3.5	^1H	broad T	cs	[1]					
Cs^+	3.0				[2]	8	295	$\theta = 54.75$	N	[18]
NH_4^+						4	280–288		X	[19]
						8	304		X	[20]
Ag^+						4	243		X	[21]
Be^{2+}	4	^1H	−20–100	pa	[3]	4	167		X	[22]
Mg^{2+}	6	^1H	−67–−90	pa	[4]	6	212		X	[23]
Ca^{2+}	6.0	^1H	broad T	cs	[1]	6	244		X	[24]
						5.5	240	$\theta = 51$	N	[25]
						8	246		E,X	[56]
Sr^{2+}	5.0	^1H	−20–100	pa	[3]	8	264		X	[26]
Ba^{2+}	5.7	^1H	−20–100	pa	[3]	9.5	290		X	[27]
Cr^{2+}						4[b]	208		E	[28]
Mn^{2+}						6	220		X	[29]
Fe^{2+}	5.6–5.8	^1H	−40–−80	cs	[5]	6	212		X	[29]
						6	213	$\theta = 32$	N	[30]
Co^{2+}	5.9	^1H	−38–−63.7	pa	[6]	6	208		X	[29]
Ni^{2+}	6	^{17}O	< 127	cs	[7]	6	204		X	[29]
						5.8	205	$\theta = 30$	N	[31]
Cu^{2+}						{4	{194		X	[29]
						2	238			
Zn^{2+}	5.9–6.4	^1H	−120	cs	[4]	6	204		X	[29]
Cd^{2+}	4.6	^1H	−20–100	pa	[3]	6	231		X	[32]
Hg^{2+}	4.9	^1H	−20–100	pa	[3]	6	241		X	[33]
Sn^{2+}						3.6	234		X	[34]
Pb^{2+}	5.7	^1H	−20–100	pa	[3]					
Al^{3+}	6.0	^1H	−35–−52	pa	[8]	6	187		X	[35]

(contd.)

Table 2 (*continued*)

Sc^{3+}	5.1	^{1}H		pa	[8]	6	187		X	[35]
Cr^{3+}	6.0	^{17}O	20	pa	[9]	6	199.7		X	[36]
						6	197		X	[57]
						5	195	$\theta = 36$	N	[37]
Fe^{3+}						6	201	$\theta = 41$	N	[38]
Ga^{3+}	5.89	^{17}O	35	pa	[10]	6	196		X	[57]
Y^{3+}						8	236		X	[39]
Rh^{3+}						6	206		X	[40]
In^{3+}	5.8–6.0	^{1}H	−89–−99	pa	[11]	6	215		X	[41]
Ti^{3+}						5	223.6		X	[42]
Bi^{3+}						8	241		X	[58]
La^{3+}	6.0–6.4	^{1}H	−105–−120	cs	[5]	9.1	258		X	[43]
						{6	{252		X	[59]
						3	264			
Ce^{3+}	4.5–5.5	^{1}H	−110	cs	[5]					
Pr^{3+}						9.2	254		X	[43]
Nd^{3+}						8.9	251		X	[43]
Sm^{3+}						8.8	247		X	[44]
Eu^{3+}						8.3	245		X	[44]
Gd^{3+}	8 or 9	^{1}H			[12]	8.0	237		X	[45]
Tb^{3+}						8.0	240		X	[39]
Dy^{3+}						7.9	239		N	[46]
Er^{3+}						8.0	236		X	[39]
Tm^{3+}						8	233	$\theta = 12$	N	[47]
Yb^{3+}						8	233	$\theta = 12$	N	[47]
Lu^{3+}						8.0	234		X	[48]
Sn^{4+}	6.0	^{1}H		pa	[8]	8.0	234		X	[48]
Th^{4+}	9.1	^{1}H		pa	[8]	7.9–8.1	248.6		X	[49]
						10	245		X	[60]
U^{4+}						7.9	250		X	[50]
						10	242		X	[60]
F^-						6	269		X	[19]
Cl^-						5.3–6.2	222–226	$\Psi = 0–6$	N	[51]
Br^-						8	320		X	[52]
I^-						8.3	358		X	[53]
NO_3^-						7.2	340[c]		X	[21]
ClO_4^-						8	380[c]		X	[54]
SO_4^{2-}						8.2	379		X	[55]

[a]broad *T*: values obtained from temperature *T* and concentration *c* dependencies of chemical shifts over a broad range of *T* and *c*; cs: determined from chemical shifts; pa: determined from peak area; X: X-ray diffraction; N: neutron diffraction; E: EXAFS; [b]$Cr^{2+}-OH_2$ bonds elongated along the axis were not observed; [c]$X-OH_2$ distance beyond the O atom directly combined with *X*

[1] Vorgin J, Knapp PS, Flint WL, Anton A, Highberger G, Malinowski ER (1971) J Chem Phys **54**: 178; [2] Creekmore RW, Reilley CN (1969) J Phys Chem **73**: 1563; [3] Swift TJ, Sayer WG (1966) J Chem Phys **44**: 3567; [4] Matwiyoff NA, Taube H (1968) J Am Chem Soc **90**: 2796; [5] Fratiello A, Kubo V, Peak S, Sanchez B, Schuster RE (1968) Inorg Chem **10**: 2552; [6] Matwiyoff NA, Darley PE (1968) J Phys Chem **72**: 2659; [7] Neely JW, Connick RE (1972) J Am Chem Soc **94**: 3519; [8] Fratiello A, Lee RE, Schuster RE (1970) Inorg Chem **9**: 391; [9] Alei M, Jackson JA (1964) J Chem Phys **41**: 3402; [10] Fiat D, Connick RE (1966) J Am Chem Soc **88**: 4754; [11] Fratiello A, Lee RE, Nishida VM, Schuster RE (1968) J Chem Phys **48**: 3705; [12] Morgan LO (1963) J Chem Phys **38**: 2788; [13] Lee H-G, Matsumoto Y,

(Table 2 ref. notes continued)

Yamaguchi T, Ohtaki H (1983) Bull Chem Soc Jpn **56**: 443; [14] Ohtomo N, Arakawa K, Takeuchi M, Yamaguchi T, Ohtaki H (1981) Bull Chem Soc Jpn **54**: 1314; [15] Newsome JR, Neilson GW, Enderby JE (1980) J Phys Chem Solid State Phys **13**: L923; [16] Caminiti R, Licheri G, Paschina G, Piccaluga G, Pinna G (1980) J Chem Phys **72**: 4522; [17] Pálinkás G, Radnai T, Haijdu H (1980) Z Naturforsch **35a**: 107; [18] Ohtomo N, Arakawa K (1979) Bull Chem Soc Jpn **52**: 2755; [19] Narten AH (1970) J Chem Phys **74**: 765; [20] Musinu H, Paschina G, Pinna G (1981) Chem Phys Lett **80**: 163; [21] Yamaguchi T, Johansson G, Holmberg B, Maeda M, Ohtaki H (1984) Acta Chem Scand **A38**: 437; [22] Yamaguchi T, Ohtaki H, Spohr E, Pálinkás G, Heinzinger K, Probst MM (1986) Z Naturforsch **41a**: 1175; [23] Caminiti R, Cerioni G, Crisponi G, Cucca P (1988) Z Naturforsch **43a**: 317; [24] Yamaguchi T, Hayashi S, Ohtaki H (1989) Inorg Chem **28**: 2434; [25] Cummings S, Enderby JE, Howe RA (1980) J Phys Chem Solid State Phys **13**: 1; [26] Caminiti R, Musinu A, Paschina G, Pinna G (1982) J Appl Crystallogr **15**: 482; [27] Albright JN (1972) J Chem Phys **56**: 3783; [28] Miyanaga T, Watanabe I, Ikeda S (1988) Chem Lett 1073; [29] Ohtaki H, Yamaguchi T, Maeda M (1976) Bull Chem Soc Jpn **49**: 701; [30] Herman GJ, Neilson GW (1992) J Phys Condens Matter **4**: 649; [31] Soper AK, Neilson GW, Enderby JE, Howe RA (1977) J Phys Chem Solid State Phys **10**: 1793; [32] Ohtaki H, Maeda M, Ito S (1974) Bull Chem Soc Jpn **47**: 2217; [33] Sandström M, Persson I, Ahrand S (1978) Acta Chem Scand **A32**: 607; [34] Yamaguchi T, Lindquist O, Claeson T, Boyce JB (1982) Chem Phys Lett **93**: 528; [35] Caminiti R, Radnai T (1980) Z Naturforsch **35a**: 1368; [36] Caminiti R, Licheri G, Piccaluga G, Pinna G (1978) J Chem Phys **69**: 1; [37] Broadbent RD, Neilson GW, Sandström M (1992) J Phys Condens Matter **4**: 639; [38] Herdman GJ, Neilson GW (1992) J Phys Condens Matter **4**: 627; [39] Johansson G, Niinisto L, Wakita H (1985) Acta Chem Scand **A39**: 359; [40] Caminiti R, Cucca D Atzei, Squintu F, Bongiovanni G (1985) Z Naturforsch **40a**: 1319; [41] Maeda M, Ohtaki H (1977) Bull Chem Soc Jpn **50**: 1893; [42] Glaser J, Johansson G (1982) Acta Chem Scand **A36**: 125; [43] Habenshuss H, Spedding FH (1979) J Chem Phys **70**: 3758; [44] Habenshuss H, Spedding FH (1980) J Chem Phys **73**: 442; [45] Steel ML, Wertz DL (1976) J Am Chem Soc **98**: 4424; [46] Cossy C, Barnes AC, Enderby JE (1989) J Chem Phys **90**: 3254; [47] Yamaguchi T, Tanaka S, Wakita H, Misawa M, Okada I, Soper AK, Howell SW (1991) Z Naturforsch **46a**: 84; [48] Habenshuss H, Spedding FH (1979) J Chem Phys **70**: 2797; [49] Johansson G, Magini M, Ohtaki H (1991) J Solution Chem **20**: 775; [50] Pocev S, Johansson G (1973) Acta Chem Scand **27**: 2146; [51] Cummings S, Enderby JE, Neilson GW, Newsome JR, Howe RA, Howells WS, Soper AK (1980) Nature **287**: 714; [52] De Barros Marques MI, Cabaco MI, Sousa Oliveira MA, Alves Marques M (1982) Chem Phys Lett **91**: 222; [53] Tamura Y, Yamaguchi T, Okada I, Ohtaki H (1987) Z Naturforsch **42a**: 367; [54] Johansson G, Wakita H (1985) Inorg Chem **24**: 3047; [55] Caminiti R (1986) J Chem Phys **84**: 3336; [56] Jalilehvand F, Spångberg D, Lindqvist-Reis P, Persson I, Sandström M (2001) J Am Chem Soc **123**: 431; [57] Lindqvist-Reis P, Muñoz-Paez M, Diaz-Moreno S, Pattnaiki S, Persson I, Sandström M (1998) Inorg Chem **37**: 6657; [58] Näslund J, Persson I, Sandström M (2000) Inorg Chem **39**: 4012; [59] Näslund J, Lindqvist-Reis P, Persson I, Sandström M (2000) Inorg Chem **39**: 4006; [60] Moll H, Denecke MA, Jelilahvand F, Sandström M, Grenthe I (1999) Inorg Chem **38**: 1759

[1] Impey RW, Maden PA, McDonald IR (1983) J Phys Chem **87**: 5071; [2] Herz HG (1973) "Water – A Comprehensive Treatise", Franks F (ed) Plenum, New York, Vol. 6, p. 301; [3] Salmon PS, Howells WS, Mills R (1987) J Phys Chem Solid State Phys **20**: 5727; [4] Connick RE, Fiat DN (1963) J Chem Phys **39**: 1349; [5] Friedman HL (1985) Chim Scripta **25**: 42; [6] Swift TJ, Connick RE (1962) J Chem Phys **37**: 307; [7] Salmon PS, Bellisent-Funel M-C, Herdman GJ (1990) J Phys Condens Matter **2**: 4297; [8] Fiat DN, Connick RE (1968) J Am Chem Soc **90**: 608; [9] Chmelnik AM, Fiat DN (1972) J Magn Reson **8**: 325; [10] Hunt J, Plane RA (1954) J Am Chem Soc **76**: 5960; [11] Salmon PS, Herdman GJ, Lindgren J, Read MC, Sandström M (1989) J Phys Condens Matter **1**: 3459; [12] Dodgen HW, Liu G, Hunt JP (1981) Inorg Chem **20**: 1002; [13] Grant M, Jordan RB (1981) Inorg Chem **20**: 55; [14] Herdman GJ, Salmon PS (1991) J Am Chem Soc **113**: 2930; [15] Rutenberg AC, Taube H (1952) J Chem Phys **20**: 825; [16] Hugi-Cleary D, Helm L, Merbach AE (1985) Helv Chim Acta **68**: 545; [17] Reuben J, Fiat D (1969) J Chem Phys **51**: 4918

distribution in hydration numbers over a relatively wide range of numbers. Evidence for the distribution of hydration numbers around an ion has been obtained from MD for many ions of large ionic size and low electronic charge, such as alkaline metal ions and halide ions. The results for many anions belong to case (*iv*), whereas lanthanide ions are believed to belong to case (*iii*).

The observed bond length between an ion and a water molecule is close to the sum of the ionic radius reported in the literature and the *van der Waals* radius of a water molecule (140 pm). According to ND results, water molecules around an ion are not directed towards the dipole of the water molecule but tilt by an angle θ. An explanation of the tilt angle in terms of the sp^3 structure of a water molecule is no more acceptable, because it is not believed that a water molecule has an sp^3 electronic configuration of the ST2 type [26], but two positive charge at the hydrogen atoms and one negative charge at a point very close to the central oxygen atom, and no electronic orbital lobes expanding in the opposite area from the hydrogen atoms beyond the oxygen atom are observed by MO calculations [27]. The three point-model for water has been widely accepted after the successes achieved by *Matsuoka, Clementi*, and *Yoshimine* [28] with various modifications for the electronic charge distribution and bond length in the water molecule.

The tilt of water molecules in the hydration shell is interpreted in terms of hydrogen bonding of the water molecule with water molecules on the second hydration shell. A non-zero tilt angle ($\theta > 0°$) allows to form three hydrogen bonds, whereas zero angle ($\theta = 0°$; the dipole of a water molecule directs to the central ion) means only two hydrogen bonds, and thus the former can form a more stable hydration sphere than the latter. MD simulations for anionic hydration show that the dipole of water molecules is almost exactly directed towards the central anion ($\Psi = 0$–$10°$). This results agrees well with the experimentally obtained tilt angle for Cl^- ($\Psi = 0$–$6°$) (see Table 2).

Dynamic Properties of Water Molecules in Hydration Shells of Ions

In contrast to static properties, *i.e.* structures, of hydrated ions studied by XD, ND, EXAFS, and NMR which usually gave energetic information of ions in water, dynamic properties, such as the residence time and rotational behavior of water molecules in the hydration shells of ions, can be found from kinetic and dynamic studies of electrolyte solutions. NMR, QENS, and MD studies may provide such dynamic information of hydrated ions. The residence times of water molecules in the first hydration shell of various ions are summarized in Table 3.

In Table 3, τ_{H_2O} denotes the averaged residence time of a water molecule in the bulk. τ_O and τ_H represent the residence time of oxygen and hydrogen atoms, respectively, of a water molecule in the first hydration shell of an ion. τ_H is usually smaller than τ_O due to the proton transfer reaction between the first hydration shell and a nearest neighbor water molecule:

$$\begin{array}{c} M\cdots\underset{\displaystyle \mathrm{H}}{\underset{|}{\mathrm{O}}}\!-\!\mathrm{H}\cdots\underset{\displaystyle \mathrm{H}}{\underset{|}{\mathrm{O}}}\!-\!\mathrm{H} \Leftrightarrow M\!-\!\underset{\displaystyle \mathrm{H}}{\underset{|}{\mathrm{O}^-}}\cdots\mathrm{H}\!-\!\underset{\displaystyle \mathrm{H}}{\underset{|}{\mathrm{O}^+}}\!-\!\mathrm{H} \end{array}$$

τ_{H_2O} is evaluated to be $(4.6 \pm 0.3) \times 10^{-12}$ s. The value of τ_O depends remarkably on the kind of ion. The value is very small for inert ions such as Cr^{3+} and Al^{3+}, but

Table 3. Residential time of water molecule in the bulk and of oxygen and hydrogen atoms of hydrated water molecules in the first hydration shell of ions (τ_{H_2O}, τ_O, and τ_H)

Ion	τ_{H_2O}/s	τ_O/s	τ_H/s	Method[a]	Ref.
Li^+	4.8×10^{-12}			MD	[1]
		3.3×10^{-11}		MD	[1]
			$\sim 3 \times 10^{-11}$	NMR	[2]
Na^+	3.8×10^{-12}	9.9×10^{-12}		MD	[1]
K^+	4.8×10^{-12}	4.0×10^{-12}		MD	[1]
		$< 10 \times 10^{-10}$		QENS	[3]
Cs^+		$< 10 \times 10^{-10}$		QENS	[3]
Be^{2+}		$\geqq 3 \times 10^{-4}$		NMR	[4]
Mg^{2+}		2×10^{-6}		NMR	[5]
			$\geqq 5 \times 10^{-5}$	QENS	[3]
Ca^{2+}		$< 10^{-10}$		NMR	[5]
			$\leqq 10 \times 10^{-10}$	QENS	[3]
Mn^{2+}		3.2×10^{-8}		NMR	[6]
Fe^{2+}		3.13×10^{-7}		NMR	[6]
Co^{2+}		8.8×10^{-7}		NMR	[6]
Ni^{2+}		3.7×10^{-5}		NMR	[6]
			$\geqq 5 \times 10^{-9}$	QENS	[3]
		1×10^{-4}		NMR	[6]
Cu^{2+}		5×10^{-9}		NMR	[6]
			$\leqq 10^{-10}$	QENS	[3]
Zn^{2+}			$> 10^{-10}$	QENS	[7]
Ti^{3+}		1.0×10^{-5}		NMR	[8]
V^{3+}		6×10^{-4}		NMR	[9]
		$1.8\text{–}2.2 \times 10^5$		NMR	[10]
Cr^{3+}			2×10^{-6}	NMR	[2]
			$\geqq 5 \times 10^{-9}$	QENS	[11]
Fe^{3+}		6×10^{-3}		NMR	[12, 13]
			$\geqq 5 \times 10^{-9}$	QENS	[14]
Co^{3+}		10^5		NMR	[15]
Al^{3+}		0.78		NMR	[16]
			$\geqq 5 \times 10^{-9}$	QENS	[14]
Ga^{3+}		5.5×10^{-5}		NMR	[8]
			$\geqq 5 \times 10^{-9}$	QENS	[14]
Nd^{3+}			$> 10^{-10}$	QENS	[3]
Tb^{3+}		$1.3\text{–}4.8 \times 10^{-8}$		NMR	[17]
Dy^{3+}		$3.1\text{–}7.1 \times 10^{-8}$		NMR	[17]
			$> 10^{-10}$	QENS	[14]
Ho^{3+}		$1.6 \times 10^{-8}\text{–}1.1 \times 10^{-7}$		NMR	[17]
Er^{3+}		$7.4 \times 10^{-9}\text{–}1.9 \times 10^{-7}$		NMR	[17]
Tm^{3+}		$1.5\text{–}3.0 \times 10^{-7}$		NMR	[17]
F^-	5.5×10^{-12}	20.3×10^{-12}		MD	[1]
			$\leqq 10^{-10}$	QENS	[3]
Cl^-	4.1×10^{-12}	4.5×10^{-12}		MD	[1]
			$\leqq 10 \times 10^{-10}$	QENS	[3]

[a] MD: molecular dynamics simulations, QENS: quasi-elastic neutron scattering

τ_O of alkali ions is close to that of bulk water. Generally speaking, among ions in the same group (in the same column of the periodic table), the larger the ionic radius, the smaller the τ_O value. The result coincides with considerations made by *Samoilov* [22]. Irregularity in the order of the magnitude of the τ_O values observed with the first transition metal ions have been explained by the ligand field theory applied to the stability of penta- or heptacoordinated reaction intermediates [29].

Studies on the dynamic behavior of hydrated water molecules have first been examined by relaxation methods such as temperature and pressure jumps as developed by *Eigen* and coworkers [25], but in recent years dynamic properties of hydrated water molecules have mostly been investigated by the NMR method. Exchange reactions of water molecules in the hydration shell of simple ions have been thoroughly investigated; the results are summarized in Table 4.

Activation enthalpies ($\Delta H^{\neq}$) and activation entropies ($\Delta S^{\neq}$) of water exchange reactions of hydrated ions can be obtained by measuring the temperature dependence of the rate constant of the reaction. Activation volumes ($\Delta V^{\neq}$) of the reaction can be determined by the pressure dependence of the rate constant.

$$(\mathrm{d}\ln k/\mathrm{d}T)_p = -\Delta H^{\neq}/RT + \Delta S^{\neq}/R \quad (1)$$

$$(\mathrm{d}\ln k/\mathrm{d}p)_T = -\Delta V^{\neq}/RT \quad (2)$$

$\Delta V^{\neq}$ is considered more reliable than $\Delta S^{\neq}$ with respect to value and even sign because the latter is obtained by an extrapolation of measured values to $1/T = 0$.

Mechanisms of exchange reactions of hydrated water molecules have been classified into three groups: Dissociative (D), Associative (A), and Interchange (I) mechanisms. Introduction of additional two mechanisms (Dissociative Interchange (I_d) and Associative Interchange (I_a)) between the two extremes allows to represent the reaction mechanisms more precisely.

Reaction kineticists often simply assign reactions with large negative values of $\Delta S^{\neq}$ or $\Delta V^{\neq}$ to the association mechanism, whereas reactions having large positive values of $\Delta S^{\neq}$ or $\Delta V^{\neq}$ should be dissociative. In the former reaction, a water molecule associatively attached to the hydration shell from the bulk should reduce the volume by *ca.* $-18\,\mathrm{cm}^3 \cdot \mathrm{mol}^{-1}$ at the formation of an intermediate hydrated ion with an increased hydration number. On the other hand, release of a water molecule from the hydration shell due to the dissociation of a water molecule from the hydrated ion should result in an increase of volume by about $+18\,\mathrm{cm}^3 \cdot \mathrm{mol}^{-1}$. In the interchange reaction, $\Delta S^{\neq}$ and $\Delta V^{\neq}$ should be very small and close to zero. However, the classification of reaction mechanisms from the activation parameters has no sound theoretical basis, and the conclusions are usually derived from the picture shown in Fig. 1.

The picture shows that in the dissociative mechanism (D-Mechanism) the total volume of the hydrated ion with a loss of one water molecule and one water molecule in the bulk should be larger than that of the original hydrated ion. The reverse might hold for the associative mechanism (A-Mechanism) provided the volume of the hydrated ion does not change during the change in the hydration number of the ion. It is a well-known fact that the length of an ion-ligand bond increases when the coordination number increases and decreases with the decrease in the coordination number for a given ion with a given ligand. Therefore, an ion

Table 4. Rate constants of water exchange reactions of hydrated cations and thermodynamic parameters of the reaction at 25°C

Ion	Hydration Number n	k/s^{-1}	$\Delta H^{\neq}/kJ \cdot mol^{-1}$	$\Delta S^{\neq}/kJ \cdot mol^{-1}$	$\Delta V^{\neq}/cm^3 \cdot mol^{-1}$	Reaction mechanism[a]	Ref.
Be^{2+}	4	7.3×10^2	59.2	8.4	−13.6	A	[1]
V^{2+}	6	8.7×10^1	61.8	−0.4	−4.1	I_a	[2]
Mn^{2+}	6	2.1×10^7	32.9	5.7	−5.4	I_a	[3]
Fe^{2+}	6	4.39×10^6	41.4	21.2	3.8	I_a–I_d	[3]
Co^{2+}	6	3.18×10^6	46.9	37.2	6.1	I_a–I_d	[3]
Ni^{2+}	6	3.37×10^4	52.3	17.2	7.1	I_d	[4]
Cu^{2+}	4	1×10^4	192	−16.8			[5]
	2	2×10^8	87.8				
Zn^{2+}	6	3×10^7					[6]
Ru^{2+}	6	1.8×10^{-2}	87.8	16.1	−0.4	I	[1]
Pd^{2+}	4	5.6×10^2	49.5	−26	−22	I_a or A	[6]
Pt^{2+}	4	3.9×10^{-4}	89.7	−9	−4.6	I_a or A	[1, 7]
Al^{3+}	6	1.29	84.7	41.6	5.7	I_d	[8]
Ga^{3+}	6	4×10^2	67.1	30.1	5.0	I_d	[9]
Ti^{3+}	6	1.8×10^5	43.4	1.2	−12.1	I_a or A	[9]
V^{3+}	6	5×10^2	49.4	−27.8	−8.9	I_a	[10]
VO^{2+}	5	7.41×10^2	55.6	−6.3			[11, 12]
Cr^{3+}	6	2.4×10^{-6}	108.6	11.6	−9.6	I_a	[13]
Fe^{3+}	6	6.8×10^4	40.3	80	−5.4	I_a	[14]
Ru^{3+}	6	3.5×10^{-6}	89.8	−48.3	−8.3	I_a	[1]
Rh^{3+}	6	2.2×10^{-9}	131.2	29.3	−4.2	I_a	[1]
Gd^{3+}	8	11.9×10^8	12.0	−30.9		I_a	[15]
Tb^{3+}	8	5.58×10^8	12.1	−36.9	−5.7	I_a	[16]
Dy^{3+}	8	4.34×10^8	16.6	−24.0	−6.0	I_a	[16]
Ho^{3+}	8	2.14×10^8	16.4	−30.5	−6.6	I_a	[16]
Er^{3+}	8	1.33×10^8	18.4	−27.8	−6.9	I_a	[16]
Tm^{3+}	8	0.47×10^8	22.7	−16.6	−6.0	I_a	[16]
Yb^{3+}	8	0.47×10^8	23.3	−21.0		I_a	[15]

[a]A: association mechanism, I: interchange mechanism, I_a: associative interchange mechanism, I_d: dissociative interchange mechanism, I_a–I_d: not distinguishable between I_a and I_d mechanisms

[1] Merbach AE, Akitt JW (1990) In NMR Basic Principles and Progress, Springer, Berlin, Vol. 24, p 191; [2] Merbach A (1987) Pure Appl Chem **59**: 161; [3] Ducommun Y, Newman KE, Merbach AE (1980) Inorg Chem **19**: 3696; [4] Ducommum Y, Earl WL, Merbach AE (1979) Inorg Chem **18**: 2754; [5] Swift TJ, Connick RE (1962) J Chem Phys **37**: 307; [6] Pittipardi F, Petrucci S (1967) J Phys Chem **71**: 3414; [7] Helm L, Elding LI, Merbach AE (1985) Inorg Chem **24**: 1719; [8] Hugi-Cleary D, Helm L, Merbach AE (1985) Helv Chim Acta **68**: 545; [9] Hugi-Cleary D, Helm L, Merbach AE (1987) J Am Chem Soc **107**: 4444; [10] Hugi D, Helm L, Merbach AE (1985) Helv Chim Acta **68**: 508; [11] Wurthrich K, Connick RE (1967) Inorg Chem **6**: 583; **7**: 1377 (1968); [12] Ruben J, Fiat D (1967) Inorg Chem **6**: 579; [13] Xu F-C, Krouse HR, Swaddle TW (1985) Inorg Chem **24**: 267; [14] Swaddle TW, Merbach AE (1981) Inorg Chem **20**: 4212; [15] Helm L, Merbach AE (1991) Eur J Solid State Inorg Chem **28**: 245; [16] Cosay C, Helm L, Merbach AE (1989) Inorg Chem **28**: 2699

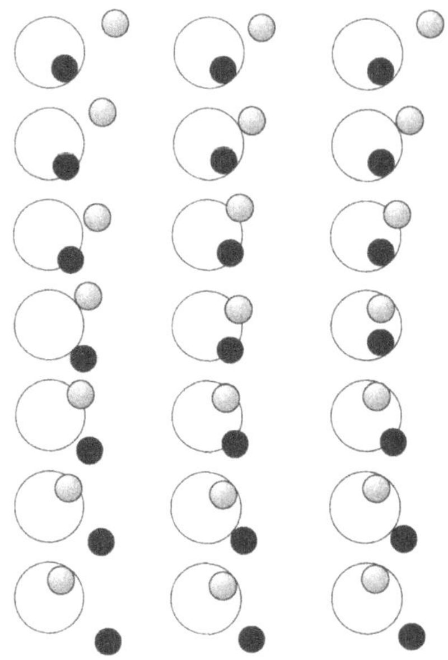

Fig. 1. Reaction mechanisms suggested from the values of activation entropies and activation volumes

with an increased hydration number formed at the intermediate step of the association mechanism in the water exchange reaction should have a larger volume than that of the original hydrated ion, and the reverse is true for an intermediate hydrated ion formed at the intermediate stage of the dissociation mechanism. The change in the volume of hydrated ions at the intermediate step of the reaction may hardly be estimated, but we cannot ignore the shrink in the volume of ions at the dissociation mechanism and inflation of the volume of an ion at the association mechanism. The volume change at the formation of an intermediate may be rather safely neglected for small ions such as hydrated multivalent transition metals, but for large ions coordinated with bulky ligands discussion of the reaction mechanisms from the data of $\Delta V^{\neq}$ should be very careful, and sometimes an erroneous conclusion may be derived.

A similar discussion can be made for the value of $\Delta S^{\neq}$. Vibrational and rotational motions of hydrated water molecules may not remain unchanged during water exchange reactions. Elongation of the metal-ligand bonds may enhance molecular motions in the hydration shell, which should increase in the entropy of the associatively formed intermediate. The loss in entropy of water molecules at the introduction to the hydration shell from the bulk should be more or less compensated by the increase in entropy due to the bond elongation. The reverse consideration can be made for the dissociation mechanism. Therefore, one should be very careful when discussing the reaction mechanism on the basis of thermodynamic quantities.

In fact, theoretical studies on water exchange reaction mechanisms of divalent and trivalent transition metal ions of the first transition period on the basis of SCF computational analysis [30, 31] lead to the conclusion that the reaction mechanism should be dissociative for Mn^{2+} and V^{2+} and trivalent ions such as Ti^{3+}, V^{3+}, Cr^{3+}, Fe^{3+}, and Ga^{3+}, which is not compatible with the associative interchange mechanism proposed for these ions on the basis of their negative activation volumes $\Delta V^{\neq}$.

Ionic Hydration under Extreme Conditions

Numerous studies have been carried out on the structure of water since 1930 after XD studies by *Mayer* [14], *Stewart* [15], and *Amaldi* [16] and the theoretical work by *Bernal* and *Fowler* [17]. The water structure at supercooled temperatures and in the glassy state has also been investigated from the viewpoints of chemistry, physics, geology, biology, and engineering [32, 33]. Water can remain liquid below 0°C at 1 atmosphere pressure and is then believed to have a similar structure like water under normal conditions. Glassy water, which has deformed hydrogen bonds, is formed by quick quenching of liquid water.

Water under supercritical conditions ($t_c = 374.2$°C, $p_c = 22.1$ MPa, $\rho_c = 0.3228\,g \cdot cm^{-3}$) has attracted much interest not only in the fields of chemistry and physics, but also geology in connection of underground waters, biology from the knowledge of evolution of life, and engineering for decomposing organohalogeno compounds and synthesis of organic and inorganic compounds.

The structure of water at high temperatures and high pressures, especially under supercritical conditions, has been extensively investigated by many authors employing various methods, and the conclusions derived from these studies are rather divergent. However, in recent years, the solution X-ray diffraction technique with a special cell arrangement has effectively been applied to structural studies at high temperatures and high pressures (HTHP), and the conclusions on the structure of water in the supercritical region seem to indicate that supercritical water consists of small clusters containing hydrogen-bonds and monodispersed gas-like water molecules. An HTHP cell used for XD studies of water is shown in Fig. 2. Recent studies on the structure of supercritical water have been summarized in some articles [34, 35].

According to XD studies [34–36], the water-water distance slightly increased with decreasing density (expansion of water due to decreasing density) from 282 pm at $1.09\,g \cdot cm^{-3}$ to 291 pm at $0.75\,g \cdot cm^{-3}$; then the water-water distance is kept practically unchanged with further decreasing water density down to $\rho = 0.71\,g \cdot cm^{-3}$. On the other hand, the number of nearest water molecules (the coordination number n) monotonously decreases with decreasing water density. The coordination number n of a water molecule is 4.4 at ambient temperature and pressure with $\rho = 1\,g \cdot cm^{-3}$, and n decreases to 1.7 at $\rho = 0.71\,g \cdot cm^{-3}$ (Fig. 3). From these results it can be concluded that water forms small clusters, in which relatively few water molecules are bound to each other through hydrogen bonds; the size of the cluster may decrease with temperature, and some other water molecules may be almost monomerly dispersed to form a gas-like phase.

It has been stated that the solubility of electrolytes decreases in supercritical water compared with that in normal water, in contrast to the increased solubility of

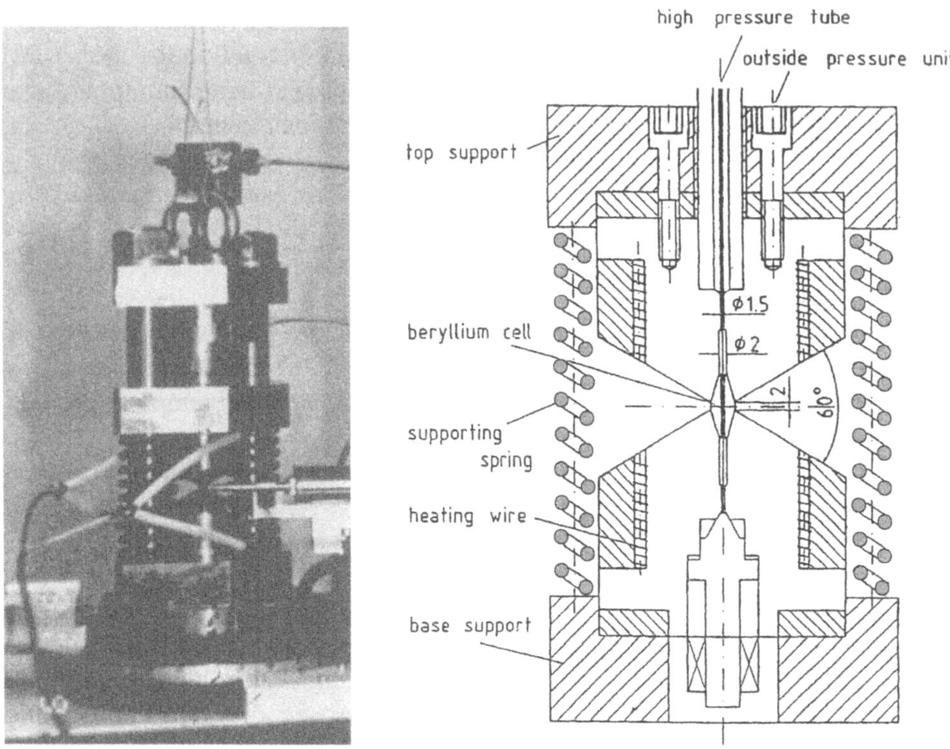

(a) (b)

Fig. 2. A high temperature-high pressure cell for X-ray diffraction experiment of liquids; (a): photograph of the cell, (b): construction and dimensions of the cell

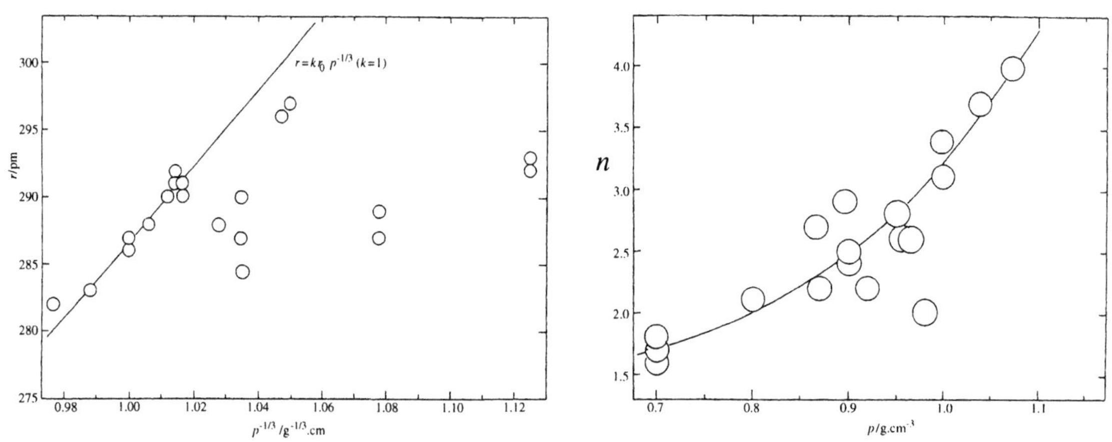

(a) (b)

Fig. 3. The nearest water-water distance r and the coordination number n at high temperatures and high pressures; (a): distance r plotted against $\rho^{-1/3}$; the straight line is drawn on the basis of the equation, $r = kr^0 \cdot \rho^{-1/3}$ where r^0 denotes the water-water distance in normal water at ambient conditions and $k = 1$; ρ represents the density of water; (b): number of water molecules at the nearest neighbor, n plotted against ρ

nonelectrolytes which hardly dissolve in water under ambient conditions. However, this expression is not clear, because the statement does not take into account the decreased density of water under supercritical conditions. It is obvious that hydration of electrolytes must be reduced in water of low density, and non-electrolytes, which are not well hydrated by water, can mix with gaseous water. Therefore, comparison between solubilities of electrolytes and non-electrolytes in supercritical water must be made at the same density of water. It is not easy to study the structure and properties of electrolyte solutions at supercritical temperatures and pressures by keeping the water density close to unity. The behaviour of hydrated ions in supercritical water has not been investigated so extensely, and a rather limited number of results have been reported. Some examples are given in the following sections.

The strontium(II) ion combines with 7.3 water molecules in aqueous 0.2 mol · dm^{-3} $Sr(NO_3)_2$ solution at room temperature, but it has only 3.5 water molecules at 385°C and 27 MPa [37]. However, it should be noted that the density of water at 385°C and 27 MPa is very low (about 0.4 g · cm^{-3}).

An EXAFS study [38] on a dilute aqueous solution of $NiBr_2$ showed that the coordination number of Ni^{2+} of 6.1 at normal conditions decreased to 2.6 at 425°C and 68.1 MPa, and a fairly large amount of Ni^{2+}–Br^- ion pairs with a Ni^{2+}–Br^- distance of 240 pm were formed. The length of the Ni^{2+}–O bonds of the hydrated nickel(II) ion in the supercritical water was found to be 209.5 pm, much longer than the Ni^{2+}–O bond length of 204 pm in a 0.1 mol · dm^{-3} solution at room temperature [10]. The density of water under these conditions is 0.65 g · cm^{-3}, and thus the concentration of water is only about 2/3 of that of normal water. A very similar result was obtained for a 0.2 mole · dm^{-3} solution of $NiBr_2$, the result indicating that the concentration dependence of the coordination number and the bond length is not significant. In the solution of $NiBr_2$ it was found that the hydration number of the nickel(II) ion decreased with increasing temperature and pressure, but the decrease in the hydration number may be more simply described in terms of the decrease in the density of water in the solution.

Ionic Solvation in Nonaqueous Solvents

Since most electrolytes are less soluble in nonaqueous solvents than in water, solvation phenomena of ions in nonaqueous solution have been much less investigated compared to those in aqueous solution. However, nonaqueous solutions of electrolytes have been interesting for physical chemists in understanding ion-molecule interactions and recognizing their importance in various industries such as material production, electrolysis, and metal plating. Furthermore, ionic solvation in nonaqueous solvents has attracted the interest of electrochemists, physical chemists, and analytical chemists as the most fundamental phenomenon of nonaqueous solution chemistry. With the development of methods for its investigation, various studies on structure and dynamics of electrolytes in nonaqueous solvents have been attempted.

Many alcohols and some nonaqueous solvents such as N,N-dimethylformamide (*DMF*), dimethyl sulfoxide (*DMSO*), or acetonitrile (*AN*) have medium relative dielectric constants of $\varepsilon_r = 35$–46 between high (water) and low (benzene and *n*-hexane) dielectric solvents, and many electrolytes can be dissolved in these solvents. The donor and acceptor properties of these medium dielectric solvents are comparable with those of water (donor numbers $D_N = 20–30$, acceptor numbers $A_N = 20–40$; water: $D_N = 18.0$, $A_N = 54.8$); thus, donor-acceptor interactions between ions and solvent molecules can be compared with those with water molecules to make the electrolyte soluble. Thermodynamic studies on electrolyte solutions in nonaqueous solvents have thus been explored rather widely [4]. Thermodynamic data are, of course, very important to understand the solution behavior of electrolytes in solvents, but they do not give a clear picture about the solvation structure. On the contrary, NMR, XD, ND, and EXAFS can provide definitive information about the solvation structure of individual ions in solution. Thermodynamic and structural data are complementary to each other, and both are indispensable for understanding the solution chemistry of electrolytes in nonaqueous solvents. However, unfortunately, compared with thermodynamic data, structural data for nonaqueous solutions are rare.

Studies the on solvation structure of ions in nonaqueous solvents have been recently attempted by using molecular dynamics simulations, but few results have been reported so far and are less reliable because the potential functions employed in the calculations were not highly polished.

The copper(II) ion forms a distorted octahedral structure in aqueous solution [23], and the ion has a similar distorted octahedral structure in *DMF* with Cu–O_{eq} and Cu–O_{ax} distances of 201–203 and 284–287 pm, respectively [39]. It has been found by an EXAFS measurement that the nickel(II) ion forms an octahedral solvation structure in various nitriles such as acetonitrile, propionitrile, butyronitrile, isobutyronitrile, valeronitrile, and benzonitrile similar to water [40]. The length of the Ni–O bond in the nitriles is 205.0–206.9 pm, which is not significantly different from the Ni–O bond length found in water (204–205 pm [40–42]).

1,3-Propanediamine (trimethylenediamine) is a bidentate ligand with two nitrogen atoms in a molecule of strong donor property. The cobalt(II) ion in 1,3-propanediamine combines with three solvent molecules to form an octahedral hexacoordinated solavte [43]. The Co–N bond length is 217 pm, which is longer than that in water in which the Co–O bond amounts to 208 pm [41, 43]. The difference in the bond lengths may arise from the bidentate character of the formation of a five-membered ring of the solvent rather than from the different atomic sizes of nitrogen and oxygen atoms in 1,3-propanediamine and water, respectively. In *n*-propylamine, the cobalt(II) ion forms an equilibrium mixture of octahedral hexacoordinated and tetrahedral tetracoordinated solvates [43]. The Co–N bond length is the same as in 1,3-propanediamine for the octahedral species, but the length in the tetrahedral one is 201 pm.

In a bulky solvent such as 1, 1, 3, 3-tetramethylurea (*TMU*) ions often cannot form an octahedral structure due to steric hindrance between ligating solvent molecules. The cobalt(II) ion forms a tetracoordinated solvate in *TMU* [44].

The solvation structure of lanthanide ions (Ln^{3+}) in *DMF* and N,N-dimethylacetamide (*DMA*) has been compared using the EXAFS method; no significant difference was observed [45].

The silver(I) ion can form various coordination structures depending on the coordinating solvent and the ligand molecules. Hard and soft properties of ligand atoms also affect the coordination structure of solvates of the silver(I) ion. The silver(I) ion forms a tetracoordinated solvate in oxygen-donating solvents such as water [46–48] and *DMF* [49] and in nitrogen-donating ones such as ammonia [50], acetonitrile [49–51], and pyridine [50, 52]. The Ag–N bond length was determined to be 224–226 pm [49, 50] in *AN* and 232 pm [52] in pyridine, which is slightly shorter than the Ag–O bond length (237–240 pm [46–48]).

The solvation structure of the silver(I) ion in *DMSO* is not conclusive at the present stage of investigation. It has been found that Ag^+ forms a tetrahedral structure in *DMSO* with four Ag–O bonds in the first coordination solvation shell by means of EXAFS [53]. On the other hand, a Ag–N bond together with three Ag–O bonds has been observed in the same solvent by XD [54]. Recently, *Persson* [55] reported that he did not find the Ag–N bond, and that the $[Ag(DMSO)_4]^+$ ion shows only the four Ag–O bonds. The structural analysis of the Ag^+ solvate in *DMSO* is presently repeated. Structural data of solvated ions in typical nonaqueous solvents are summarized in Table 5.

Preferential Solvation of Ions in Mixed Solvents

Mixed solvents have widely been used in thermodynamic, dynamic, and kinetic studies on ionic interactions in solution with changing solvent properties of the reaction medium. Dielectric constants and viscosities of solvents are changed by varying solvent composition of the mixtures. Theories proposed for explaining the variation of the thermodynamic, dynamic, and kinetic data usually assumed a homogeneous continuum with specific bulk properties, and theories which can be successfully applied to a neat medium usually fail in mixed solvent systems. The *Born* equation is not valid in other solvents than water for the calculation of solvation energies of ions, and activity coefficients in nonaqueous and mixed solvents calculated by using the *Debye-Hückel* equation do not agree with those determined experimentally. Conductivities of an ion in isoviscous mixed solvents are usually not constant. The disagreement of theoretical values with experimental ones has been traced to the inadequacy of the simple continuum assumption. The different solvent composition in the solvation shell from that in the bulk may be another important factor for causing the discrepancies.

It can easily be expected that in a mixed solvent, in which one solvent has a larger solvation affinity to an ion than the other, the former solvent may be preferentially solvated with the ion than the latter. However, there is no proper method to determine the solvent composition in the neighbourhood of an ion. The XD method can be a suitable tool to determine the solvent composition of the first solvation shell of ions. The EXAFS method, if it could be applied to mixed solvent systems of ions in an accurate and reliable way, could be useful. *Raman* spectroscopy is another powerful tool for the investigation of preferential solvation of ions.

Acetonitrile (*AN*; *Gutmann*'s donor number: $D_N(AN) = 14.1$; *Gutmann-Mayer*'s acceptor number: $A_N(AN) = 19.3$) and N,N-dimethylformamide (*DMF*; $D_N(DMF) = 26.6$, $A_N(DMF) = 16.0$) are both aprotic solvents of medium dielectric constants. In *AN-DMF* mixtures, the silver(I) ion forms $[Ag(AN)_4]^+$, $[Ag(AN)(DMF)_3]^+$, $[Ag(AN)_2(DMF)_2]^+$, $[Ag(AN)_3(DMF)]^+$, and $[Ag(DMF)_4]^+$ species depending on the solvent composition [49]. The Ag–O (*DMF*) bond length

Table 5. Selected values of ion-ligand atom distances (r_{M-X}) and number of solvent molecules (n) coordinated the ion in various solvents

Ion	*DMSO* r/pm	n	*DMF* r/pm	n	*DMA* r/pm	n	*AN* r/pm	n	*DMTF* r/pm	n	*DMPU* r/pm	n	P_y r/pm	n	*PDA* r/pm	n
Ca^{2+}	232	6	229	6												
Sr^{2+}	254	6									255	6				
Ba^{2+}	276	6									275	6				
Cu^{+}	209	4					199	4	233[m]	4						
Ag^{+}	224[a] 257[b]	4	233	4			226	4	255[m]	4	241	4				
Au^{+}							219	4	228[m]	4			216	4		
Mn^{2+}							220	6			210	5				
Fe^{2+}	210	6					214	6			202	4				
Co^{2+}	209.5	6	213	6			211.6	6			200	5			217	6
Ni^{2+}	206.3	6					206.7	6			200	5				
Cu^{2+}	{198[d] {–	6[c]	{201[d] {284	6[c]			{199[d] {–	6[c]			192	4[i]				
Zn^{2+}	212	6					208	6	236[m]	4	195	4				
Cd^{2+}	229	6	230	6					270[m]	6	229	6				
Hg^{2+}	239	6[e]							257.2[m]	4						
Hg_2^{2+}	222[f]	4					221[k]	2[l]			222	4				
Sn^{2+}									267 –	4[g]						
Pb^{2+}									289	6						
Pd^{2+}	204[h] 223	4[i]														
Pt^{2+}	207[h] 221.5	4[i]					200	4[i]	230	4[i]						
Sc^{3+}	209	6														
Fe^{3+}	202	6														
Y^{3+}	236	8[j]									224	6				

Bi^{3+}	241	8[j]					279	6	232	8
La^{3+}	250	8[j]	248.6	7.3	247.5	6.5	244	~7		
Ce^{3+}			247.4	8.1	245.2	6.6				
Pr^{3+}			244.7	7.4	242.0	6.9				
Nd^{3+}			243.8	7.4	240.2	6.1				
Sm^{3+}			241.6	8.9	236.7	8.4				
Eu^{3+}			238.5	7.7	234.7	6.7				
Gd^{3+}			238.5	7.5	233.3	7.2				
Tb^{3+}			236.9	7.5	231.9	7.3				
Dy^{3+}			236.0	7.7	230.5	7.0				
Ho^{3+}			234.6	7.8	229.3	6.9				
Er^{3+}			233.6	7.2	227.3	7.6				
Tm^{3+}			232.5	7.5	226.7	6.8				
Yb^{3+}			230.9	7.5	223.8	6.3				
Lu^{3+}			229.8	7.9	222.1	6.0				
Ga^{3+}	196	6					224	4	192	5
In^{3+}	214	6					261	6	214	6
Bi^{3+}	241	8[j]					279	6	233	6

DMSO: dimethyl sulfoxide; *DMF*: N,N-dimethylformamide; *DMA*: N,N-dimethylacetamide; *AN*: acetonitrile; *DMTF*: N,N-dimethylthioformamide; *DMPU*: 1,3-dimethylpropylene urea; *Py*: pyridine; *TMU*: tetramethylurea; *PDA*: 1,3-propanediamine; [a]three *M*–O bonds; [b]one *M*–S bond; [c]*Jahn-Teller* distorted octahedron; [d]four equatorial Cu–O bonds; [e]second-order *Jahn-Teller* distorted octahedron; [f]Hg–Hg bond length: 250 pm; [g]two bonds are longer than the other two; [h]two *M*–O bonds and two *M*–S bonds; [i]square-planar. [j]square antiprism; [k]Hg–Hg bond length: 254 pm; [l]linear; [m]*M*–S bond; [n]square pyramid; [o]bidentate ligand

Table 6. Preferential solvation of ions in binary **A**–**B** mixtures

Ion	Mole fraction of solvent **A** in the bulk	Main species in solution	Average metal-ligating atom distance r/pm	Mole fraction of solvent **A** in the first solvation shell
	AN-DMF Mixture [1] ($\mathbf{A} = AN, \mathbf{B} = DMF$)			
Ag$^+$	0.00	$[Ag(DMF)_4]^+$	Ag–O: 233	0.00
	0.15	$[Ag(AN)(DMF)_3]^+$	Ag–O: 235 Ag–N: 226	0.2(5)
	0.25	$[Ag(AN)_2(DMF)_2]^+$	Ag–O: 237 Ag–N: 224	0.5(0)
	0.33	$[Ag(AN)_2(DMF)_2]^+$	Ag–O: 237 Ag–N: 224	0.5(0)
	0.50	$[Ag(AN)_2(DMF)_2]^+$(~ 50%) $[Ag(AN)_3(DMF)]^+$(~ 50%)	Ag–O: 237 Ag–N: 224 Ag–O: 241 Ag–N: 224	0.6(3)
	0.67	$[Ag(AN)_3(DMF)]^+$	Ag–O: 241 Ag–N: 224	0.7(5)
	1.00	$[Ag(AN)_4]^+$	Ag–N: 226	1.00
	DMSO-DMF Mixture [2] ($\mathbf{A} = DMSO, \mathbf{B} = DMF$)			
Ca^{2+}	0.00	$[Ca(DMF)_6]^{2+}$	Ca–O: 229	0.00
	0.25[a]	$[Ca(DMF)_4(DMSO)_2]^{2+}$ and/or $[Ca(DMF)_3(DMSO)_3]^{2+}$	Ca–O: 228–229	0.3(3)–0.5(0)
	0.50	$[Ca(DMF)_2(DMSO)_4]^{2+}$	Ca–O: 230	0.6(7)
	0.75	$[Ca(DMF)(DMSO)_5]^{2+}$	Ca–O: 232	0.8(3)
	1.00	$[Ca(DMSO)_6]^{2+}$	Ca–O: 232	1.00

[a]Not distinguishable between the two species, probably approximately 1:1 mixture to be formed
[1] Ozutsumi K, Kitakaze A, Iinomi M, Ohtaki H (1997) J Mol Liquids **73/74**: 385; [2] Ozutsumi K, Ikuno N, Suzuki T, Ohtaki H (1999) 49th National Conference on Coordination Chemistry, September, Sapporo, Japan

was determined to be 233–241 pm, whereas the Ag–N (*AN*) bond was 224–226 pm, and the Ag–O bond is gradually elongated by the stepwise addition of acetonitrile molecules to the solvation shell. The solvent composition of the first solvation shell of the main species formed in various *AN-DMF* mixtures, together with the lengths of individual Ag–O and Ag–N bonds, are summarized in Table 6. The accuracies of the results are not too high, but the variation of the solvent composition of the solvation shell with the composition in the bulk can be recognized.

It shall now be attempted to introduce a quantitative measure of preferential solvation of ions. When the solvent composition of an **A**-**B** mixture in the bulk is defined to be $[\mathbf{A}]_B$ and $[\mathbf{B}]_B$, where $[\ ]_B$ denotes the concentration of a solvent in the bulk (*e.g.* mole fraction), and the concentrations of solvents **A** and **B** in the first

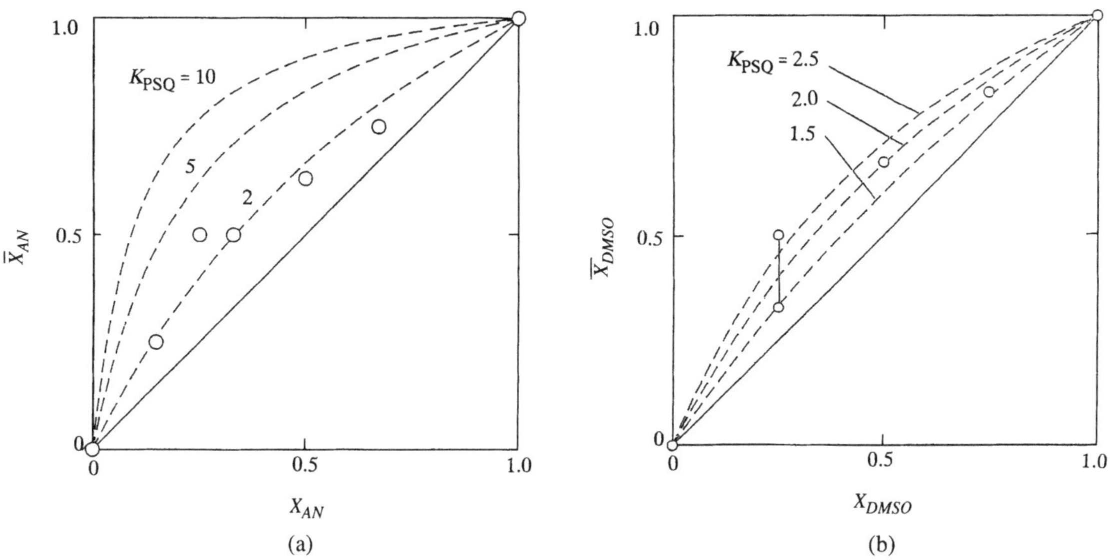

Fig. 4. Plots of mole fraction of a solvent **A** in the solvation shell, $\bar{x}_{\mathbf{A}}$, against mole fraction of solvent **A** in the bulk $x_{\mathbf{A}}$; (a): Ag(I) in *AN-DMF* mixtures; (b): Ca^{2+} in *DMSO-DMF*

solvation shell are denoted by $[\mathbf{A}]_{SS}$ and $[\mathbf{B}]_{SS}$, respectively, the preferential solvation quotient K_{PSQ} can be defined as follows:

$$K_{PSQ} = \frac{[\mathbf{A}]_{SS}/[\mathbf{B}]_{SS}}{[\mathbf{A}]_B/[\mathbf{B}]_B} = \frac{\bar{x}_{\mathbf{A}}/(1-\bar{x}_{\mathbf{A}})}{x_{\mathbf{A}}/(1-x_{\mathbf{A}})} \tag{3}$$

$\bar{x}_{\mathbf{A}}$ and $x_{\mathbf{A}}$ denote the mole fraction of solvent **A** in the first solvation shell and in the bulk, respectively. When K_{PSQ} is larger than unity, the ion is preferentially solvated with solvent **A**.

A plot of $\bar{x}_{AN}$ in the first solvation shell of the Ag^+ ion against x_{AN} in the bulk solvent is given in Fig. 4a. All experimental points fall on the line drawn of K_{PSQ} being about 2. If it is assume that the preferential solvation occurs due to the difference in the *Gibbs* energies of solvation of the two solvents towards the Ag^+ ion, the preferential solvation quotient can be given as the difference in the *Gibbs* free energies of solvation of *AN* and *DMF*, and the difference, $\Delta\Delta G^0_{solv}$, can be evaluated from the *Gibbs* free energies of transfer of the Ag^+ ion from water to *AN* and from water to *DMF*. Thus,

$$-RT\ln K_{PSQ} = \Delta\Delta G^0_{solv} = \Delta G^{tr}_{Ag}(H_2O \to AN) - \Delta G^{tr}_{Ag}(H_2O \to DMF) \tag{4}$$

Insertion of the reported values of $\Delta G^{tr}_{Ag}(H_2O \to AN)$ and $\Delta G^{tr}_{Ag}(H_2O \to DMF)$ (-23.2 and $-20.8\,kJ\cdot mol^{-1}$, respectively [4]) into Eq. (4) leads to $\Delta\Delta G^0_{solv} = -2.4\,kJ\cdot mol^{-1}$. The value of $-RT\ln K_{PSQ}$ at $K_{PSQ} = 2$ is $-1.7\,kJ\cdot mol^{-1}$, which is fairly close to the value of $\Delta\Delta G^0_{solv}$.

The calcium ion is a typical divalent cation. The solvation number of Ca^{2+} has been reported to be six or eight depending on solvent, temperature, and pressure. In *DMSO-DMF* mixtures the Ca^{2+} ion is preferentially solvated with *DMSO* over *DMF* (see Table 6 and Fig. 4b). The K_{PSQ} value estimated from the plot is about 2. Therefore, $-RT\ln K_{PSQ} = \Delta\Delta G^0_{solv}$ can be evaluated to be $-1.7\,kJ\cdot mol^{-1}$. The

Gibbs free energy of transfer of Ca^{2+} from *DMF* to *DMSO* has not been reported in the literature, but it may be roughly estimated from the values of alkali metal ions to be about -2–$-3\,kJ \cdot mol^{-1}$ (-5.0 [Li^+], -3.8 [Na^+], -2.7 [K^+], -0.7 [Rb^+], and $-2.2\,kJ \cdot mol^{-1}$ [Cs^+]), which is again close to the value evaluated from the plot of Fig. 4b ($-1.7\,kJ \cdot mol^{-1}$). Therefore, it can be concluded that preferential solvation occurs due primarily to the difference in the *Gibbs* energies of solvation of the two solvents under examination [56].

Solvent-solvent interaction is another important factor controlling preferential solvation of ions. 1, 1, 3, 3-Tetramethylurea ($D_N(TMU) = 31$) is a much stronger donor solvent than water ($D_N(H_2O) = 18$), and thus *TMU* is expected to be preferentially solvated with cations over water. However, in H_2O-*TMU* mixtures with a larger proportion of water, the Co(II) ion preferentially solvated with water. The results can be explained in terms of strong *TMU*-H_2O ($A_N(H_2O) = 54.8$) intermolecular interactions in the bulk, which result in the disappearance of free *TMU*. In mixtures containing *TMU* as a major portion, Co(II) is preferentially solvated with *TMU* [57]. Specrtophotometric evidence shows that Co(II) forms a tetrasolvated $[Co(TMU)_4]^{2+}$ complex in neat *TMU* [57].

Yokoyama et al. [58] have determined the solvent composition in the solvation shell of Co^{2+} in various amide-water mixtures by XD, the plots of $\bar{x}_{amide}$ against x_{amide} being shown in Fig. 5. The *Gibbs* free energy of solvation or *Gutmann*'s donor number of these amides cannot be significantly different, but they have different abilities for hydrogen-bonding with water molecules due to different numbers of

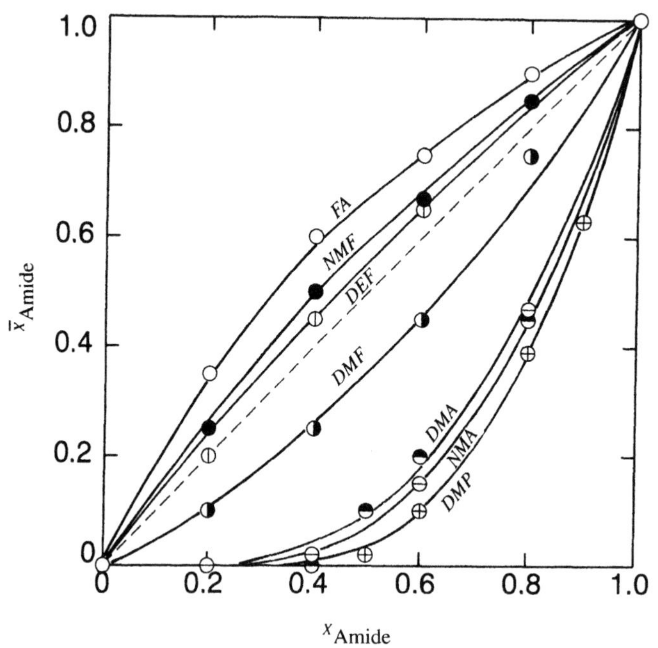

Fig. 5. Plots of mole fraction of an amide in the solvation shell, $\bar{x}_{Amide}$, of Co(II) against mole fraction of the amide in the bulk x_{Amide}; ○: formamide (*FA*), ●: N-methylformamide (*NMF*), ⦶: N,N-diethylformamide (*DEF*), ◑: N,N-dimethylformamide (*DMF*), ◓: N,N-dimethylacetamide (*DMA*), ⊖: N-methylacetamide (*NMA*), ⊕: N,N-dimethylpropionamide (*DMP*); quoted from Ref. [57]

free amino groups in their molecules. Moreover, they have different substituents which result in different volumes of bulkiness of the molecules; this may cause a steric hindrance in the rather crowded solvation sphere of the ion. They explained their results primarily in terms of different solvation abilities of the amides, which may be more or less proportional to *Gutmann*'s D_N. The preferential solvation of formamide $(FA, D_N = 24)$ over water $(D_N = 18)$ can be evidence thereof. When amino protons are substituted with alkyl groups, the amino group in the molecule attracts a lower number of water molecules in the bulk, so that the concentration of free water molecules increases. Thus, Co^{2+} can be solvated with more water than amide molecules. This explanation is similar to that given for the behaviour of the cobalt(II) ion in *TMU*-H_2O mixtures [57]. The different volumes of solvent molecules with bulky substituents may be an additional reason for causing different degrees of preferential solvation of the amides. The reason why N,N-diethylformamide (*DEF*) is advantageous in preferential solvation over water compared with N,N-dimethylformamide (*DMF*) (see Fig. 5) is not clear, but hydrophobicity of the longer alkyl chains attached to the nitrogen atom in the former may play a significant role.

Raman spectroscopy can be another useful technique for the study of preferential solvation of ions. *Umebayashi et al.* [59] have applied the method to the determination of the total solvation number of various ions together with the composition of the solvent molecules in the first solvation shell. They determined the total number of solvated molecules in the first coordination sphere, as well as the numbers of individual solvent molecules of Mn(II), Ni(II), Cu(II), and Zn(II) in *DMF-DMA* and *DMF-TMU* mixtures. In Fig. 6 their results for Mn(II) in *DMF-DMA* and *DMF-TMU* (Figs. 6a and 6b, respectively) and for Ni(II) in *DMF-DMA* mixtures (Fig. 6c) are quoted [59].

It should be noted in that Mn(II) shows no preferential solvation in *DMF-DMA* mixtures in spite of the larger donor number of *DMA* $(D_N = 28.7)$ compared to *DMF* $(D_N = 26.6)$ (Fig. 6a). The result suggests that the energy lost due to the steric hindrance of bulky *DMA* in the first solvation shell may approximately be

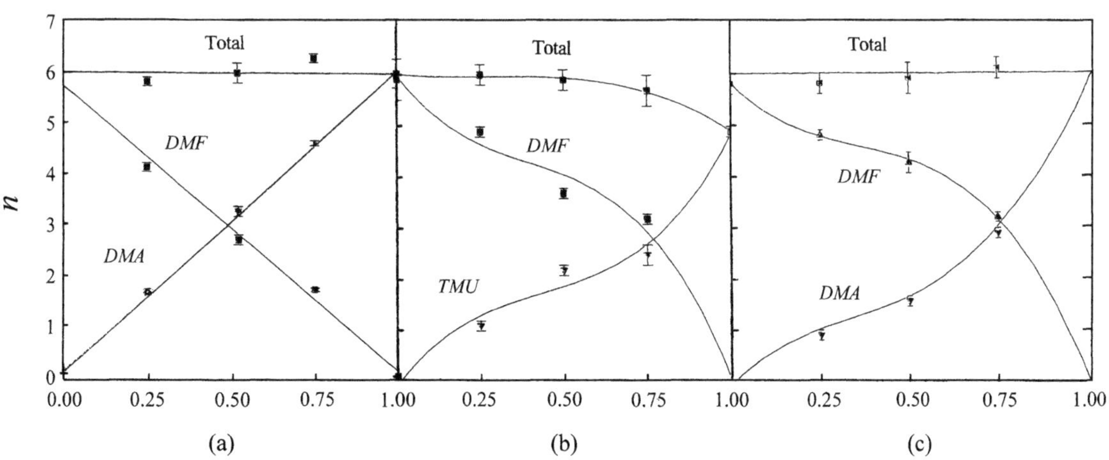

Fig. 6. Solvation of Mn(II) and Ni(II) in binary mixtures; (a): Mn(II) in *DMF-DMA*, (b): Mn(II) in *DMF-TMU*, (c): Ni(II) in *DMF-DMA*; quoted from Ref. [58]

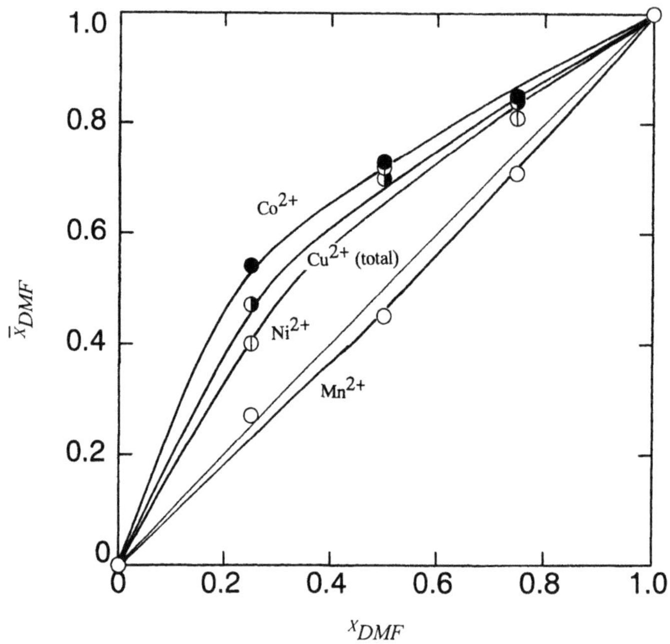

Fig. 7. Plots of mole fraction of *DMF* in the solvation shell, $\bar{x}_{DMF}$, of divalent transition ions against mole fraction of *DMF* in the bulk, x_{DMF}; Co^{2+} (●), Cu^{2+} (total) (◑), Ni^{2+} (⦶), and Mn^{2+} (○); quoted from Ref. [61]

compensated by an energy proportional to the difference in the donor properties of the two solvents ($2.1\,\mathrm{kcal \cdot mol^{-1}} = 5.0\,\mathrm{kJ \cdot mol^{-1}}$). The steric effect caused by the bulky methyl group in *DMA* may hinder the rotation of the *DMA* molecule in the solvation sphere along the -C–N- axis, which may result in a decrease in the entropy of solvation which is more pronounced for *DMA* than for *DMF*. Therefore, $\Delta\Delta G^0_{\mathrm{solv}}$ of the Mn(II) ion should be much less than the expected value from the difference in D_N of *DMA* and *DMF*. The bulkier *TMU* ($D_N = 31$) is less preferentially solvated to the Mn(II) ion than *DMF* in spite of a larger donor property in the former (Fig. 6b). The bulkiness of the substitutents in *TMU* causes a decrease in the total solvation number of Mn(II). In the case of Ni(II) (see Fig. 6c), the total solvation number of Ni(II) remains unchanged over the whole composition range, and *DMF* is preferentially solvated to the Ni(II) ion than *DMA*. Again, *DMA* with its larger donor number is less preferred by the Ni(II) ion due to its bulky volume.

Plots of $\bar{x}_{DMF}$ *vs.* x_{DMF} in *DMF-DMA* mixtures recalculated from the data reported by *Umebayashi et al.* [59] for Co^{2+}, Cu^{2+}, Ni^{2+}, and Mn^{2+} are shown in Fig. 7. All these ions have an octahedral structure, although the structure of the Cu^{2+} ion is distorted by the *Jahn-Teller* effect, an effect which will be discussed in more detail later. The order of preferential solvation of *DMF* over *DMA* is $Co^{2+} > Ni^{2+} \gg Mn^{2+}$. *Umebayashi* and coworkers [59] have explained the smallest preferential solvation of Mn^{2+} (ionic radius: 81 pm) in terms of the large size of the ion. However, Co^{2+} (79 pm) has a larger ionic size than Ni^{2+} (70 pm).

The behavior of the zinc(II) ion is unique in *DMF-DMA* mixtures. It has been observed that the coordination number of the Zn(II) ion decreases from six in neat *DMF* to four in neat *DMA* (Fig. 8a), which shows that the Zn(II) ion prefers *DMF*

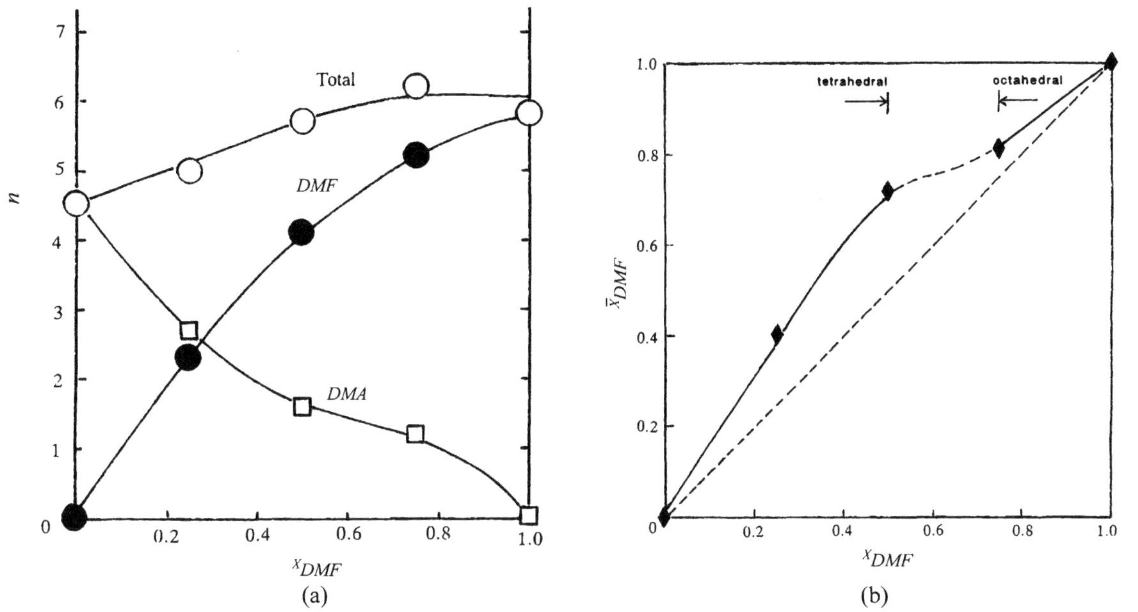

Fig. 8. Solvent composition in the solvation shell of Zn(II) ion in *DMF-DMA* mixtures; (a): total number of solvent molecules in the first solvation shell and the numbers of individual solvents; (b): $\bar{x}_{DMF}$ *vs.* x_{DMF}; quoted from Ref. [61]

over *DMA* in spite of a larger donicity of *DMA* [59]. A plot of $\bar{x}_{DMF}$ *vs.* x_{DMF} (Fig. 8b) gives a curve with an inflexion point around $x_{DMF} = 0.5$–0.7, probably due to the change in the coordination structure from octahedral to tetrahedral.

The enthalpy of transfer, $\Delta H^{tr}_{Mn}(DMF \rightarrow DMA)$, of Mn(II) from *DMF* to *DMA* at 25°C [60] is negative ($-6.9\,kJ \cdot mol^{-1}$), but the values of other divalent ions are positive (7.9 for Co(II), 17.4 for Ni(II), and $21.8\,kJ \cdot mol^{-1}$ for Zn(II)). The results show that divalent metal ions, except for Mn(II), are stabilized in *DMF* better than *DMA*, which is contrary to a result expected from the solvation abilities of *DMF* and *DMA*.

Solvation of Cu(II) changes in a rather complicated way in *DMF-DMA* [61]. Since Cu(II) has a distorted octahedral structure with two elongated axial and four shortened equatorial solvent molecules, replacement of one solvent with another occurs differently at the axial and equatorial positions, the latter showing a stronger ion-solvent bonding than the former. According to their result, *DMF* is preferentially solvated at the equatorial position, whereas *DMF* less favorably solvates Cu(II) than *DMA* at the axial position, where less steric hindrance among solvated molecules is expected. In the average, *DMF* is preferentially solvated with Cu(II) in *DMF-DMA* mixtures (Fig. 9).

Concluding Remarks

The study of ionic solvation is traditional but still modern, and it is one of the most fundamental subjects in solution chemistry. The extension of studies in aqueous solution to nonaqueous and mixed solvents leads to a widened view for chemistry in aqueous solution itself. Expansion of solvent composition from single-component to

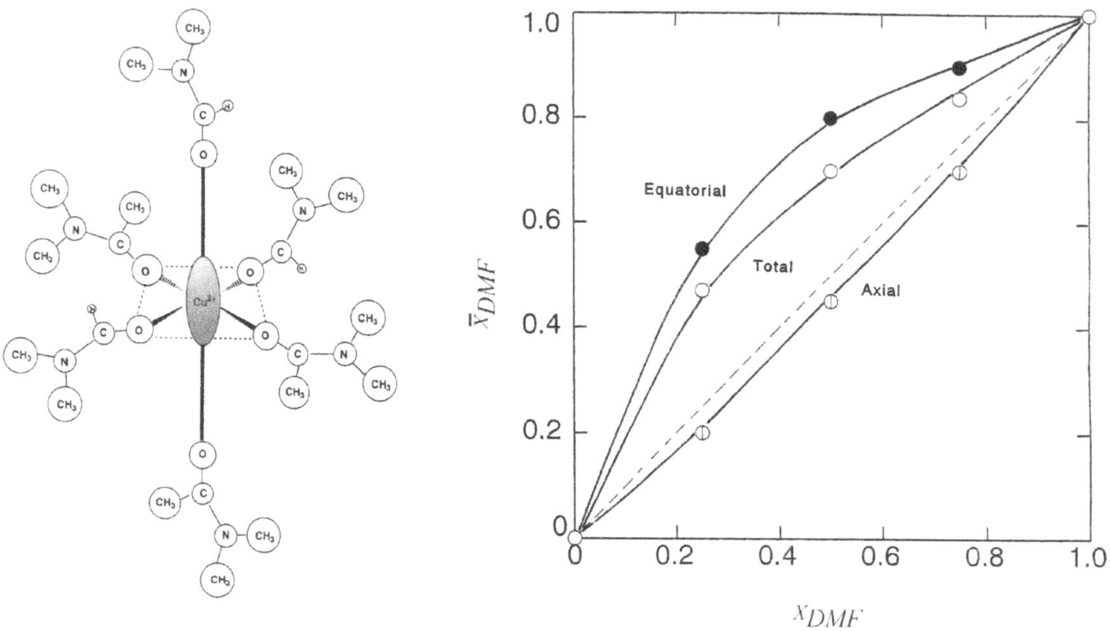

Fig. 9. Preferential solvation of the Cu(II) ion in *DMF-DMA* mixtures; (a): a structural model of a mixed solvated, distorted octahedral Cu(II) ion; (b): plots of $\bar{x}_{DMF}$ *vs.* x_{DMF} for the total composition of Cu(II), solvent composition at the equatorial and axial positions of the Cu(II) ion; quoted from Ref. [61]

multi-component systems brings new interesting results. Ionic solvation is a subject of not only solution chemistry but also analytical chemistry, physical chemistry, electrochemistry, and even industrial and environmental chemistry. Studies on ionic solvation should be much more deeply investigated over wide ranges of temperature, pressure, and solvent composition in the future to present interesting results.

Acknowledgements

The author expresses his deepest thanks to Prof. *Kazuhiko Ozutsumi* of Ritsumeikan University and Dr. *Tamas Radnai* of the Hungarian Academy of Sciences, Budapest, for their long lasting cooperation and support. He thanks his colleagues at the periods of his academic lives at the Tokyo Institute of Technology, Institute for Molecular Science, and Ritsumeikan University. He is especially indebted to Prof. *Ingmar Persson* of the Swedish University of Agricultural Sciences, Uppsala, Sweden, for his helpful contribution for the tabulation of structural data of solvated ions in nonaqueous solutions (Table 5). He also thanks Prof. *Haruhiko Yokoyama* of Yokohama City University and Prof. *Shin-ichi Ishiguro* and Dr. *Yasuhiro Umebayashi* of Kyushu University for providing unpublished data.

References

[1] Burgess JH (1978) Metal Ions in Solution. Ellis Horwood, New York
[2] Conway BE (1981) Ionic Hydration in Chemistry and Biophysics. Elsevier, Amsterdam
[3] Burgess J (1988) Ions in Solution: Basic Principles of Chemical Interactions. Ellis Horwood, Chichester, UK
[4] Marcus Y (1988) Ion Solvation. Wiley, Chichester, UK

[5] Bockris JO'M, Reddy AKN (eds) (1998) Modern Electrochemistry, Ionics, vol 1, 2nd edn. Plenum, New York
[6] Ohtaki H (1985) Yoeki Kagaku (Solution Chemistry) Shokabo, Tokyo (in Japanese)
[7] Ohtaki H (1987) Yoeki no Kagaku (Chemistry of Solutions) Dai-Nippon Tosho, Tokyo (in Japanese)
[8] Ohtaki H (1990) Ion no Suiwa (Hydration of Ions) Kyoritsu Shuppan, Tokyo (in Japanese)
[9] Marcus Y (1993) Chem Rev **88**: 1475
[10] Ohtaki H, Radnai T (1993) Chem Rev **93**: 1157
[11] Chemical Society of Japan (ed) (1976) Ion to Yobai (Ions and Solvents) Kagaku Sosetsu No. 11, Gakkai Shuppan Center (in Japanese)
[12] Chemical Society of Japan (ed) (1995) Yoeki no Bunshiron-teki Byozo (Molecular Pictures of Solutions) Kikan Kagaku Sosetsu No. 25. Gakkai Shuppan Center (in Japanese)
[13] Arrhenius S (1887) Z Phys Chem **1**: 631
[14] Mayer AW (1930) Ann Phys **5**: 701
[15] Stewart GW (1930) Phys Rev **35**: 1462; (1931) **37**: 9
[16] Amaldi W (1931) Phys Zeitschrift **32**: 914
[17] Bernal JD, Fowler RH (1933) J Chem Phys **1**: 515
[18] Debye P, Hückel E (1923) Phys Z **24**: 185, 305
[19] Bjerrum N (1926) Det Kgl Danske Videnskab Math-Fys Medd **7**: 1
[20] Born M (1920) Z Physik **1**: 45
[21] Frank HS, Wen W-Y (1957) Disc Faraday Soc **24**: 133
[22] Samoilov OY (1957) Disc Faraday Soc **24**: 141
[23] Ohtaki H, Maeda M (1974) Bull Chem Soc Jpn **47**: 2197
[24] Enderby JE (1985) 7th International Symposium on Solute-Solute-Solvent Interactions, Reading, July 15–19
[25] Eigen M, DeMaeyer L (1955) Z Elektrochem **59**: 986
[26] Stillinger FH, Rahman A (1974) J Chem Phys **60**: 1545
[27] Kern CW, Karplus M (1972) In: Franks F (ed) The Water Molecule in Water – A Comprehensive Treatise, vol I. Plenum, New York
[28] Matsuoka O, Clementi E, Yoshimine M (1976) J Chem Phys **64**: 1351
[29] Basolo F, Pearson RG (1958) Mechanisms of Inorganic Reactions. Wiley, New York
[30] Akesson R, Pettersson LGM, Sandström M, Siegbahn PEM, Wahlgren U (1993) J Phys Chem **97**: 3765
[31] Akesson R, Pettersson LGM, Sandström M, Wahlgren U (1994) J Am Chem Soc **116**: 8705
[32] Franks F (1982) Water-A Comprehensive Treatise, vol 7. Plenum, New York, p 215
[33] Angel CA (1983) Ann Phys Chem **34**: 593
[34] Ohtaki H, Radnai T, Yamaguchi T (1997) Chem Soc Rev 41
[35] Nakahara M, Yamaguchi T, Ohtaki H (1997) Recent Res Devel Phys Chem **1**: 17
[36] Radna T, Ohtaki H (1996) Mol Phys **87**: 103
[37] Pfund DM, Darab JG, Fulton JL, Ma Y (1994) J Phys Chem **98**: 13102
[38] Wallen SL, Palmer BJ, Fulton JL (1998) J Chem Phys **108**: 4039
[39] Ozutsumi K, Ishiguro S, Ohtaki H (1988) Bull Chem Soc Jpn **61**: 945
[40] Inada Y, Funahashi S (1997) Anal Sci **13**: 373
[41] Ohtaki H, Yamaguchi T, Maeda M (1976) Bull Chem Soc Jpn **49**: 701
[42] Soper AK, Neilson GW, Enderby JE, Howe RA (1977) J Phys Chem Solid State Phys **10**: 1793
[43] Aizawa S, Iida S, Matsuda K, Funahashi S (1996) Inorg Chem **35**: 1338
[44] Ozutsumi K, Asaka A, Iinomi N, Ohtaki H (1996) Rits SR Center Activity Report, Ritsumeikan Univ, p 39
[45] Ishiguro S, Umebayashi Y, Kato K, Takahashi R, Ozutsumi K (1998) J Chem Soc Faraday Trans **94**: 3607
[46] Yamaguchi T, Lindqvist O, Boyce JB, Claesson T (1984) Acta Chem Scand **A38**: 423

[47] Yamaguchi T, Johansson G, Holmberg B, Maeda M, Ohtaki H (1984) Acta Chem Scand **A38**: 437
[48] Sandström M, Neilson GW, Johansson G, Yamaguchi T (1985) J Phys Chem Solid State Phys **18**: L1115
[49] Ozutsumi K, Kitakaze A, Iinomi M, Ohtaki H (1997) J Mol Liquids **73/74**: 385
[50] Yamaguchi T, Wakita H, Nomura M (1988) J Chem Soc Chem Commun 433
[51] Nilsson K, Oskarsson A (1984) Acta Chem Scand **A38**: 79
[52] Nilsson K, Oskarsson A (1982) Acta Chem Scand **A36**: 605
[53] Tsutsui Y, Sugimoto K, Wasada H, Inada Y, Funahashi S (1997) J Phys Chem A **101**: 2900
[54] Ozutsumi K, Ohtaki H (1999) 26th International Conference on Solution Chemistry, July 26–31, Fukuoka, Japan
[55] Persson I (private communication)
[56] Ozutsumi K, Ikuno N, Suzuki T, Ohtaki H (1999) 49th National Conference on Coordination Chemistry, September 24–26, Sapporo, Japan
[57] Ikuno N, Asaka T, Ueohmae M, Ozutsumi K, Ohtaki H (1998) 21st National Symposium on Solution Chemistry, Sapporo, October 8–10
[58] Yokoyama H, Tada Y (1998) 65th Annual Meeting of Electrochemical Society of Japan, Tokyo, April 1–3, Tokyo; Yokoyama H, Tada Y, Ohtsu H, Tanabe M, Hirano R (1998) 21st National Symposium on Solution Chemistry, Sapporo, October 8–10; 26th International Conference on Solution Chemistry, July 26–31, 1999, Fukuoka, Japan
[59] Umebayashi Y, Matsumoto K, Watanabe M, Katoh K, Ishiguro S (2001) Anal Sci **17**: 15; Umebayashi Y, Watanabe M, Ishiguro S (1999) 26th International Conference on Solution Chemistry, July 26–31, Fukuoka, Japan
[60] Koide M, Ishiguro S (1995) J Solution Chem **24**: 511
[61] Umebayashi Y, Ishiguro S (private communication)

Received January 16, 2001. Accepted January 31, 2001

Invited Review

Spin Equilibrium in Solutions

Hans Toftlund[*,a]

Department of Chemistry, University of Southern Denmark, DK-5230 Odense, Denmark

Summary. The newer literature on spin equilibria of Fe(II) complexes in solution is reviewed. The thermodynamics as well as the kinetics in different solvents are discussed.

Keywords. Spin crossover; Equilibrium; Kinetics; Liquid state.

Introduction

The spin crossover phenomenon has been studied very intensely during the last decade. The main focus has been on the solid-state behaviour [1], but some important investigations on spin change in solution have also been performed. Since most of the older work in this field has been covered in a review in 1989 [2], the present paper will mainly describe recent accomplishments.

Octahedral iron(II) complexes showing equilibrium between $^1A_{1g}$ and $^5T_{2g}$ are by far the most intensively studied examples of the spin crossover phenomenon. There are many reasons for that; some of them will be explained below. With this in mind it is surprising that the oldest known compounds showing the crossover phenomenon are actually a series of iron(III)dithiocarbamate complexes. This type of complexes was discovered by the French chemist *Marcel Delépine* in 1908 [3], but their anomalous magnetic properties were not described until 1931 [4]. The first example of an iron(II) spin crossover system is usually assumed to be [Fe(*phen*)$_2$(NCS)$_2$], which was investigated by *Madeja* and *König* [5] in 1963. The phenomenon was, however, predicted by *Orgel* [6] seven years before the first system was actually characterized. A more careful examination of the older literature shows that Fe(II) compounds nowadays known to be spin crossover compounds have been prepared as early as 1938. *Paul Pfeiffer*, the well-known student and collaborator of *Alfred Werner*, reported on *tris*-(2-methyl-1,10-phenanthroline)-iron(II) and pointed out that the absorption spectrum of this complex is much weaker than that of its unsubstituted 1,10-phenanthroline iron(II) analogue [7]. He even noticed the solvatochromism of this system: whereas a methanolic solution of the perchlorate salt is orange, a solution in nitromethane is

* E-mail: hto@dou.dk

[a] Dedicated to the memory of Prof. *Christian K. Joergensen*

yellow. Later, *R. J. P. Williams* noticed the unusual high magnetic moment of these complexes, and he actually suggested spin crossover as a possible explanation for this phenomenon [8]. These early observations have since been confirmed to be due to spin crossover [9].

Thermodynamics

The requirement for the observation of the spin equilibrium phenomenon is that the potential minima of the low-spin (LS) and high-spin (HS) surfaces are so close in energy that both can be thermally populated. In the case of Fe(II), the two relevant states are $^1A_{1g}$ and $^5T_{2g}$. Due to the large difference in the number of available electronic and vibrational levels in these two states, the thermal transformation from $^1A_{1g}$ to $^5T_{2g}$ is to a large extend entropy driven, and the equilibrium is most clearly observed if the singlet state is the real ground state. The strategy usually used in creating new crossover systems is to fine-tune the ligand field strength of a ligand system known to be close to the crossover point by chemical modifications. Considering the large number of Fe(II) spin crossover systems known today it is clear, from pure statistical reasons, that many equilibrium systems with the quintet state as the true ground state must also exist. Because of the experimental difficulties in detecting such a system, practically no compounds of this type are described in the literature. One simple way to test if a given system belongs to this category is to change the relative energy of the two states by pressure. A short metal-ligand distance favours the LS state; pressure will therefore stabilize the LS state, resulting in a pressure induced HS-to-LS transformation if the energies of the two states are sufficiently close to each other at ambient pressure.

The thermodynamic parameters for a series of iron(II) complexes for which the $^1A_{1g} \rightleftharpoons {}^5T_{2g}$ equilibrium has been studied in solution are listed in Table 1. The data are typically obtained from UV/Vis, magnetic susceptibility, or NMR data. In contrast to the solid-state behaviour, solvent and counter ion effects are rather modest in dilute solutions. Since no cooperativity is present in solutions, all reported transition curves are gradual. *Linert et al.* have analyzed the possible effects of non-ideality on the shape of the transition curve. Effects from the internal pressure are predicted to result in non-gradual spin crossover for concentrated solutions [33]. In a few cases, dramatic solvent effects have been reported even for dilute solutions. Typically, one of the solvents is very polar, and substitution reactions where some of the ligands are replaced by solvent molecules cannot be ruled out. *Linert et al.* have investigated [Fe(*pyben*H)$_3$]$^{2+}$ [20] and [Fe(*bispybim*)]$^{2+}$ [28] in a number of different solvents (for abbreviations, see Appendix).

In these cases, interactions between the benzimidazole NH function and the different solvents seem to modify the ligand field strength and thereby change the ratio between HS and LS in different solutions. A set of data on [Fe(*pyben*H)$_3$]$^{2+}$ in *DMF* from an other group [21] shows unreasonably large ΔH^0 and ΔS^0 values; in this case ligand, dissociation is likely to have occurred. Obviously, data of this latter type should not be considered as true spin crossover interconversions. Ligand dissociation and replacement reactions are more likely to occur for complexes of

Table 1. Thermodynamic parameters for some Fe(II) spin crossover systems in solution

Complex	Solvent	ΔH^0/kJ · mol^{-1}	ΔS^0/J · mol^{-1} · K^{-1}	Ref.
[Fe(*pyim*H)$_3$]$^{2+}$	$(CH_3)_2CO$	15.9	48.3	[19]
[Fe(*pyben*H)$_3$]$^{2+}$	$(CH_3)_2CO$	19.7	78.0	[19]
[Fe(*pyben*H)$_3$]$^{2+}$	*DMF*	42.7	140.9	[21]
[Fe(*bispybim*)]$^{2+}$	$(CH_3)_2CO$	22.5	68.1	[33]
[Fe(HB(*pz*)$_3$)$_2$]	$(CH_3)_2CO$	16.1	47.7	[22]
[Fe(HC(*pz*)$_3$)$_2$]$^{2+}$	MeCN	18	53	[14]
[Fe(HC(3,5-Me$_2$*pz*)$_3$)$_2$]$^{2+}$	*DMF*	20	58	[13]
[Fe(*tacn*)$_2$]$^{2+}$	D_2O	23	67	[23]
[Fe(*tp*[10]*ane*N$_3$)]$^{2+}$	EtCN	23.6	84	[24]
[Fe(*lpp*[9]*ane*N$_3$)]$^{2+}$	MeOH	17.1	59	[25]
[Fe(*tpt*Me*tame*)]$^{2+}$	EtCN	19.4	85	[11]
[Fe(*btpa*)]$^{2+}$	MeOH	27.6	89	[26]
[Fe((NH$_2$)$_2$*sar*)]$^{2+}$		12	30	[27]
[Fe Fe(*L*)$_3$]$^{4+a}$	MeCN	24	72.9	[36]
[Fe Fe(*L*)$_3$]$^{4+b}$	MeCN	41	104.6	[36]

[a] First step; [b] Second step

uni- and bidentate ligands, but even for multidentate ligands, replacement of a single chelate arm has been observed [10].

It is obvious that the accessible temperature range is limited by the boiling and freezing points of the solvents; therefore, certain selections of systems with optimal parameters have been performed. The spin transition in solution is so gradual that it will extend over more than hundred degrees in the accessible temperature range. Even if the critical temperature T_c (the temperature where the HS and LS concentrations are equal) is situated in the middle of the chosen temperature range it is often difficult to cover the complete spin transition. This usually results in a loss of accuracy in the determinations of the equilibrium constants. Nevertheless, quite good and straight *van't Hoff* plots are obtained in most cases. The quality of the data ought to be reflected in good linear isokinetic plots (ΔH^0 *vs.* ΔS^0). Such a plot based on the data of Table 1 shows quite a large scatter. However, if the systems are arranged according to the nature of ligands, much better correlations are obtained. Systems with bidentate and tridentate ligands show large reaction enthalpy values and define a line with a slope of 385 K. Systems based on hexadentate ligands generally show small reaction enthalpies, and the slope of this plot is 223 K. Within this type of systems some still deviate from the rest. An extreme case is [Fe(*tpt*Me*tame*)]$^{2+}$, which combines an unusual low reaction enthalpy with high reaction entropy. It is believed that this anomaly reflects the stiffness of the system [11]. It is remarkable that the cage system [Fe((NH$_2$)$_2$*sar*)]$^{2+}$ fits into the plot even though both the enthalpy and the entropy changes are small [27]. In this system, all donor atoms are secondary nitrogen functions, whereas in the *tpt*Me*tame* system they are all tertiary nitrogen functions.

There are not enough thermodynamic studies of other spin crossover systems. Iron(III) and cobalt(II) crossover complexes have rarely been studied in solution. Recent systematic deuterium NMR studies on *bis*-(cyclopentadienide)-manganese(II) complexes [12] reveal common features, which are reflected in a very well defined isokinetic plot with a slope 328 K. In this case, both HS and LS signals were observed at low temperature. It is very unusual that spin crossover phenomena are studied in this straightforward manner. Previous investigations using ^{1}H NMR also turned out to have been misinterpreted. A very recent study on *bis*-[*tris*-(pyrazolyl)-methane]iron(II) in *DMF* where it is claimed that HS and LS signals are observed at temperatures below 283 K [13] is clearly in contradiction with previous laser flash studies where the relaxation time for this system in MeCN is found to be about 25 ns at room temperature [14]. Obviously, what has been assigned to signals of an LS iron(II) complex are just signals from the free ligand obtained as a result of dissociation. On the other hand, with higher magnetic fields and therefore increased NMR time scales at hand, maybe some of the classical systems considered be too fast for the technique ought to be reinvestigated.

Interpretations of minor differences between solid-state and solution data are probably not meaningful. There seems to be a tendency towards stabilization of LS in the solid phase compared to solution, but there are exceptions to this rule. Dramatic differences might be related to changes in the molecular structures in the different phases. $[Fe(tacn)_2]Br_2 \cdot 4H_2O$ is LS in the solid phase but becomes paramagnetic when the crystallization water is removed under vacuum [29]. In aqueous solution this compound is a spin equilibrium system [23]. Apparently, hydrogen bonding to the NH functions is important in controlling the ligand field strength in this case. The corresponding N-methylated ligand Me_3*tacn* forms a HS *bis*-complex with Fe(II), but recently *Hagen et al.* have shown that the combination of three acetonitrile molecules and one Me_3*tacn* moiety in the coordination sphere of Fe(II) gives a crossover complex [30]. In this case, the T_c of the acetonitrile solution is about 100 K lower than that of the neat solid. A detailed NMR study, however, indicates that rapid exchange with the triflate counter ions might be responsible for the change in behaviour. 1,4,7-Triazacyclononane functionalized with 2-pyridylmethyl and 6-methyl-2-pyridylmethyl groups provides an environment around Fe(II) which is in the crossover region. $[Fe(lpp[9]aneN_3)]$ shows a decrease in T_c of 90° for the methanol solution compared to the solid perchlorate [25]. A comparison with similar systems suggests that the bulky 6-methyl-2-pyridylmethyl group might be replaced by a methanol molecule upon dissolution of the complex salt.

In recent years a couple of oligomeric spin crossover systems have been investigated in solution. *Williams et al.* have prepared an interesting series of homo- and heterodimetallic podates with helicate structures [34, 35]. A comparison of thermal spin crossover transitions of a solution of a Fe(II) dinuclear triple helicate with an analogous mononuclear complex shows the helicate transition to occur at higher temperature in a two-step process, implying a negative cooperativity [36].

Lehn et al. have reported an even more complex supramolecular structure [37]. The introduction of sterically demanding substituents in certain dinucleating pyridyl ligands enabled them to assemble $[Fe(II)_4L_4]^{8+}$ grid structures which show the crossover phenomenon both in solid state and in solution. ^{1}H NMR spectra at variable temperature in an acetone/acetonitrile mixture show a rather complicated behaviour with both HS and LS signals, suggesting an apparent slow HS $\rightleftharpoons$ LS interconversion.

Dynamics of Spin Crossover Reactions

The interchange between HS and LS states of a metal ion can be considered as simple intramolecular electron transfer processes, and an understanding of the detailed mechanism of these fundamental reactions is therefore highly desirable. Since 1989 the kinetics of many new systems have been investigated. The laser flash technique is now established as the most commonly used for this type of study in solution at temperatures down to 180 K. Activation parameters for some recently reported systems are collected in Table 2.

Variable pressure studies of spin crossover complexes [15] have provided molar activation volumes that in general imply the volume of the transition state to be placed about midway between the HS and LS relaxed states, so contraction of the solvated complex occurs on going from HS to the transition state. The solution activation parameters for most of the systems in Table 2 fall in the usual range for iron(II) spin crossover compounds. In view of the large change in coordination distances (approximately 0.2 Å) it is frequently assumed that the reaction coordinate is primarily of radial nature [16]. However, the importance of other conformational changes and twisting modes has also been stressed [2, 16].

It has been suggested that the variation of the activation energy, as *e.g.* represented by $\Delta H^{\ddagger}$ can be linked to the ligand structure [2]. In this pictures ligands that more readily accommodate twisting modes should reach their transition state more easily, and consequently have lower activation barriers than complexes with ligands which prevent easy twisting, thereby necessitating a more radially expanded transition state with a higher barrier. In Table 2, especially the low values of $\Delta H^{\ddagger}$ for $[Fe(dpa)_2]^{2+}$ as compared to the hexadentate ligand systems have been suggested as evidence that the activation process consists largely of a twist mode [16]. The facial coordinating tridentate *dpa* ligand poses no restraints on a trigonal twist motion. The relatively large value obtained for the new complex $[Fe(lpp[9]aneN_3)]^{2+}$ might be in favour of this model. The increased barrier could be considered to result from the fact that a twisting mode is less favourable for this complex due to steric interactions between the methyl group in position six of one of the pyridine groups and the two other pyridine moieties.

Table 2. Activation parameters for the HS to LS process for some spin crossover complexes in solution

Complex	Solvent	$\Delta H^{\ddagger}$/kJ · mol^{-1}	$\Delta S^{\ddagger}$/J · mol^{-1} · K^{-1}	Ref.
$[Fe(dpa)_2]^{2+}$	MeOH	2.14		[26]
$[Fe(tp[10]aneN_3)]^{2+\,a}$	EtCN	28.49	−18.3	[24]
$[Fe(lpp[9]aneN_3)]^{2+}$	MeOH	9.4	−63.5	[25]
$[Fe(tptMetame)]^{2+}$	MeOH	5.6	−75	[11]
$[Fe(btpa)]^{2+}$	MeCN/PrCN	15.5^{b}		[17]
$[Fe(btpa)]^{2+\,a}$	MeCN/PrCN	27.5^{b}		[17]
$[Fe(tpchxn)]^{2+}$	MeOH	9.72	−68.1	[31]
$[Fe(tpchxn)]^{2+\,a}$	MeOH	26.39	−8.3	[31]
$[Fe(tpen)]^{2+}$	MeOH	7.23	−76.5	[16]

^{a} Slow process; ^{b} E_a from an *Arrhenius* plot

It is of interest to investigate whether special structural features of the ligands can be used to design systems which can be trapped in a certain conformation and thereby slow the relaxation rate considerably. *Toftlund et al.* have pursued such a strategy. Some of the highest barriers observed up to now have been obtained following this strategy. However, a common feature of these new systems is the occurrence of biphasic relaxation kinetics. The fast decay processes show barriers typical for other crossover systems, whereas the slow decay processes have activation barriers that are almost twice as large as those of the fast systems. Such high values indicate that bigger geometrical rearrangements of the coordination sphere take place than simple lengthening of the metal-ligand bonds. In the case of $[Fe(btpa)]^{2+}$ the crystal structure analysis of the PF_6^- salt might provide a clue [18]. The Fe(II) complexes occupy two crystallographic non-equivalent lattice sites. One site has an almost octahedral $[FeN_6]$ coordination sphere with mean Fe-N lengths of 2.00 Å. The other site consists of a highly distorted $[FeN_6]$ unit with mean Fe-N bond lengths of 2.23 Å and a twisting of the bipyridyl backbone along the 2,2′-C,C bond. One of the two non-coordinating pyridyl arms points towards the Fe centre with a Fe-N distance of only 3.16 Å. No HS $\rightleftharpoons$ LS interconversion is observed in the solid state, probably because the lattice packing does not allow large conformational changes to occur.

In order to explain the biphasic kinetics in solution it has been concluded that two metastable HS states with distinctly different equilibrium geometries and lifetimes exist side by side [11]. One denoted HS_A differs from the LS state

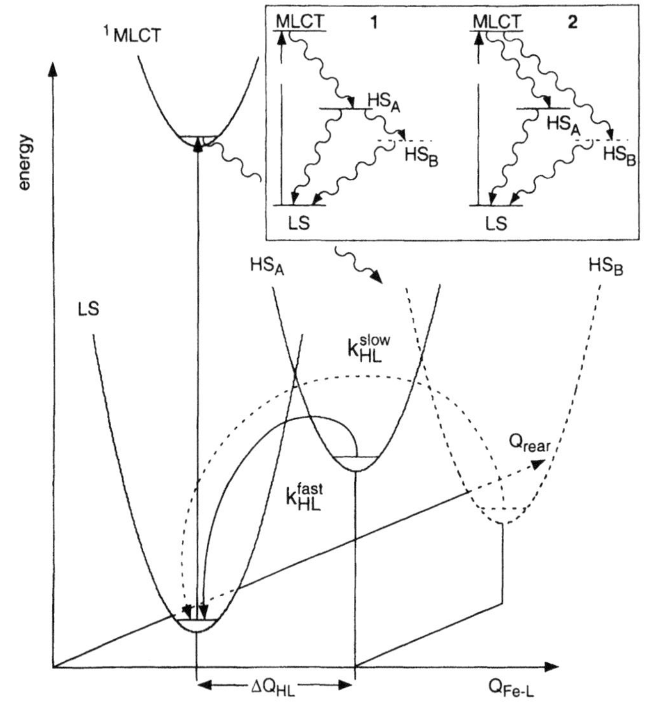

Fig. 1. Configurational coordinate diagram with the LS, MLCT, HS_A, and HS_B potential wells schematically illustrating the biphasic HS $\rightleftharpoons$ LS relaxation process following pulsed laser excitation of $[Fe(btpa)]^{2+}$ [17]

basically by substantially longer Fe-L bonds. For the other one, denoted HS_B, an additional rearrangement of the pyridyl arms is involved. From kinetics, HS_B is inferred to be at lower energy than HS_A. This is summarized in the two-dimensional configurational coordinate diagram of Fig. 1. The two different pathways resulting in transient populations of the two HS states after pulsed laser excitation into the $^1A_1$ to ^1MLCT transition is sketched in the insert of Fig. 1. They cannot be distinguished on the basis of the laser flash photolysis experiments alone; however, combination with the finding by *McCusker* [32] that HS states generally are populated a few picoseconds after the ^1MLCT transition it seems that the consecutive decay pathway (mechanism 1) is the most likely in this case.

Conclusions

The development of advanced materials and devices for nanotechnology requires systems that form switchable molecules or supermolecular systems which enable highly efficient information storage. The light-induced excited spin state trapping of crossover compounds provides promising examples of materials for this application. Most studies in this area have been performed on crystalline solids [1], but recent studies on polymeric films doped with $[Fe(btpa)]^{2+}$ have shown that room temperature trapping of a metastable HS form is possible [38]. This is an encouraging result as it demonstrates that constructing ligands with very specific and restrictive demands on the molecular structural dynamics may substantially increase the intersystem crossing barrier.

Acknowledgments

The author would like to thank a number of coworkers who have been involved in the reported research. He thanks the following authors for allowing him to quote from papers not yet published: Prof. *Alan F. Williams*, Prof. *John J. McGarvey*, Dr. *Alan Hazell*, and Dr. *Kenneth B. Jensen*.

Appendix

Abbreviations:

bispybim:	2,6-*bis*-(benzimidazol-2-yl)-pyridine
btpa:	N,N,N′,N′-*tetrakis*-(2-pyridyl)-6,6′-*bis*-(aminomethyl)-2,2′-bipyridine
$(NH_2)_2$*sar*:	dimethylsarcofagine
dpa:	*bis*-(2-pyridylmethyl)-amine
$HB(pz)_3^-$:	hydro-*tris*-(1-pyrazolyl)-borate
$HC(pz)_3$:	*tris*-(1-pyrazolyl)-methane
$HC(3,5\text{-}Me_2pz)_3$:	*tris*-(3,5-dimethyl-1-pyrazolyl)-methane
L:	*bis*-(5-(1-methyl-2-(6′-methyl-2′-pyridyl)-benzimidazol-2-yl)-methane
*pyim*H:	2-(2′-pyridyl)-imidazole
*pyben*H:	2-(2′-pyridyl)-benzimidazole
lpp[9]*ane*N_3:	1-(6-methyl-2-pyridylmethyl)-4,7-*bis*-(2-pyridylmethyl)-1,4,7-triazacyclononane
tp[10]*ane*N_3:	1,4,7-*tris*-(2-pyridylmethyl)-1,4,7-triazacyclodecane
tacn:	1,4,7-triazacyclononane

Me_3*tacn*:	1,4,7-trimethyl-1,4,7-triazacyclononane
*tpt*Me*tame*:	1,1,1-*tris*-((N-(2-pyridylmethyl)-N-methylamino)-methyl)-ethane
tpen:	N,N,N′,N′-*tetrakis*-(2-pyridylmethyl)-1,2-ethanediamine
tpchxn:	N,N,N′,N′-*tetrakis*-(2-pyridylmethyl)-1*R*,2*R*-cyclohexanediamine

References

[1] Gütlich P, Hauser A, Spiering H (1994) Angew Chem Int Ed **33**: 2024
[2] Toftlund H (1989) Coord Chem Rev **94**: 67
[3] Delépine M (1908) Bull Soc Chim Fr **3**: 643
[4] Cambi L, Cagnasso A (1931) Atti Accad Naz Lincei **13**: 809
[5] Madeja Y, König E (1963) J Inorg Nucl Chem **25**: 377
[6] Orgel LE (1956) Quelques Problèmes de Chimie Minérales. 10ême Conseil de Chimie, Bruxelles, p 289
[7] Pfeiffer P, Christeleit W (1938) J prakt Chem NF **151**: 127
[8] Williams RJP (1953) Proc Symp on Coordination Chemistry. Copenhagen, note p 56
[9] Goodwin HA, Sylva RN (1968) Aust J Chem **21**: 83
[10] Toftlund H *et al.* J Chem Soc Dalton (to be published)
[11] Al-Obaidi AHR, Jensen KB, McGarvey JJ, Toftlund H, Jensen B, Bell SEJ, Carrol JG (1996) Inorg Chem **35**: 5055
[12] Köhler FH, Scleisinger B (1992) Inorg Chem **31**: 2853
[13] Reger DL, Little CA, Rheingold AL, Lam M, Liable-Sands LM, Rhagitan B, Concolino T, Mohan A, Long GJ, Briois V, Grandjean F (2001) Inorg Chem **40**: 1508
[14] McGarvey JJ, Toftlund H, Al-Obaidi AHR, Taylor KP, Bell SEJ (1993) Inorg Chem **32**: 2469
[15] McGarvey JJ, Lawthers I, Heremans K, Toftlund H (1990) Inorg Chem **29**: 252
[16] McCusker JK, Rheingold AL, Hendrickson DN (1996) Inorg Chem **35**: 2100
[17] Schenker S, Stein PC, Wolny JA, Brady C, McGarvey JJ, Toftlund H, Hauser A (2001) Inorg Chem **40**: 134
[18] Hazell A, Toftlund H, J Chem Soc Dalton Trans (to be published)
[19] McGarvey JJ, Lawthers I, Heremans K, Toftlund H (1984) J Chem Soc Chem Comm 1575
[20] Boca R, Baran P, Dlhán L, Sima J, Wiesinger G, Renz F, El-Ayaan U, Linert W (1997) Polyhedron **16**: 47
[21] Addison AW, Burman S, Wahlgren CG, Rajan OA, Rowe TM, Sinn E (1987) J Chem Soc Dalton Trans 2621
[22] Jesson JP, Trofimenko S, Eaton DR (1967) J Am Chem Soc **89**: 3158
[23] Turner JW, Schultz FA (1999) Inorg Chem **38**: 358
[24] Al-Obaidi AHR, McGarvey JJ, Taylor KP, Bell SEJ, Jensen KB, Toftlund H (1993) J Chem Soc Chem Comm 536
[25] Koikawa M, Hazell A, Jensen KB, McGarvey JJ, Pedersen JZ, Toftlund H, J Chem Soc Dalton Trans (to be published)
[26] Jensen KB (1997) Mono and Dinuclear Iron Complexes. PhD Thesis. Department of Chemistry, Odense University
[27] Martin LL, Hagen KS, Hauser A, Martin RL, Sargeson AM (1988) J Chem Soc Chem Commun 1313
[28] Linert W, Konecny M, Renz F (1994) J Chem Soc Dalton Trans 1523
[29] Wiegardt K, Küppers HJ, Weiss J (1985) Inorg Chem **24**: 3067
[30] Blakesley D, Payne SC, Hagen KS (2000) Inorg Chem **39**: 1979
[31] McCusker JK, Toftlund H, Rheingold AL, Hendrickson DN (1993) J Am Chem Soc **115**: 1797
[32] McCusker JK, Walda KN, Dunn RC, Simon JD, Magde D, Hendrickson DN (1993) J Am Chem Soc **115**: 298

[33] Koudriavtsev AB, Linert W (2001) Monatsh Chem **132**: 235
[34] Charbonnière LJ, Williams AF, Piguet C, Bernardinelli G, Riverra-Minten E (1998) Chem Eur J **4**: 485
[35] Edder C, Piguet C, Bernardinelli G, Mereda J, Bochet CG, Bünzli J-CG, Hopfgartner G (2000) Inorg Chem **39**: 5059
[36] Williams AF (personal communication)
[37] Breuning E, Ruben M, Lehn J-M, Renz F, Garcia Y, Ksenofontov V, Gütlich P, Wegelius E, Rissanen K (2000) Angew Chem **39**: 2504
[38] McGarvey JJ, Toftlund H (to be published)

Received June 11, 2001. Accepted (revised) June 20, 2001

Invited Review

Thermochromism and Solvatochromism in Solution

Usama El-Ayaan[1,2], **Fumiko Murata**[2], and **Yutaka Fukuda**[2,*]

[1] Department of Chemistry, Faculty of Science, Mansoura University, Mansoura 35516, Egypt
[2] Department of Chemistry, Faculty of Science, Ochanomizu University, Tokyo 112-8610, Japan

Summary. The thermochromic behaviour of various coordination compounds in solution is discussed, with a special focus on cyclic diamine chelates. Thermally induced spin-crossover phenomena of iron(II) complexes are also considered. The solvatochromic behaviour of mixed-ligand complexes is presented in detail.

Keywords. Thermochromism; Spin-crossover; Solvatochromism; Solvent effect; Magnetic properties.

Introduction

The colours of coordination compounds originate from (*i*) d–d transitions, (*ii*) CT transitions (*ML*- and *LM*-CT), (*iii*) intervalence transitions, and (*iv*) ligand π–π^* transitions. The change of chemical and physical conditions (temperature, pressure, solvent, *etc.*) strongly influences the colours of the complexes. Colour change induced by temperature is known as thermochromism. Solvatochromism is another property of many coordination compounds; in this case, colour change is a consequence of solvent polarity.

Thermochromism

The term thermochromism is defined as the change of the colour of a substance, or a system of substances, when heated to a certain temperature. The change can be reversible, *i.e.* the original colour returns upon cooling, or irreversible.

Thermochromism on thermally induced spin-crossover iron(II) complexes

Thermally induced spin-state transitions between the low-spin (LS) and the high-spin (HS) states of transition metal complexes with d^4 to d^7 electronic configuration are observed when the ligand field splitting energy ($\Delta_0 = 10\,Dq$) is approximately

* Corresponding author. E-mail: usama@cc.ocha.ac.jp

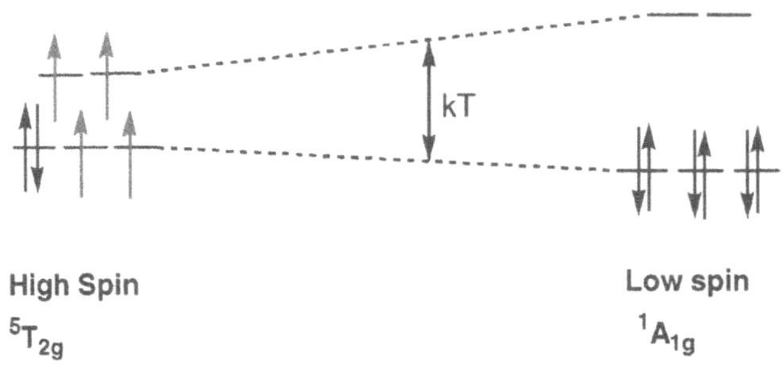

Fig. 1. High-spin and low-spin states in iron(II) compounds

outweighed by the spin-pairing energy P [1, 2]. In other words, a spin-crossover phenomenon is observed when the energy difference between the HS and LS states is of the order of kT (Fig. 1). In the course of such a spin-crossover, temperature dependent changes of the magnetic moment are observed.

A thermally induced LS $\leftrightarrows$ HS transition is characterized by an $x = f(T)$ curve, where x is the molar fraction of HS molecules and $(1 - x)$ that of LS molecules. A method to obtain such a curve is to measure the temperature dependence of the magnetic susceptibility. χT is constant in the temperature range where all molecules are in the same spin state ($(\chi T)_{LS}$ and $(\chi T)_{HS}$, respectively). Therefore, the $x = f(T)$ curve can be deduced from the experimental $\chi T = f(T)$ curve according to Eq. (1).

$$x = \chi T - (\chi T)_{LS}/(\chi T)_{HS} - (\chi T)_{LS} \quad (1)$$

Spin-crossover phenomena ($S = 0$ to $S = 2$ transitions) have been well documented for various iron(II) complexes [3, 4]. Among them, [Fe(*phen*)$_2$(NCS)$_2$], [Fe(*phen*)$_2$(NCSe)$_2$], and [Fe(*bpy*)$_2$(NCS)$_2$] have been studied using different experimental techniques. It should be pointed out that [Fe(*phen*)$_3$]$X_2 \cdot n$H$_2$O complexes keep their spin state unchanged when the temperature is varied, *i.e.* without showing any spin-crossover phenomena. They are either of low-spin $(^1A_1, S = 0)$, intermediate-spin $(^3T_1, S = 1)$, or high-spin $(^5T_2, S = 2)$ nature [5]. Generally, some asymmetry in the ligand field sphere can produce an appropriate balance between the ligand field strength and the spin pairing energy, thus leading to the conditions necessary for the spin-crossover.

The Fe(II) complex [Fe(*ptz*)$_6$](BF$_4$)$_2$ (Fig. 2; *ptz* = 1-propyltetrazole) is known to exhibit a spin transition at 135 K with a spectacular change of colour from purple (LS) to white (HS) [6, 7]. Its UV/Vis spectrum at room temperature is typical of a $^5T_{2g}$ ground state with a $^5T_{2g} \leftrightarrows {}^5E_g$ spin-allowed transition at 12250 cm^{-1}, whereas that at 10 K corresponds to a $^1A_{1g}$ ground state with $^1A_{1g} \rightarrow {}^1T_{1g}$ and $^1A_{1g} \rightarrow {}^1T_{2g}$ spin-allowed transitions at 18400 and 26700 cm^{-1}.

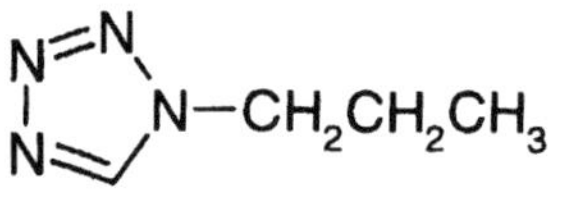

Fig. 2. Structure of 1-propyltetrazole

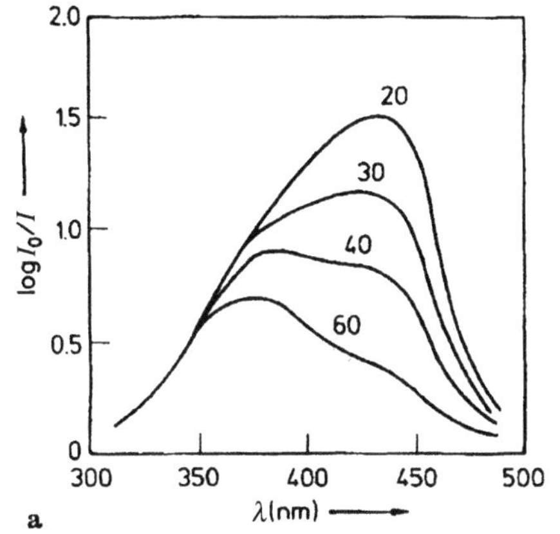

Fig. 3. Absorption spectra of $[Fe(amp)_3]^{2+}$ at different temperatures; (a) 4×10^{-4} *M* in CH_3CN/H_2O, (b) 1.9×10^{-4} *M* in H_2O; the temperature (°C) is given on each curve

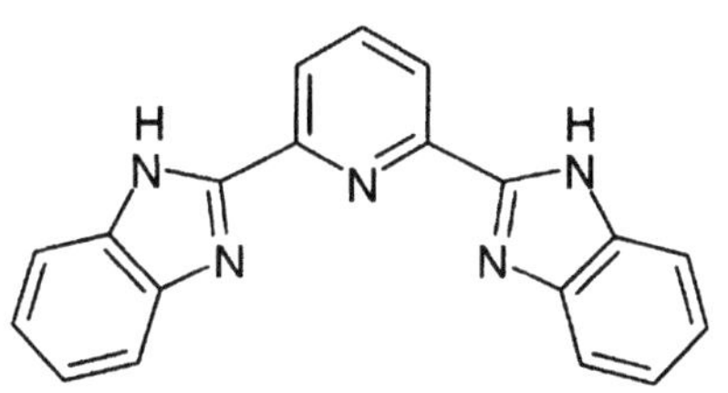

Fig. 4. Structure of 2,6-*bis*-(benzimidazol-2′-yl)-pyridine

Temperature-dependent electronic spectra of $[Fe(amp)_3]^{2+}$ (*amp* = 2-amino-methyl-pyridine) exhibit two intense absorption bands at 440 and 380 nm in water or water/acetonitrile [8]. The relative intensities of these bands change sharply with temperature (Fig. 3).

The iron(II) complex $[Fe(bzimpy)_2](ClO_4)_2$ (Fig. 4; *bzimpy* = 2,6-*bis*-(benzi-midazol-2′-yl)-pyridine) has been shown to exhibit a pronounced spin-crossover $(^1A_{1g}(O_h) \leftrightarrows {}^5T_{2g}(O_h))$ in solution [9].

Temperature-dependent UV/Vis spectroscopic measurements of $[Fe(bzimpy)_2]$-$(ClO_4)_2$ in ethanol:methanol = 1:1 from 300 down to 15 K and in a polyvinyl alcohol (*PVA*) matrix have been carried out [10]. The CT bands located in the region of 316–362 nm were found to be sensitive to the spin transition (Fig. 5). Upon cooling they increase in intensity because of the higher extinction coefficient in the low-spin species due to π-back donation. Also, the peak located at 570 nm shows both LS and HS properties [11]. The increase in intensity upon cooling indicates a raising amount of LS, *i.e.* the extinction of the LS state is higher (Fig. 5).

4-*X*-2,6-*bis*-(benzimidazol-2′-yl)-pyridine (*X* = H, Cl, OH) can coordinate as a tridentate ligand to metal ions of d^4–d^7 electronic configuration in distorted octahedral geometry [12, 13]. Among these complexes, $[Fe(X\text{-}bzimpy)_2]^{2+}$ exhibits a pronounced spin-crossover behaviour both in solution and in the solid state [14, 15]. Thermally induced spin-crossover in solution has been observed in

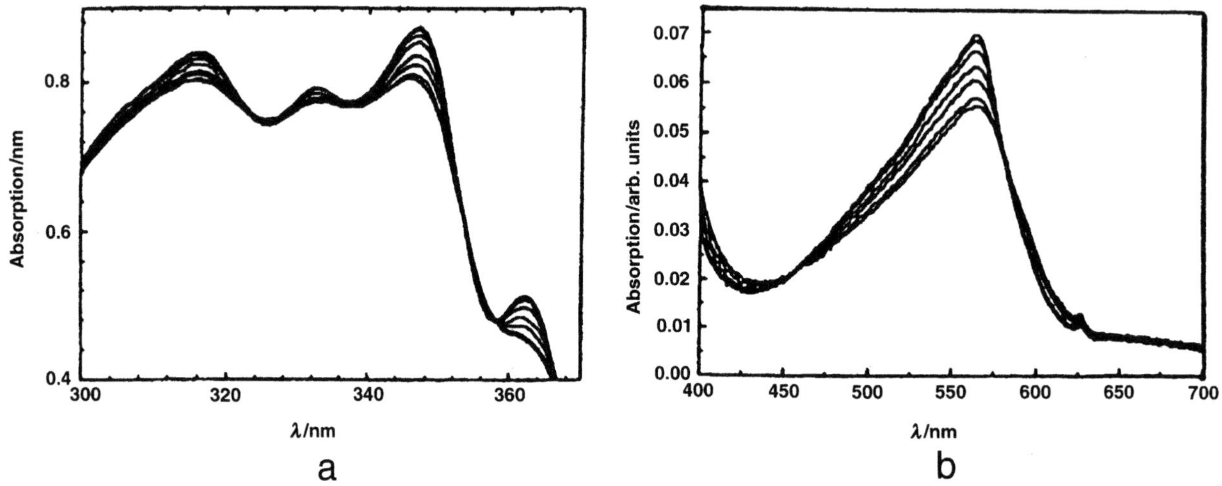

Fig. 5. Temperature dependent UV/Vis spectra of $[Fe(\textit{bzimpy})_2](ClO_4)_2$ in *PVA* in the wavelength ranges (a) 300–370 and (b) 400–700 nm; temperatures (K) from bottom to top: 300, 265, 225, 195, 150, 100, 50, 15

Table 1. Magnetic parameters of $[Fe(\textit{X-bzimpy})_2]^{2+}$ and $[Fe(\text{H-}\textit{bzimpy}H_{-1})_2]$ in MeOH and thermodynamic parameters derived thereof

T/K	$[Fe(\text{H-}\textit{bzimpy})_2]^{2+}$		$[Fe(\text{H-}\textit{bzimpy}H_{-1})_2]$	
	μ/μ_B	K_{sc}	μ/μ_B	K_{sc}
240	2.32	0.22	1.65	0.10
250	3.14	0.48	2.62	0.29
260	3.63	0.77	3.21	0.52
270	4.12	1.28	3.70	0.83
280	4.36	1.69	4.18	1.37
290	4.62	2.40	4.44	1.87
300	4.82	3.31	4.52	2.08
310	4.95	4.26	4.78	3.09
320	5.03	5.11	5.03	5.11
ΔH (kJ/mol)	2.77 ± 0.6		25.6 ± 1.2	
ΔS (J/K · mol)	85.1 ± 2.1		92.8 ± 4.3	

$[Fe(\text{H-}\textit{bzimpy})_2]^{2+}$ and its deprotonated form $[Fe(\text{H-}\textit{bzimpy}H_{-1})_2]$ [16]. The magnetic moments of both complexes in methanolic solution were derived by applying the *Evans* method [17] at different temperatures (Table 1).

$[Fe(\textit{pybzim})_3]X_2 \cdot nH_2O$ (Fig. 6; *pybzim* = 2-(2′-pyridyl)-benzimidazol) is a spin-crossover system which has been studied by various techniques [18–21]. NMR measurements on the perchlorate complex $[Fe(\textit{pybzim})_3](ClO_4)_2 \cdot H_2O$ at different temperatures in methanol, acetonitrile, acetone, nitromethane, and dimethylsulfoxide have been reported (Table 2) [22]. The temperature dependence of the effective magnetic moment (evaluated by the *Evans* method) allows to conclude that spin-crossover occurs in these solutions.

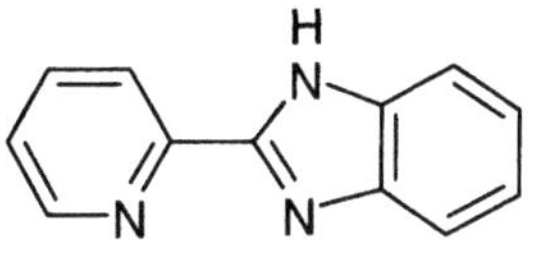

Fig. 6. Structure of 2-(2′-pyridyl)-benzimidazole

Table 2. Temperature dependent magnetic moments of $[Fe(pybzim)_3](ClO_4)_2 \cdot H_2O$ in various solvents using *Evans*' method

T/K	μ/μ_B				
	MeOH	*Ac*	*AN*	*NM*	*DMSO*
325	–	–	–	–	5.37
320	4.74	4.96	3.56	4.76	5.33
315	–	–	–	–	5.29
310	4.83	4.95	3.24	4.69	5.24
300	4.90	4.87	3.19	4.44	
290	4.80	4.79	3.13	4.36	
280	4.68	4.71	3.08	3.94	
270	4.57	4.46	3.02	3.88	
260	4.45	4.20			
250	4.19	3.56			
240	3.92	2.85			
230	3.64	1.97			
220	3.21	1.60			

In methanol, the spin transition occurs at 220 K and culminates at 300 K. Even better resolution of the spin-crossover was observed in acetone. In contrast, in acetonitrile only a small variation of the effective magnetic moment was recorded from 270 to 320 K. In nitromethane and dimethylsulfoxide, only the high-spin state is stabilized within the accessible temperature range.

It is interesting to note that in the solid state the spin-crossover behaviour of such complexes varies remarkably with the counter anion, *e.g.* $X = BPh_4^-$, BF_4^-, PF_6^-, NO_3^-, Br^-, I^-, and ClO_4^- [23–25]. Also, the conditions of the preparation may cause different magnetic properties [26]. Additionally, in solution the nature of the solvent affects the spin transition [27].

Potential applications of thermally induced spin-crossover materials

Smart materials respond to their environment as *e.g.* illustrated by photochromic eyeglasses which darken upon exposure to ultraviolet light. Hence, materials with fast reversible responses to environmental stimuli are sought as components of smart systems. Similar to photochromic materials, thermochromic materials reversibly respond to heat and exhibit substantial colour changes upon small temperature variations. Spin-crossover materials are a class of thermochromic materials exhibiting fast and reversible colour changes amenable to display, memory, and switching devices. These colour changes also can be induced by light (photochromism) or pressure (piezochromism). Due to the nature of the mechanism of the thermo-, photo-, or piezochromic responses (*i.e.* redistribution of the electron density at a metal ion site within the molecule), in addition to reversibility these changes occur extremely fast. As a consequence of their favourable properties (fast colour change, strong contrast between colours, long-term colour stability, and lack of moving parts), these materials are completely recyclable and amenable to fast, low power-consuming, high data density display and storage devices for future systems [28].

Thermochromism of cyclic diamine complexes

In the solid state, cyclic diamine complexes are known to exhibit thermochromism. For example, the orange complex $[Cu(daco)_2](NO_3)_2$ (*daco* = 1,5-diazacyclooctane) with a square-planar geometry around the copper(II) centre shows a reversible thermochromism. When heated above *ca.* 90°C the colour changes quite abruptly from orange to violet. Upon cooling, it gradually becomes orange again [29]. These observations can be explained by the hydrogen-bond breaking and reforming between the ligand amino proton and the nitrate anion.

In solution, thermochromism of Ni complexes with cyclic diamines is also interesting. In aqueous solution, the nickel(II) complex $[Ni(dachda)(OH_2)_2]$ (*dachda* = 1,4-diazacycloheptanediacetate) and its alkyl derivatives (Fig. 7) show a pronounced thermochromism [30].

In case of the *dachda* complex, heating shifts the equilibrium of Eq. (2) to the right, resulting in a change of the geometry around the nickel(II) centre from octahedral to square-planar.

$$\underset{\text{green (room temp)}}{[Ni(dachda)(H_2O)_2]} \leftrightarrows \underset{\text{orange (high temp)}}{[Ni(dachda)] + 2H_2O} \quad (2)$$

The UV/Vis spectra of $[Ni(dachda)(H_2O)_2]$ at different temperatures show that the peak intensity at $21 \times 10^3\,cm^{-1}$ gradually increases as the temperature increases (Fig. 8), favouring the square-planar structure.

In case of all other Ni complexes discussed, the equilibrium is shifted from a 5-coordinated to a square-planar geometry (Eq. (3)); in other words, the alkyl substituent stabilizes the planar species, so that the thermochromic change involving

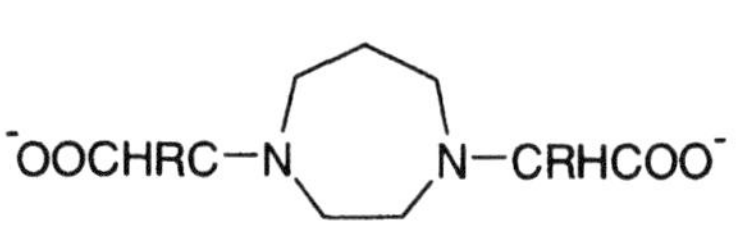

R	Abbreviation
H	*dachda*
Me	*dachdma*
Et	*dachdea*
n-Pr	*dachdpa*

Fig. 7. Structure of 1,4-diazacycloheptanedicarboxylate

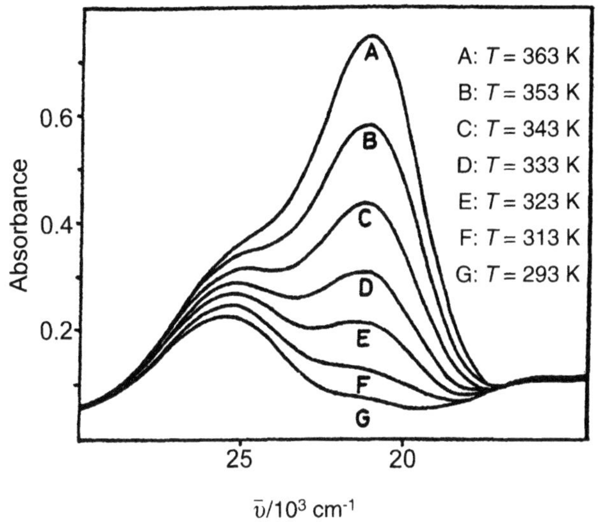

Fig. 8. Electronic absorption spectra of [Ni(*dachda*)$(H_2O)_2$] at different temperatures (2.3×10^{-2} *M* in H_2O)

its formation by heating occurs at much lower temperatures than for the *dachda* complex.

$$\underset{\text{5-coordinated}}{[\mathrm{Ni}(\mathit{dachdRa})(\mathrm{H_2O})]} \underset{\text{cool}}{\overset{\text{heat}}{\rightleftharpoons}} \underset{\text{square-planar}}{[\mathrm{Ni}(\mathit{dachdRa})]} + \mathrm{H_2O} \quad (3)$$

It is interesting to note that when a large excess of $NaClO_4$ is added to a green solution of [Ni(*dachda*)$(H_2O)_2$], the colour changes from green to orange at room temperature. This effect of colour change induced by $NaClO_4$ addition is known as the inert salt effect and is explained by decreasing the coordination ability of the free water molecules due to the presence of a large amount of the inert salt in solution. The addition of $NaClO_4$ is sometimes important to improve thermochromism, like in the case of $[\mathrm{Ni}(\mathit{diam})_2(\mathrm{H_2O})_2]^{2+}$ (*diam* = ethylenediamine or propylenediamine). Although the ethanolic solution of this complex is thermochromic and changes its colour reversibly from blue-violet to orange upon heating (Eq. (4)), the aqueous solution is not [31].

$$\underset{\text{blue-violet}}{[\mathrm{Ni}(\mathit{diam})_2(\mathrm{EtOH})_2]^{2+}} \underset{\text{cool}}{\overset{\text{heat}}{\rightleftharpoons}} \underset{\text{orange}}{[\mathrm{Ni}(\mathit{diam})_2]^{2+}} + 2\mathrm{EtOH} \quad (4)$$

However, the addition of a large amount of $NaClO_4$ makes the aqueous solution notably thermochromic due to the inert salt effect (Eq. (5)).

$$[\mathrm{Ni}(\mathit{diam})_2(\mathrm{H_2O})_2]^{2+} \underset{\text{cool}}{\overset{\mathrm{NaClO_4}\ \text{heat}}{\rightleftharpoons}} [\mathrm{Ni}(\mathit{diam})_2]^{2+} + 2\mathrm{H_2O} \quad (5)$$

The thermochromic properties of the nickel complex of 1,4-*bis*-(3-aminopropyl)-piperazine (*bapp*) have also been studied (Fig. 9) [32]. Interestingly, the piperazine ligand itself can only coordinate as a bidentate ligand in the chair form, but not in the boat form. When two (3-aminopropyl) groups are introduced to the piperazine

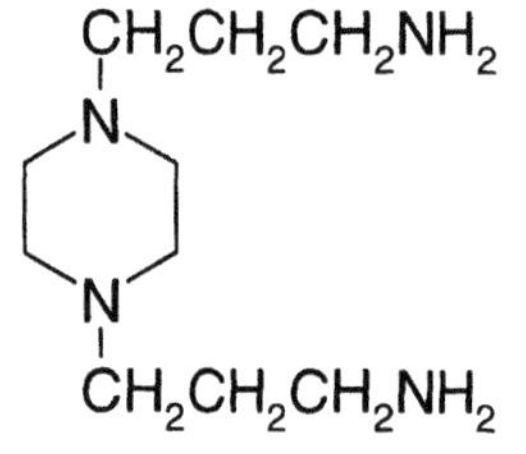

Fig. 9. Structure of 1,4-*bis*-(3-aminopropyl)-piperazine (*bapp*)

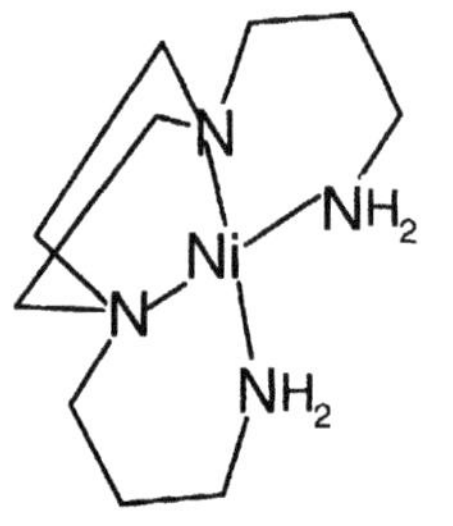

Fig. 10. Structure of $[Ni(bapp)]^{2+}$

moiety, the resulting tetradentate *bapp* ligand can stabilize the piperazine skeleton in the boat conformation through coordination to a metal center.

In certain organic solvents (*DMF* and acetonitrile), the orange colour of $[Ni(bapp)]^{2+}$ (Fig. 10) becomes blue upon cooling to *ca.* −50°C, indicating the formation of an octahedral species $[Ni(bapp)(solv)_2]^{2+}$ at lower temperatures. When Cl^- or Br^- ions are present in such solutions, another thermochromic equilibrium (Eq. (6)) between the halide-containing 5-coordinated species and the original square-planar structure has been confirmed [33].

$$\underset{\text{square planar}}{\underset{\text{orange}}{[Ni(bapp)]^{2+}}} + Cl^- \leftrightarrows \underset{\text{5-coordinaed}}{\underset{\text{blue-violet}}{[Ni(bapp)Cl]^{+}}} \qquad (6)$$

Solvatochromism

Solvatochromism is defined as the pronounced change in position and sometimes intensity of an electronic absorption or emission band accompanying a change in the polarity of the medium. This can be even detected visually by the change of solution colour when going from one solvent to another.

Solvatochromism of mixed-ligand complexes

Mixed-ligand complexes of copper(II) or nickel(II) ions show a pronounced solvatochromism in solution. Extensive studies have been focused on mixed-ligand complexes containing N-alkylated diamines (*diam*) and β-diketonates (*dike*) of the general formula $[M(diam)(dike)]^+$ (Fig. 11) [34–38]. An interesting property of these complexes is their high solubility in various organic solvents due to increasing hydrophobic properties of the groups surrounding the central metal ion, *i.e.* the alkyl substituent groups on the *diam* and the *dike* ligands. The study has been extended to cover a series of different substituents on both ligands (Tables 3, 4).

According to different interactions between the solvent and the studied solute complex, solvatochromic behaviour can be classified into the following categories.

Perchlorate and tetraphenylborate mixed-ligand complexes (solute-solvent interactions)

Compounds of the type $[M(diam)(dike)]X (M = Cu(II)$ or $Ni(II)$, $X = ClO_4^-$ or BPh_4^-) are of square-planar structure. These anions act only as counter ions and

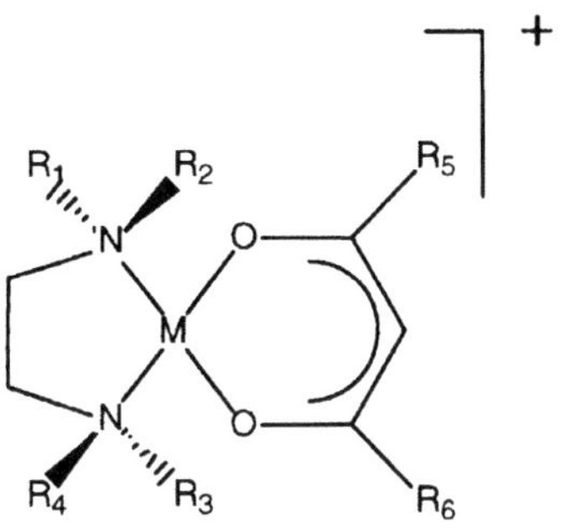

Fig. 11. Mixed-ligand complexes $[M(dike)(diam)]X$

Table 3. Diamine ligands with different substituents R^1–R^4

Ligand	Abbreviation	R^1	R^2	R^3	R^4
N,N,N′,N′-Tetramethylethylenediamine	*tmen*	Me	Me	Me	Me
N,N,N′,N′-Tetraethylethylenediamine	*teen*	Et	Et	Et	Et
N,N′-Dimethylethylenediamine	*sym-dmen*	Me	H	Me	H
N,N-Dimethylethylenediamine	*unsym-dmen*	Me	Me	H	H
N,N′-Diethylethylenediamine	*sym-deen*	Et	H	Et	H
N,N-Diethylethylenediamine	*unsym-deen*	Et	Et	H	H

Table 4. β-Diketonate ligands with different substituents R^5–R^6

Ligand	Abbreviation	R^5	R^6
Acetylacetonate	*acac*	Me	Me
Benzolylacetonate	*bzac*	Me	Ph
Dibenzoylmethanate	*dibm*	Ph	Ph
Trifluoroacetylacetonate	*tfac*	Me	CF_3
Hexafluoroacetylacetonate	*hfac*	CF_3	CF_3
Dipivaloylmethanate	*dipm*	*tert*-Bu	*tert*-Bu

do not interact with any solute (complex cation) in the solid state and in solution [39]. Therefore, the complex structure depends only on the donor properties of the solvent used (*Gutmann*'s donor number *DN*), and only solute-solvent interactions can be considered [40]. In a typically non-coordinating solvent (low *DN*) such as 1,2-dichloroethane (*DCE*), solutions of [Ni(*diam*)(*dike*)]*X* are diamagnetic and red in colour, whereas in highly coordinating solvents (high *DN*), such as *DMSO* or *DMF*, paramagnetic green or blue solutions are obtained (Eq. (7)).

$$\underset{\text{red}}{[\mathrm{Ni}(diam)(dike)]^+} + 2\,solv \leftrightarrows \underset{\text{blue or green}}{[\mathrm{Ni}(diam)(dike)(solv)_2]^+} \qquad (7)$$

Increasing donor ability of solvents shifts the equilibrium to the right. Substitutions on both $diam(R^1\text{-}R^4)$ and $dike(R^5\text{-}R^6)$ have a pronounced effect on the above equilibrium (Eq. (7)). Electron-donating substituents on *dike* shift the equilibrium to the left, whereas electron-withdrawing substituents favour the solvent coordination to the metal center and, consequently, shift the equilibrium to the right hand side. In other words, when R^5 and R^6 are electron-donating, the ligand field strength of *dike* is increased. The coordination sphere around the metal ion becomes electron-rich, which makes it difficult for a solvent molecule to approach the axial centres. The equilibrium constant K for the above reaction (Eq. (7)) was found to increase in the following order:

$$dipm < acac < bzac < dibm < tfac < hfac$$

Spectroscopic data show that the equilibrium constant decreases with increasing number and size of the alkyl groups on *diam*, obviously because of steric hindrance. The observed order of decreasing K is as follows [41]:

$$sym\text{-}dmen > unsym\text{-}dmen > sym\text{-}deen > tmen > unsym\text{-}deen > teen$$

Spectroscopic studies on [Cu(*diam*)(*dike*)]ClO_4 complexes in many non-aqueous solvents showed a good correlation between their d–d absorption maxima and the donor strength of solvent used [42]. Absorption peaks of complexes in inert solvents like *DCE* (low *DN*) were shifted to higher frequencies than in solvents with high *DN* such as *DMSO*.

Halide and pseudohalide mixed-ligand complexes (solute-solute-solvent interactions)

The behaviour of halide and pseudohalide complexes in various solvents is very different from that of the perchlorate compounds. The copper(II) complex [Cu(*diam*)(*dike*)*X*] is 5-coordinated (square-pyramidal), whereas the nickel(II) complex is a dinuclear, octahedral complex bridged by NCS^- ligands: [(*tmen*)(*acac*)Ni$(NCS)_2$Ni(*acac*)(tmen)]. In this review, the solution behaviour of the copper(II) complexes [Cu(*diam*)(*dike*)*X*] where $X = Cl^-$, Br^-, I^-, or NCS^- is discussed. In this case, both the anion X and the solvent molecule can coordinate to the copper(II) centre, and a competition of these two types of ligands at the axial sites of the chelate cation takes place (solute-solute-solvent interactions, Fig. 12). In

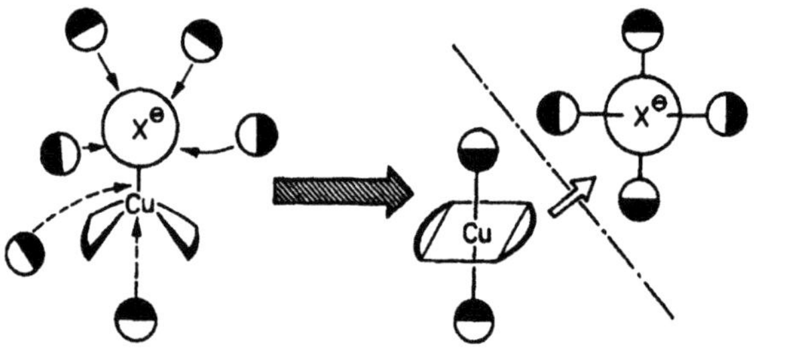

Fig. 12. Schematic representation of the dissociation of a 5-coordinated complex in a polar solvent; the solvent molecules are represented by balls; their white and black halves correspond to the positive and negative ends of the molecules, respectively

a polar solvent, the negative centre of the 5-coordinated complex, *i.e.* X, attracts the positive ends of solvent molecules. The Cu-X bond is weakened by this interaction, and finally the X anion, which is rather strongly solvated, dissociates off [38]. The main driving force of this reaction is the solvation power of the solvent as expressed by the acceptor number (*AN*) [43–45].

With two solvents of similar *AN*, that with higher *DN* will be favoured with respect to coordination to the metal centre. Solvent molecules pull out the anion X as acceptors, whereas as donors they solvate the complex cation. Therefore, a new polarity parameter [39, 46] of the solvent, the polarity factor P_f, is introduced as a linear combination of *AN* and *DN* (Eq. (8)).

$$P_f = a \times (AN) + b \times (DN) \tag{8}$$

Our results confirm the ease of anion (X) dissociation with increasing P_f in the order $I^- > Br^- > Cl^-$. In [Cu(*tmen*)(*dike*)X] complexes, the increasing strength of the Cu-X bond results in a stabilization of the 5-coordinated species which in turn is reflected in the decrease of its ν_{max}. In solvents of very low P_f (*DCE* or acetone), where the 5-coordinated species predominate, ν_{max} values increase following the order $I^- < Br^- < Cl^-$, thus indicating increasing stabilization of 5-coordinated species in the same order. The ν_{max} diagram of Cl- and Br-complexes with different kinds of *dike* in various solvents (Fig. 13) has been reported. In these complexes, ν_{max} changes in the order MeOH > EtOH > *DMSO* > *DMF* > *NM* > CH_3CN >

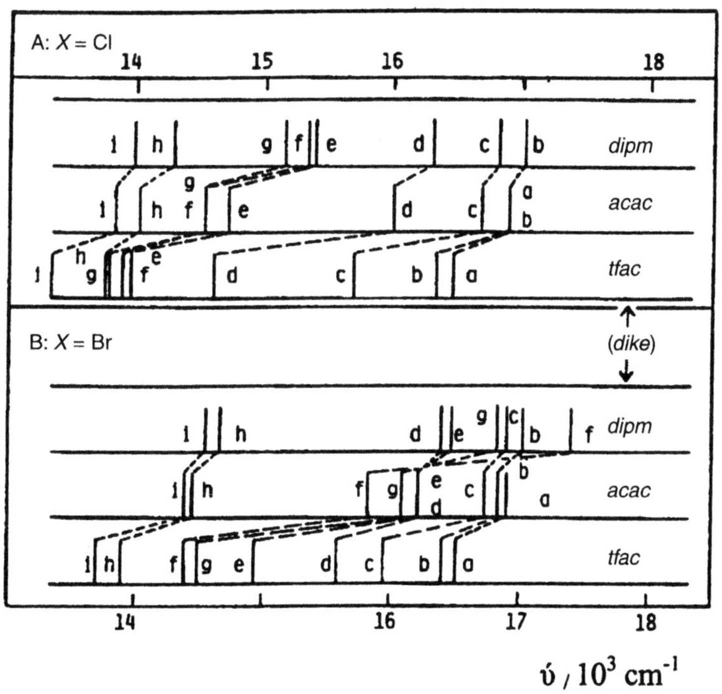

Fig. 13. ν_{max} values of the complexes [Cu(*tmen*)(*dike*)X] in various solvents; A: [Cu(*tmen*)(*dike*)Cl], B: [Cu(*tmen*)(*dike*)Br]; solvents from (a) to (i): H_2O (a), CH_3OH (b), C_2H_5OH (c), *DMSO* (d), *DMF* (e), CH_3NO_2 (f), CH_3CN (g), *DCE* (h), CH_3COCH_3 (i)

$DCE \cong$ acetone which is nearly the same order as that of P_f. In the same solvent, the change of ν_{max} with different *dike* moieties follows the order *dipm* > *acac* > *tfac* > *hfac*. This is in accordance with the ligand field strength of the *dike* ligands. The anion effect and substituent effect on *dike* may be called Axial Ligand Field Variation Effect and Equatorial Ligand Field Variation Effect, respectively. Owing to the latter effect, the tetragonal structure is stabilized, especially in the case of *dipm* chelates.

Nitrato and bis-(β-diketonato) mixed-ligand complexes

Two kinds of mixed-ligand complexes with octahedral geometry are involved in this case. The first complex contains a bidentate nitrato anion and is characterized by the general formula $[Ni(diam)(dike)NO_3]$. In the second one, $[Ni(diam)(dike)_2]$, an additional *dike* ligand is introduced. In solvents of low *DN*, the nitrato complexes dissolve without structural changes to form non-conducting solutions. Solvents with higher *DN* displace the NO_3^- ligands, and the quilibrium described by Eq. (9) is established [47] which is accompanied by only minor spectroscopic variations [48, 49] due to the unchanged geometry.

$$\underset{\text{octahedral}}{[Ni(diam)(dike)NO_3]} + 2\,\text{solv} \leftrightarrows \underset{\text{octahedral}}{[Ni(diam)(dike)(solv)_2]^+} + NO_3^- \qquad (9)$$

The octahedral $[Ni(diam)(dike)_2]$ complexes dissolve in various kinds of organic solvents without structural and spectroscopic changes. The introduction of the additional *dike* shields the central metal cations effectively by three organic ligands with eight alkyl groups. As a result, these chelates are very soluble in non-polar organic solvents. The strong shielding effect also prevents solute-solute interactions, thus allowing to obtain pure crystals of such chelates by sublimation. A list of donor and acceptor numbers together with some other polarity parameters are collected in Table 5 [43, 44].

The behaviour of nickel complexes of the general formula $[Ni(diam)(dike)X]$ ($X = BPh_4^-, ClO_4^-, NO_3^-, dike$) in various organic solvents is summarized in the following schematic diagram (Fig. 14) showing all possible structural changes induced by the polarity of the solvent (solvatochromism).

Mixed-ligand complexes as colour indicators for solvent polarity: an application of solvatochromism

The mixed-ligand complex $[Cu(acac)(tmen)]ClO_4$ is known to be solvatochromic owing to the structure changes caused by the solvation processes [34, 38]. The coordination centre which determines the position of the d–d band is extremely sensitive to nucleophilic solvent attack. This fact led to its use as a colour indicator for solvent polarity for *DN* values within the range of 5 to 30. However, anomalies appear in very weak or very strong donor solvents (Table 6). The λ_{max} value of this complex decreases with decreasing *DN* from pyridine to nitromethane; when the *DN* gets still smaller, however, it increases notably again. In such weak donor solvents λ_{max} becomes slightly concentration dependent [50]. Such problems have

Table 5. Donor numbers (*DN*), acceptor numbers (*AN*), dielectric constants (*D*), and dipole moments (μ/Debye) of various solvents at 25°C

Solvent	*DN*	*AN*	*D*	μ
Acetic acid (AcOH)		52.9	6.2	1.75
Acetic anhydride (Ac_2O)	10.5		20.7 (19°C)	2.82
Acetone (*Ac*)	17.0	12.5	20.7	2.86
Acetonitrile (*AN*)	14.1	18.9	36.0	3.44
Acetyl chloride	0.7		15.8	2.71
Benzene (*Bz*)	0.1	8.2	2.3	0.0
Benzonitrile (*BN*)	11.9	15.5	25.2	4.05
Benzyl cyanide	15.1		18.4 (27°C)	3.50
Chloroform ($CHCl_3$)	0.0	25.1	4.7	1.15
1,2-Dichloroethane (*DCE*)	0.0	16.7	10.1	1.75
Diethyl ether (Et_2O)	19.2	3.9	4.2	1.25
Dimethylformamide (*DMF*)	26.6	16.0	36.7	3.86
Dimethylsulfoxide (*DMSO*)	29.8	19.3	46.7	3.90
1,4-Dioxane (*diox*)	14.8	10.8	2.2	0.45
Ethanol (EtOH)	20.0	37.1	24.3	1.70
Ethyl acetate (EtOAc)	17.1	9.3	6.0	1.88
Hexane	0.0	0.0	1.9	0.0
Methanol (MeOH)	19.0	41.3	32.6	1.70
Nitrobenzene (*NB*)	4.4	14.8	34.8	4.03
Nitroethane (*NE*)			28.5	
Nitromethane (*NM*)	2.7	20.5	36.7	3.57
Pyridine (*Py*)	33.1	14.2	12.3	2.37
Tetrahydrofuran (*THF*)	20.0	8.0	7.4	1.75
Triethylamine (Et_3N)	61.0	1.4	2.4	0.79
Trimethyl phosphate (*TMP*)	23.0	16.3	20.6 (20°C)	
Water (H_2O)	18.0	54.8	78.4	1.84

been overcome by using $[Cu(acac)(tmen)]BPh_4$. UV/Vis spectra of this complex (Fig. 15) in various solvents of $DN > 5$ are very similar to those of the perchlorate, but the anomalies in very weak donor solvents are absent. The λ_{max} value continues to decrease with decreasing solvent basicity; this can be expressed in terms of *Schmid*'s DN_B [51, 52]. The deviation of the perchlorate complex at low DN_B is due to ion-pair formation, *i.e.* weak coordination of the perchlorate anion to the chelate plane. In case of tetraphenylborate, no effective ion-pair formation will occur since BPh_4^- is sterically demanding and inert. The regular spectroscopic trend therefore continues also in very weak donor solvents with no anomaly. In addition to the d–d band, the UV/Vis spectra of $[Cu(acac)(tmen)]BPh_4$ show a continuous background absorption in the blue part which seems to be the tail of a strong CT band in the near ultraviolet. With the decrease of *DN* or DN_B, this absorption is shifted towards red and overlaps remarkably with the d–d band in very weak donor solvents. The λ_{max} values of the d–d band in these solvents were corrected by assuming that d–d band has the same ν-symmetrical, *pseudo-Gaussian* shape as in other solvents.

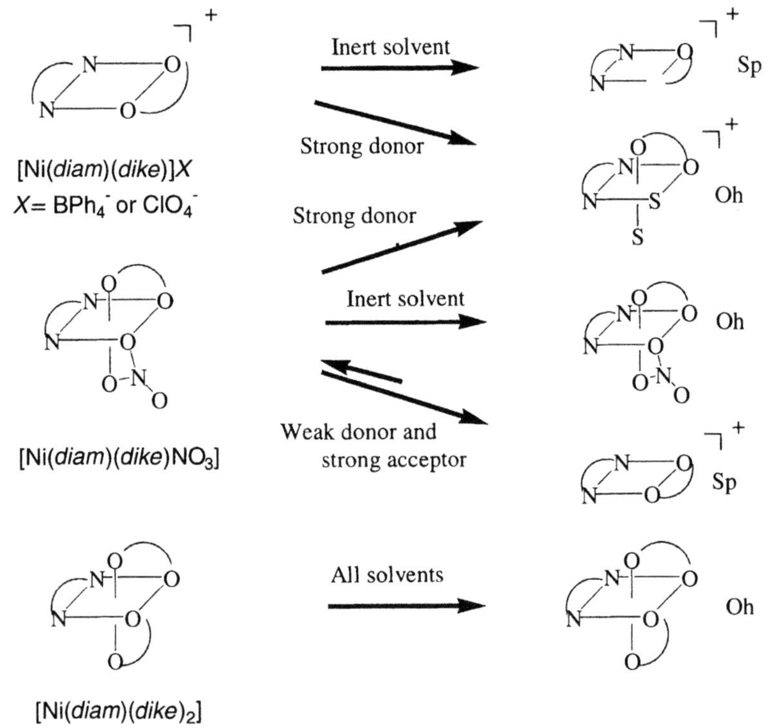

Fig. 14. Schematic diagram showing all possible structural changes of [Ni(*diam*)(*dike*)*X*] complexes resulting from solute-solute-solvent interactions

Table 6. Absorption maxima of the d–d band of $[Cu(acac)(tmen)]X$ in various solvents

Solvent	λ_{max}/nm			DN/DN_B^a
	$X = ClO_4^-$	$X = BPh_4^-$		
		Observed	Corrected	
Dichloromethane ($10^{-2}\,M$)	550[c]	488	515	(1.2)
Chlorobenzene ($10^{-3}\,M$)	572[c]	480	520	(1.2)
1,2-Dichloroethane ($10^{-2}\,M$)	543[c]	498	520	(2.9)
Nitrobenzene ($10^{-3}\,M$)	533	490	522	(3.7)
Nitromethane	532	527	530	(4.8)
Acetonitrile	578	573		14.1
Acetone	571	569		17.0
Dimethylformamide	603	602		26.6
Pyridine ($10^{-1}\,M$)	624[b]	623[b]	613	33.1
2-Picoline ($10^{-2}\,M$)	624[b]	620[b]		

[a] DN_B in parentheses; [b]strongly concentration dependent; [c] slightly concentration dependent

Conclusions

Thermochromism of various nickel(II) complexes with cyclic diamines or their derivatives in solution is always accompanied by structural changes. Thermochromic colour changes caused by changes in the spin state are sometimes observed

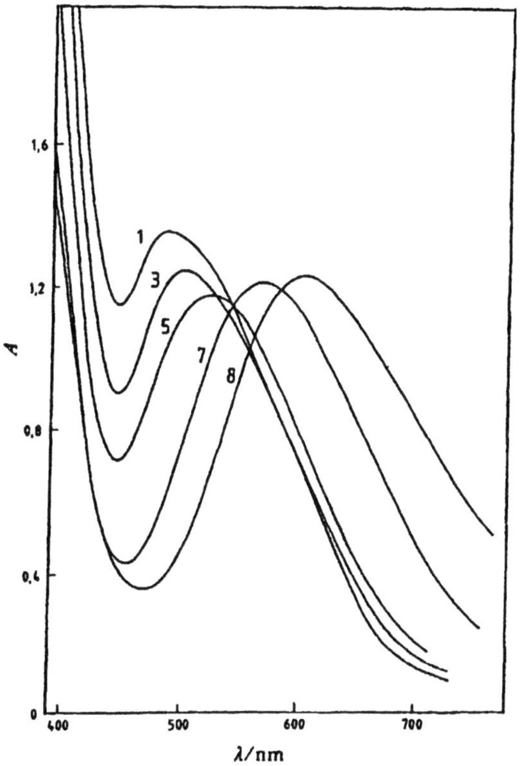

Fig. 15. UV/Vis spectra of [Cu(*acac*)(*tmen*)]BPh_4 in dichloromthane (1), 1,2-dichloroethane (3), nitromethane (5), acetone (7), and *DMF* (8)

in iron(II) complexes. Investigations on the solvatochromism of different chelates show how the colour changes can be used for the understanding of various polarity concepts in addition to studying their coordination structure.

Acknowledgements

This work was partly supported by a Grant-in-Aid (Scientific Research No. 11304046) from the Ministry of Education, Science, Culture, and Sports of Japan.

References

[1] Kahn O (1993) Molecular Magnetism. VCH, New York
[2] Toftlund H (1989) Coord Chem Rev **94**: 67
[3] Gütlich P, Hauser A, Spiering H (1994) Angew Chem **106**: 2109
[4] König E (1991) Struct Bonding **76**: 51
[5] Gütlich P (1981) Struct Bonding **44**: 83
[6] Deccurtins S, Gütlich P, Köhler CP, Spiring H, Hauser A (1984) Chem Phys Lett 1
[7] Deccurtins S, Gütlich P, Hasselbach KM, Hauser A, Spiring H (1985) Inorg Chem **24**: 2174
[8] Chum HL, Vanin JA, Holanda MID (1982) Inorg Chem **21**: 1146
[9] Addison AW, Burman S, Wahlgren CG (1987) J Chem Soc Dalton Trans 2621
[10] Renz F, Hasegawa M, Hoshi T, El-Ayaan U, Linert W, Fukuda Y (1999) Mol Cryst Liq Cryst **335**: 531

[11] Renz F, El-Ayaan U, Linert W, Fukuda Y (1999) Mol Cryst Liq Cryst **335**: 521
[12] Strauß B, Gutmann V, Linert W, Jameson RF (1992) Monatsh Chem **123**: 537
[13] Strauß B, Gutmann V, Linert W, Jameson RF (1993) Monatsh Chem **124**: 515
[14] Linert W, Konecny M, Renz F (1994) J Chem Soc Dalton Trans 1523
[15] Enamullah M, Linert W (1995) J Coord Chem **35**: 325
[16] Enamullah M, Renz F, El-Ayaan U, Wiesinger G, Linert W (1997) Vibr Spect **335**: 521
[17] Evans DF (1959) J Chem Soc 2003
[18] Beattie JK, McMahon KJ (1988) Aust J Chem **41**: 1315
[19] McGarvey JJ, Lawthers I, Heremans K, Toftlund H (1990) Inorg Chem **29**: 252
[20] McGarvey JJ, Lawthers I (1982) J Chem Soc Chem Commun 906
[21] Dose EV, Hoselton MA, Sutin N, Tweedle MF, Wilson LJ (1978) J Am Chem Soc **100**: 1141
[22] Bocá R, Baran P, Dlhán L, Šima J, Wiesinger G, Renz F, El-Ayaan U, Linert W (1997) Polyhedron **16**: 47
[23] Goodgame DML (1972) Bull Soc Chem Fr 3
[24] Sams JR, Scott JC, Tsin TB (1973) Chem Phys Lett **18**: 451
[25] Sams JR, Tsin TB (1976) J Chem Soc Dalton Trans 488
[26] Sasaki Y, Shigematsu T (1973) Bull Chem Soc Jpn **46**: 3438
[27] Reeder KA, Dose EV, Wilson LJ (1978) Inorg Chem **17**: 1071
[28] Miller JS (2000) Encyclopedia of Smart Materials. Wiley, New York
[29] Yamaki S, Fukuda Y, Sone K (1982) Chem Letters 269
[30] Fukuda Y, Miyamae H, Yamagata K, Sone K (1984) Chem Letters 1309
[31] Sone K, Fukuda Y (1987) Inorganic Thermochromism, Inorganic Chemistry Concepts, vol 10. Springer, Berlin Heidelberg
[32] Iwasaki N, Sone K, Fukuda Y (1975) Z Anorg Allg Chem **412**: 170
[33] Gibson JG, McKenzie ED (1971) J Chem Soc **A**: 1029
[34] Fukuda Y, Sone K (1972) Bull Chem Soc Jpn **45**: 465
[35] Fukuda Y, Shimura A, Mukaida M, Fujita E, Sone K (1974) J Inorg Nucl Chem **36**: 1265
[36] Fukuda Y, Miura Y, Sone K (1977) Bull Chem Soc Jpn **50**: 154
[37] Fukuda Y, Sato N, Hoshino N, Sone K (1981) Bull Chem Soc Jpn **54**: 428
[38] Sone K, Fukuda Y (1983) Ions and Molecules in Solution, Studies in Physical and Theoretical Chemistry, vol 27. Elsevier, Amsterdam, p 251
[39] Fukuda Y, Yasuhira M, Sone K (1985) Bull Chem Soc Jpn **58**: 3518
[40] Gutmann V (1967) Coord Chem Rev **2**: 239
[41] Nga NT, Fukuda Y, Sone K (1977) Bull Chem Soc Jpn **50**: 154
[42] Soukup RW, Schmid R (1985) J Chem Edu **62**: 459
[43] Gutmann V (1978) The Donor-Acceptor Approach to Molecular Interactions. Plenum Press, New York London
[44] Gutmann V (1968) Coordination Chemistry in Non-Aqueous Solutions. Springer, Vienna
[45] Mayer U (1983) Ions and Molecules in Solutions. Elsevier, Amsterdam, p 219
[46] Schmid R, Gutmann V (1982) Coord Chem Rev **43**: 150
[47] Fukuda Y, Sone K (1972) J Inorg Nucl Chem **34**: 2315
[48] Fukuda Y, Sone K (1975) J Inorg Nucl Chem **37**: 455
[49] Fukuda Y, Morishita R, Sone K (1976) Bull Chem Soc Jpn **49**: 1017
[50] Soukup RW, Sone K (1987) Bull Chem Soc Jpn **60**: 2286
[51] Schmid R (1983) J Sol Chem **12**: 135
[52] Schmid R, Sapunov VN (1982) Non-Formal Kinetics. VCH, Weinheim

Received April 2, 2001. Accepted (revised) April 19, 2001

Invited Review

Recent Advances in the Description of the Structure of Water, the Hydrophobic Effect, and the Like-Dissolves-Like Rule

Roland Schmid*

Institute of Inorganic Chemistry, Technical University of Vienna, A-1060 Vienna, Austria

Summary. Following a critical survey of the vast recent literature, the state of the art may be summarized as follows:

(A) Water structure. The key is appreciating the next-nearest neighbour aspect. Thus, liquid water may be conceived as a fluctuating mixture of broadly two groups of structure elements: (*i*) an open ice-I_h-type outer neighbour O···O bonding at about 4.5 Å and (*ii*) a dense ice-II-type outer neighbour O···O bonding at about 3.4 Å. On the other hand, the nearest neighbour O···O distance of about 2.8 Å and the number of these neighbors (4) is very similar in the solid and liquid state. The characterization of the two states may be directed either by the geometry of the H-bonds (more linear H-bonds in (*i*) and more bent H-bonds in (*ii*) or by the bonding forces operating (H-bonding favours the ordered open state (*i*), oxygen–oxygen interactions favour the random dense state (*ii*). Basically, the nature of liquid water can be understood in terms of a competition between H-bond (*Coulomb*) and dispersion (*van der Waals*) forces. Since the bonding characteristics in crystalline phases carry over to the liquid state, any molecular dynamics (MD) model of the liquid would have first of all to reproduce well the ice polymorph structures under appropriate thermodynamic conditions.

(B) Hydrophobic effect. The two classic approaches, *i.e.* the clathrate cage model and the cavity-based model, appear to be just different perspectives on the same physics. The particular features of water are (*i*) the small molecular size or, more specifically, the small size of the space between water molecules and the low expansibility, and (*ii*) the structure of the water molecule with the same number of donor and acceptor sites arranged tetrahedrally. Due to (*i*), cavity formation is particularly demanding, and this is the main contributor to the hydrophobic effect. This is mitigated by the capability of water, due to (*ii*), to form a cage around a nonpolar solute without sacrificing much of the H-bonding; rather, H-bonding networks are stabilized by the presence of guest molecules. In view of the tangential orientation of the first-sphere waters, such a cage can be compared with an elasticated net effecting strong solute–solvent dispersive interactions, rendering the solubility of nonpolar gases exothermic at room temperature. Furthermore, cavity formation largely determines the excess entropy, whereas dispersive forces determine the excess enthalpy. This gives rise to compensation behaviour when the solute size varies. Whereas an increase in solute size enhances the cavity formation energy, polarizability is also increased, and this leads to stronger solute–water interaction. Unfortunately,

* E-mail: rschmid@mail.zserv.tuwien.ac.at

present models of cavity formation predict positional entropies that are far in excess of the experimental entropies so that orientational contributions due to cage formation are hard to accommodate.

(C) Like-dissolves-like rule. The number of exceptions is dramatically reduced if the term polarity is given a broader meaning. Instead of identifying it solely with dipolarity, it should also include higher multipolar properties, in particular quadrupolarity. Quadrupolar solvent effects on solvation and reactivity are receiving increasing attention, particularly in low dielectric solvents.

Keywords. Water structure; Hydrophobic effect; Cavity formation; Solvent reorganization energy; Compensation effect; Quadrupolar solvents.

Introduction

Hydrophobicity is, without doubt, one of the most widely discussed features of contemporary solution chemistry and biochemistry. It is invoked in treating phenomena as different as the cleaning action of soaps and detergents, the influence of surfactants on surface tension, the immiscibility of nonpolar substances with water [1], the formation of biological membranes and micelles [2, 3], the folding of biological macromolecules in water [4], clathrate hydrate formation [5], and the binding of a drug to its receptor [6]. Besides, hydrophobic interactions are considerably involved in self-assembly, leading to the aggregation of nonpolar solutes or, equivalently, to the tendency of nonpolar oligomers to adopt chain conformations in water relative to a nonpolar solvent [7].

Due to these wide span of applications, the hydrophobic effect has often been separated into two categories: hydrophobic hydration describing structural and dynamic changes of water around a nonpolar solute, and hydrophobic interaction referring to the tendency of nonpolar solutes to aggregate in aqueous solution. In these terms, the latter category can be viewed as the reverse of the former.

Historically, the concept of hydrophobicity arose in the context of the low solubility of nonpolar compounds in water. Thus, unlike simple organic solvents, the insertion of a nonpolar solute into water at room temperature is (1) strongly unfavourable though slightly favoured by enthalpy, but (2) strongly opposed by a large, negative change in entropy, and (3) accompanied by a large positive heat capacity. An example is given in Table 1 showing the thermodynamic properties of methane dissolved in water and in carbon tetrachloride. In general, the solubility of nonpolar gases in water is 1 to 3 orders of magnitude smaller than in common hydrocarbon liquids [8]. Formerly unaware of the importance of dispersion (*van der Waals* (*vdW*), nonbonded) interactions, these thermodynamic characteristics were interpreted in terms of the formation of an ordered structure of water (an

Table 1. Solution thermodynamics of methane in water and carbon tetrachloride at 25°C; data are taken from *Lazaridis T, Paulaitis ME* (1992) J. Phys. Chem. **96**: 3847 and Ref. [67]

	Water	CCl_4
ΔG/kJ · mol^{-1}	+8.4	+0.9
ΔH/kJ · mol^{-1}	−10.9	−1.2
ΔS/J · mol^{-1} · K^{-1}	−64.6	−7.1
298ΔS/kJ · mol^{-1}	−19.2	−2.1
ΔC_p/J · mol^{-1} · K^{-1}	217.5	(0 to 42)

Table 2. Change in the solution thermodynamics of methane in water with temperature; data are taken from Ref. [142b]

	25°C	65°C	160°C
$\Delta G/\mathrm{kJ \cdot mol^{-1}}$	8.4	10.5	17.4
$\Delta H/\mathrm{kJ \cdot mol^{-1}}$	−10.9	−2.3	+11.3
$\Delta S/\mathrm{J \cdot mol^{-1} \cdot K^{-1}}$	−64.6	−37.7	−14.1
$T\Delta S/\mathrm{kJ \cdot mol^{-1}}$	−19.2	−12.8	−6.1

iceberg) around the nonpolar solute, which, though exothermic, is entropically unfavourable. It should be emphasized, however, that (due to the large positive heat capacity change involved) both heat and entropy of solvation are strongly temperature dependent. As is shown in Table 2, the large positive free energy of mixing of hydrocarbons with water is dominated by entropy only at ambient temperatures, whereas it is dominated by enthalpy at higher temperatures where the disaffinity is enlarged. Paradoxically, where hydrophobicity is strongest, entropy plays a minor role. Consequently, models that focus on water at around room temperature miss much of the thermodynamics of the oil/water solvation process. In addition, since the hydration of any solute – polar, nonpolar, or ionic – is accompanied by a decrease in entropy [9], the negative entropy of hydration is arguably not the main characteristic feature of hydrophobicity. The qualitative similarity in the hydration entropy behaviour of polar and nonpolar groups contrasts sharply with the opposite sign of the heat capacity change in polar and nonpolar group hydration. Nonpolar solutes have a large positive heat capacity of hydration, whereas polar groups have a smaller negative one. Thus, the large heat capacity increase might be what truly distinguishes the hydrophobic effect from other solvation phenomena [10].

Indeed, thermodynamic observables reflect bulk or averaged quantities. Thus, with the advent of computer simulation techniques (molecular dynamics (MD), Monte Carlo (MC)) every endeavour has been made to decipher features that are not directly measurable. After three decades of such studies it appears that simulation of chemical systems is a powerful way of gaining information on the atomic scale. However, the prolific literature devoted to simulations of water and aqueous solutions is increasingly hard to survey. In addition, several theoretical approaches to solvation exist, and the correspondence between the various terms appearing in different theories is not clear. Furthermore, establishing a connection between theoretical terms and the many intuitive ideas about solvation has proved to be difficult. Consequently, different qualitative conclusions have frequently been drawn as to the physical origin of hydrophobicity.

Another source of confusion is that because solvation is complex, insights are commonly obtained by studies of simpler reference fluids. The conclusions drawn from these studies are then extrapolated to real fluids. Since there is no unique way of doing this, the results may differ depending on the scheme of extrapolation adopted [11]. For all these reasons, there are currently many contradictory papers on the subject. Thus, in a quite recent communication [12] it is claimed that water should not be blamed for the hydrophobic effect. Instead, the formation of molecular aggregates is not limited to aqueous solutions, but it is a universal phenomenon associated with appreciably non-ideal solutions. The history of the study of the hydrophobic effect is in fact a vivid example of a tenet given by *Bertrand Russell* saying: "Unless we can know something, without knowing everything, it is obvious that we never know something."

It is therefore appropriate to review the subject with emphasis placed on publications that have appeared since the review by *Blokzijl* and *Engberts* in 1993 [13]. The review of the present author may also be noted [14]. It should be stressed here that the vast literature on the topic is virtually impossible to survey comprehensively. In what follows only a few papers (and references therein) that have paved the way to the present state of the art will be cited. It is not intended just to give an overview of existing ideas, but to sort out facts and fiction, both experimental and theoretical. It should be mentioned that this study is restricted to the phenomenon of hydrophobic hydration, thus omitting hydrophobic interaction.

Structure of Liquid Water

Crucial for all aspects of aqueous chemistry is an understanding of the structure of water itself. In a liquid, molecules are in constant motion and do not have fixed positions and orientations. Therefore, any structure of the liquid must be described in terms of probabilities and averages. The numerous models for water structure proposed so far fall into two main classes: continuum and mixture models, differing basically in their perception of the hydrogen bonds. The former suggests almost completely H-bonded molecules in a continuous network where distortion of the H-bonds results in a smooth distribution of H-bond distances, angles, and energies. In this case, there should be a one-peak distribution of H-bond strengths and geometries. *Giguère*, therefore, termed this model 'uniformist' [15]. However, this scheme is incompatible with both the physical properties of water and its behaviour manifested in (properly disentangled) high-resolution *Raman*, near IR, and neutron scattering spectra. For the former, the (roughly ten) thermodynamic and kinetic anomalies, such as the density maximum at 4°C, and the compressibility minimum at 45°C, cannot be explained. From spectroscopic results, there is evidence of an equilibrium between broadly two different groups of water structural elements as follows.

Raman studies on water as a solvent go back at least to 1965 when *Wall* and *Hornig* obtained a single broad OD band displaying only slight asymmetry [16]. Lack of evidence for two or more distinct peaks caused them to favour an interpretation involving a continuum distribution of hydrogen bond properties. *Walrafen's* subsequent work at higher resolution displayed an asymmetry more clearly [17]. In addition, he obtained a frequency of temperature independent intensity, that is, an isosbestic crossing of isotherms. Such a behaviour is indicative of a system composed of two distinct classes of scatters as has subsequently been supported by *d'Arrigo* and coworkers [18], *Giguère* [19], *Walrafen* and coworkers [20], and *Hare* and *Sørenson* [21].

Further evidence of a two-state system stems from mid-infrared spectroscopy, *i.e.* absorption spectroscopy of fundamental ($0 \rightarrow 1$) vibrational transitions, and this may be considered one of the most powerful techniques in the study of H-bonds. However, the strong absorbance of the fundamental OH stretching gives rise to intense broad bands, prohibiting the derivation of reliable quantitative structural information from this region with standard infrared techniques. Fortunately, new techniques such as attenuated total reflection (ATR) allow to avoid saturation effects. Such ATR spectra recorded by *Maréchal* as a function of temperature can be decomposed into two spectra, a low temperature one and a high temperature one [22]. The same result has been derived from ATR infrared measurements resolved by evolutionary curve resolution [23].

A particularly intriguing case is the advancement of the interpretation of X-ray and neutron diffraction experiments for obtaining atom-atom radial distribution functions (RDFs) $g(r)$ in water. Such measurements can reveal the local co-ordination structures of molecules in liquids. According to a classical integration procedure, the O–O distribution has a well-defined peak at 2.975 Å consisting of *ca.* 4.5 oxygen atoms out to a distance of 3.3 Å from an oxygen atom at the origin [24]. These numbers and distances suggest that the near neighbor coordination of water molecules is well defined and roughly tetrahedral at any instant in time but that a substantial number of molecules (in the order of 10%) are to be found in other configurations. The location of the first minimum in the O–O RDF around 3.3 Å is usually used to define the hydrogen bond. Thus, two water molecules are commonly considered to be H-bonded if their O–O distance is less than or equal to this distance. There was a long-standing debate as to whether the local pair density in water at interparticle separations of about 3.5 Å originates from large fluctuations in the separation of first H-bonded neighbours, from indirect second-neighbor correlations in small four- and five-membered rings, or from the distinct so-called interstitial fraction of water molecules. In fact, any RDF is the one-dimensional representation of a three-dimensional structure (and as such spatially averaged) and therefore carries a limited amount of structural information. Consequently, more recent simulation [25, 26] and experimental studies [27] employ orientation resolved spatial distribution functions (SDFs) to clarify the structural behaviour. In another paper [28], molecular orientations have been defined using *Eulerian* angles.

Notwithstanding, a look at the X-ray RDFs for liquid water suggests that the first peak cannot fit well a single *Gaussian* curve. The asymmetry towards longer distances increases with temperature, pointing at another overlapped $O \cdots O$ peak, and, therefore, a more complex first coordination sphere of the water molecule [29]. That feature had already been considered in 1962 by *Némethy* and *Scheraga* [30] on the basis of the X-ray diffraction patterns of amorphous ice, where the two peaks are well resolved. A final corroboration of this idea was provided by *Bosio*, *Chen*, and *Teixeira* [31] who introduced, for the study of liquid heavy water, a new approach. Making use of the water's maximum in the temperature-density curve (at 11.2°C), measurements of the X-ray structure factor can be performed at pairs of temperatures around the maximum at which the density is a constant. By subtracting the respective RDFs they obtained isochoric temperature differentials (ITD). In this way the temperature dependence of the structural parameters can be obtained much more precisely, since many density dependent corrections are circumvented. It was then shown [32] that these ITDs display the remarkable property that only the heights of the maxima and minima, but not their positions, depend on temperature. This is interpreted to mean that the water molecules exist in two distinct configurations characterized by either four or five neighbours. A summary of the various types of isosbestic points found for the structural changes of water with variable temperature and pressure, which is strong evidence for a two-state mixture model, has been given very recently [33].

In structural terms, the results from the ITD measurements reveal that it is not adequate to consider the coordination number of water as 4.5. Rather, it is essential to differentiate between first-neighbour and second-neighbour coordination. There are a nearest-neighbour O···O distance of about 2.8 Å and at least two outer-neighbour O···O distances, one of about 3.3 Å, whose intensity grows with increasing temperature, and one near 4.5 Å, which decreases with temperature. The interpretation of these distances in terms of chemical bonding is straightforward through their close relationship with the radial distribution functions of the various forms of ice in combination with the corresponding crystal structures. In fact, as pointed out by *Kamb* and referred to by *Robinson* and coworkers [34], the RDFs of the liquid can be created by a superposition of ice-I_h, -II, and -III RDFs (the subscript h means hexagonal). In other words, the bonding characteristics in crystalline phases carry over to the liquid state, a reasonable feature indeed. Most intriguing is the 3.3 Å distance which does not occur at all in ice-I structures but is very close to prominent features in ice-II, -III, -V, and -VI.

Ice-II, formed from ice I by raising the pressure to about 3 kbar at temperatures around −60°C, is taken as a case study. According to *Kamb* [35, 36], the structure of ice-II can be derived from ice-I as follows: detach the hexagonal columns of ice-I from one another, move them relatively up and down parallel to the long hexagonal axis so as to give a rhombohedral stacking sequence for equivalent six-rings, rotate them by *ca.* 30° around this axis, and finally re-link them together in a more tightly fitting way than in ice-I. As a result, each oxygen is bonded to four nearest neighbours at 2.77–2.82 Å and has, in addition, a next-nearest neighbour at 3.24–3.60 Å, which is not H-bonded to the central water molecule. This is shown in Fig. 1. In ice-I_h, in comparison, there is the same nearest neighbour O···O distance of 2.75 Å, whereas the next-nearest neighbour is 4.5 Å away from the center. Another difference between ice-I_h and ice-II is the O···O···O angle

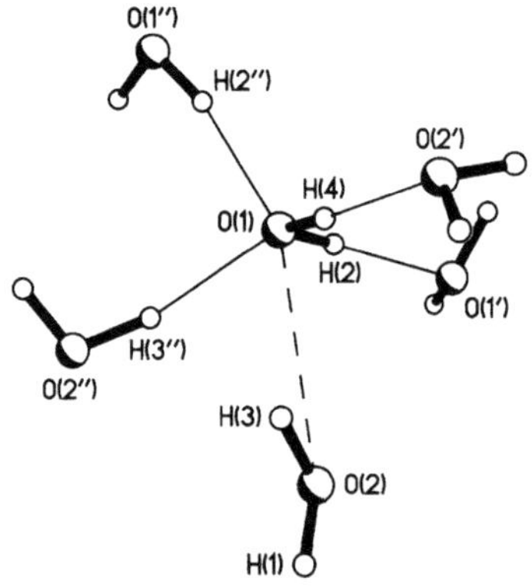

Fig. 1. Structure element of ice-II showing the next neighbours of a water molecule. Each of the four nearest oxygen atoms, at distances of 2.77–2.84 Å, is linked by a hydrogen bond. The fifth oxygen atom is non-H-bonded with an O(1)–O(2) distance of 3.24 Å. Data are taken from Ref. [36]

(*i.e.* O···H–O–H···O), which is 109.5° in the former but 93.8° (average) in the latter. Thus, H-bonds are linear in ice-I, but appreciably bent in ice-II. In liquid water, both features are simultaneously present.

It should be stressed, however, that water should not be conceived as some sort of mixture of ices, but rather is a rapidly (on picosecond timescales) fluctuating mixture of the intermolecular bonding types found in the polymorphs of ice. This two-state mixture model differs fundamentally from the traditional two-species mixture model, which considers liquid water as a mixture, or solution, of at least two molecular species, such as tetramers and octamers [37], distinguishable by the number of H-bonds. This latter model implies the concept of broken H-bonds (whose nature has never been defined) and estimates of their concentration in water, based on various criteria, diverge over an excessively wide range [38]. In fact, the presentation of diffraction data by orientational correlation functions shows that there are a range of local orientations compatible with the neutron data. This argues against the notion of water forming short lived, ice-like clusters at ambient temperature and pressure [27]. Furthermore, according to more recent *Raman* spectroscopic work [39], the concentration of free OH groups is only of the order of 1–2% by comparison with the spectra of the hydroxyl ion in aqueous solution. Therefore, some authors introduced the concept of the bifurcated H-bond, for which a hydrogen atom is bonded to three oxygen atoms [19]. This is another kind of molecular interaction, intermediate between the linear H-bond and the free OH group. It should be noted, however, that there is no suggestion of bifurcation of any of the H-bonds in ice-II.

Summing up, water may well be interpreted as a mixture of ice-I_h and ice-II type (representing all the other ice polymorphs) bonding, with average compositions depending on temperature and pressure. With increasing temperature or pressure, a transformation on the average occurs from the capacious, more rigid I_h-type outer-neighbour O···O bonding at 4.5 Å to the dense, more fragile II-type outer-neighbour O···O bonding near 3.4 Å. This implies that the bonding differences between the two states are not at the nearest-neighbour level but occur instead in the outlying non-hydrogen bonded next-nearest neighbour O···O structure. As in the polymorph structures, the transformation takes place by a bending of the inner-neighbour H-bonds without appreciably changing neither the O···O nearest neighbour distances of about 2.8 Å, nor the number of these neighbours (four).

The characterization of the two states may be directed either by the nature of the H-bonds or by the bonding forces operating. For the former, the dense and the capacious states involve linear and bent H-bonds, respectively. The latter definition is more fundamental, referring to a competition between O–O *vdW* interactions that favour random dense states and H-bonding that favours ordered open states. In this way the peculiarities of water are adequately captured: they are not simply due to H-bonding interactions, but rather to the admixing of trivial nondirectional *vdW* forces holding otherwise together the regular or unstructured liquids. It is intriguing to speculate that *vdW* interactions are necessary to offset energy losses caused by H-bonds bending to form the much more compact O···O···O angles [40]. Note that the low density of ice is due to the fact that H-bonding is stronger than *vdW* interactions. Optimal H-bonding is incommensurable with the tighter packing that would be favoured by *vdW* interactions. Ice melts when the thermal energy is sufficient to disrupt and disorder the hydrogen bonds, broadening the distribution of H-bond angles and lengths. Now among this broadened H-bond distribution, the *vdW* interactions favour those conformations of the system that have higher density. Hence, liquid water is denser than ice. In this framework all anomalies of water can be accommodated [40]. The density maximum, for instance, occurs due to the superposition of the open → dense transformation and normal thermal expansion when the temperature is raised.

These features are reflected by potential models used for the molecular dynamics simulation of liquid water of the form of Eq. (1) [28, 41], combining a *Lennard-Jones* (*LJ*) interaction between oxygen atoms with an electrostatic interaction modeled by interactions of point charges (q_i is the number of elemental charges in the ith site). On the other hand, it has been shown that a solute-hydrogen potential function need not be included [42].

$$V = 4\varepsilon\left(\left(\frac{\sigma}{r_{\text{oo}}}\right)^{12} - \left(\frac{\sigma}{r_{\text{oo}}}\right)^{6}\right) + \frac{1}{4\pi\varepsilon_0}\sum_{i,j}\frac{q_i q_j e^2}{r_{ij}} \qquad (1)$$

Hydrophobic Hydration

Ever increasing theoretical work within the last years has been devoted towards lifting the veil of secrecy about the molecular details of the hydrophobic effect, a subject of vigorous debate. Specifically, the scientific community would eagerly like to learn about the relative contributions of water-water and water-solute correlations to the thermodynamics of solution of nonpolar substances in water. Overall, there are two concepts: the older one clathrate cage model reaching back to the iceberg hypothesis of *Frank* and *Evans* [43], and the newer cavity based model.

The clathrate cage model assigns the responsibility of the low solubility of nonpolar solutes in water solely to the hydrogen bonding of water. In these terms, the structure of water is strengthened around a hydrophobic solute in order to avoid wasting hydrogen bonds. This causes a large unfavourable entropic effect, because the surrounding water molecules adopt only a few orientations (low entropy), with all water configurations fully H-bonded (low energy). Experimental evidence of structure strengthening is claimed to stem from NMR and FT-IR studies [44], NMR relaxation [45], dielectric relaxation [46], and HPLC [47].

In the cavity based model, the hard core of water molecules is more important to the hydrophobic effect than H-bonding of water. The process of solvation is dissected into two components: the formation of a cavity in the water to accommodate the solute and the interaction of the solute with the water molecules. The creation of a cavity reduces the volume of the translational motion of the solvent particles. This causes an unfavourable entropic effect. Thus, large perturbations in water structure are not required to explain hydrophobic behaviour. This conclusion arose from the surprising success of the scaled particle theory (SPT) which is a hard-sphere fluid theory, to account for the free energy of hydrophobic transfers. Since the theory only uses the molecular size, density, and pressure of water as input and does not explicitly include any special features of H-bonding of water, the structure of water is arguably not directly implicated in the hydration thermodynamics; however, the effect of the H-bonds of water is implicitly taken into account through the size and density of water. The proponents of this hypothesis argue that the entropic and enthalpic contributions arising from the structuring of water molecules largely compensate each other. In fact, there are several papers propagating exact enthalpy-entropy compensation of solvent reorganization [48–51]. Likewise, recent simulations [52, 53] and neutron scattering data [54–56] suggest that solvent structuring might be of much lower extent than previously believed. Also, a recent MD study report [57] stated that the structure of water is preserved, rather than enhanced, around hydrophobic groups.

Since there is strong evidence for both models, the importance of water structure enhancement has been rather equivocal. The reason for this is at least twofold. Firstly, theoretical models have many adjustable parameters, so their physical bases are not always clear; additionally, MD simulations usually do not take into consideration the aforementioned next-nearest neighbour aspect of liquid water. This raises serious questions about the realism of the results. Any MD model of the liquid would have first of all to reproduce well the ice polymorph structures under appropriate thermodynamic conditions. Secondly, the free energy alone masks the underlying physics in the absence of a temperature dependence study because of the entropy-enthalpy compensation.

Notwithstanding these difficulties, it appears that the two approaches, the clathrate cage model and the cavity based model, are just different perspectives of the same physics with different diagnostics reporting consequences of the same shifted balance between H-bonds and *vdW* interactions. Actually, in a very recent paper, a unified physical picture of hydrophobicity based on both the hydrogen bonding of water and the hard-core effect has been put forward [58]. Thus, hydrophobicity features an interplay of several factors. In the following the state of the art of cavity formation, solute-water correlations, and entropy-enthalpy compensation will be delineated. First, the different reference states used in the literature and their conversions will be briefly outlined, since from that side the issue has been unnecessarily obscured.

Standard states

There are numerous ways of expressing the solubility of a gas in a liquid. Of these, the *Ostwald* solubility *L*, or gas-liquid partition coefficient, is an especially useful measure. It is defined as the ratio of the concentration of gas molecules B dissolved at equilibrium in the liquid solvent to their concentration in the gas phase. In other words, *L* is the ratio of the number densities

$$L = \frac{[\mathrm{B}_s]}{[\mathrm{B}_g]} = \frac{\rho_\mathrm{B}^l}{\rho_\mathrm{B}^g} = \frac{\textit{volume of absorbed gas}}{\textit{volume of solvent}} \tag{2}$$

where ρ_B^l and ρ_B^g are the number densities of the solute B in the liquid (l) and the gaseous (g) phases, respectively. The standard states implied in this case are

$$\mathrm{B}(\text{ideal gas}, C_\mathrm{g} = 1) \rightarrow \mathrm{B}(\text{ideal solution}, C_\mathrm{s} = 1) \tag{I}$$

and

$$-RT\ln L = \Delta G_\mathrm{s} \tag{3}$$

where ΔG, henceforth, is equivalent to ΔG^* in *Ben-Naim*'s notation [59]. It is perhaps ironic to find that the directly measured quantity has the physically reasonable standard state involved, but formerly many authors proceeded to convert the primary data into less informative quantities. A frequently used standard state refers to gas at 1 atm and to a solution of unit mole fraction [60],

$$\begin{array}{ccc} \mathrm{B(ideal\ gas, 1\ atm)} & \rightarrow & \mathrm{B(ideal\ solution}, x_\mathrm{B} = 1) \\ G_g = G_g^\circ + RT\ln(p/p_0) & & G_l = G_l^\circ + RT\ln x_\mathrm{B} \end{array} \tag{II}$$

and

$$\underbrace{G_l - G_g}_{\Delta G} = \underbrace{G_l^\circ - G_g^\circ}_{\Delta G^\circ} + RT\ln(x_\mathrm{B} \cdot p_0/p)$$

For $p = p_0$,

$$\Delta G = \Delta G^\circ + RT\ln x_\mathrm{B} \tag{4}$$

The additional approximation of Eq. (5) is now applied,

$$x_\mathrm{B} = \frac{n_\mathrm{B}}{n_\mathrm{B} + n_\mathrm{A}} \approx \frac{n_\mathrm{B}}{n_\mathrm{A}} \tag{5}$$

where n_A is the mole fraction of the solvent. Now taking the volume per mole of ideal gas RT/p for n_A, and the molar liquid volume V_l for n_A, one obtains $\Delta G = \Delta G^\circ + RT\ln(RT/pV_\mathrm{l})$, where p is the external pressure. Since, however, the latter term is included in ΔG°, it has to be subtracted in order to arrive at ΔG, *i.e.*

$$\Delta G = \Delta G^\circ + RT\ln(pV_\mathrm{l}/RT) \tag{6}$$

Thus, for water at 25°C,

$$\Delta G = \Delta G^\circ - 17.87\,\mathrm{kJ} \tag{7a}$$

$$\Delta S = \Delta S^\circ + 68.27\,\mathrm{J} \tag{7b}$$

Furthermore, since $\partial(RT\ln(pV/RT)/\partial T = R\ln(pV/RT) - R$, it follows that $\Delta H = \Delta H^\circ + RT$, *i.e.*

$$\Delta H = \Delta H^\circ + 2.48\,\mathrm{kJ} \tag{7c}$$

It should also be realized that the concept of the standard state (II) is not internally consistent since Eq. (5) implies a highly dilute solution, whereas the standard state refers to the neat solute [61].

Another outmoded standard state is [62]

$$\mathrm{B(ideal\ gas, 1\ atm)} \quad \rightarrow \quad \mathrm{B(ideal\ solution}, C_\mathrm{s} = 1) \tag{III}$$

In converting reference state (III) into (I), the gas phase has to be compressed from 1 mol gas present in 24.46 dm^3 (at 298 K) to 1 mol in 1 dm^3. This requires an entropy of compression as high as $-26.58\,\mathrm{J \cdot K^{-1} \cdot mol^{-1}}$. Therefore, at 25°C,

$$\Delta G = \Delta G^\circ - RT\ln 24.46 = \Delta G^\circ - 7.93\,\mathrm{kJ} \tag{8}$$

The following considerations refer to the standard state (I), *i.e.* unit density number. This corresponds to Monte Carlo and molecular dynamics calculations performed in the isothermal-isobaric (*T,P,N*) ensemble, *i.e.* with fixed temperature, pressure, and number of molecules [63, 64].

Cavity formation

Since the venerable view of *van der Waals*, an intermolecular potential composed of repulsive and attractive parts is a fundamental ingredient of modern theories of the liquid state. Whereas the attractive interaction potential is not well known, the repulsive part, because of sharp changes with distance, is treatable by a common formalism in terms of the packing density η, which is the fraction of space occupied by the liquid molecules. The packing fraction is a key parameter in liquid state theories and is related in a simple way to the hard sphere (HS) diameter σ in a spherical representation of the molecules comprising the fluid, $\eta = \pi\rho\sigma^3/6 = \rho \cdot V_{HS}$, where ρ is the number density N/V = number of particles per unit volume and V_{HS} is the hard-sphere volume. Thus, by means of the molar liquid volume the packing fraction can be transformed into the HS diameter *via* $\eta = \pi\sigma^3 N_A/(6V_l)$. When σ is in Å and V_l in cm^3, we have

$$\eta = \frac{0.3153\sigma^3}{V_l} \tag{9}$$

The energy of dissolution of an inert gas in a molecular liquid is the result of the compensation of a positive cavity formation energy, which is the reversible work required to open up a cavity in the fluid by pushing against the repulsive solvent hard cores, and the negative stabilization energy due to attractions:

$$\Delta G_{sol} = \Delta G_{cav} + \Delta G_{att} \tag{10}$$

Since the cavity formation is a substantial part of the overall change in energy, much effort has been made towards an accurate calculation. HS theories only use two input parameters: d, which is the solute-solvent diameter ratio (σ/σ_s), and the solvent packing fraction η. The first equation was proposed from the scaled particle theory (SPT) [65, 66]:

$$\frac{\Delta G_{cav}}{RT} = \frac{2\eta^2 d^3}{(1-\eta)^3} + \frac{\eta d^2(2\eta d + 9/2)}{(1-\eta)^2} + \frac{\eta d(3 - 3d/2)}{1-\eta} + \eta d^3 - \ln(1-\eta) \tag{11}$$

The terminus SPT theory of fluids has been coined because it makes use of a coupling parameter which measures the size of a molecule and its potential field rather than the amplitude of this potential. A molecule is coupled formally to a fluid by letting it grow in size until it achieves the scale of its neighbours [65]. It is important to note that Eq. (10) is at variance with one frequently used, *e.g.* by *Graziano* [67, 68]. Thus, whereas *Graziano* calculated ΔG_{cav} for the solution of Ar in water to be 18.5 kJ/mol, Eq. (10) results in 30.9 kJ/mol, the latter being in agreement with the predictions of the equations given below.

The *Boublik-Mansoori-Carnahan-Starling-Leeland* (*BMCSL*) mixed hard-sphere equation of state has recently been shown to be somewhat superior to the SPT model in predicting the chemical potential of solvation in a HS fluid. The *BMCSL* cavity formation energy has the form [69, 70]

$$\frac{\Delta G_{cav}}{RT} = 2\frac{\eta d^3}{(1-\eta)^3} + 3\frac{\eta d^2}{(1-\eta)^2} + 3\frac{\eta d(-d^2 + d + 1)}{(1-\eta)} + (-2d^3 + 3d^2 - 1)\ln(1-\eta) \tag{12}$$

Recently, this equation has been further improved by *Matyushov* and *Ladanyi* [71], rendering it also accurate for high densities of the liquid and large sizes of the solute:

$$\frac{\Delta G_{\text{cav}}}{RT} = \frac{3\eta}{(1-\eta)} d + \frac{3\eta(2-\eta)(1+\eta)}{2(1-\eta)^2} d^2 + \frac{\eta(1+\eta+\eta^2-\eta^3)}{(1-\eta)^3} d^3 - \ln(1-\eta) \tag{13}$$

In the application of these equations, the liquid HS diameter is a crucial parameter. For its determination, the most direct method is arguably that based on inert gas solubility data [72, 73]. However, in view of the arduousness involved and the uncertainties in both the extrapolation procedure and the experimental solubilities, it is natural to look out for alternatives. From the various suggestions [74, 75] a convenient way is to adjust σ such that the computed value of some selected thermodynamic quantity related to σ is consistent with experiment. This is, in fact, the best method up to now [76]. To diminish effects of attraction, the property chosen should probe primarily repulsive forces rather than attractions. Since the low compressibility of the condensed phase is due to short-range repulsive forces, the isothermal compressibility $\beta_T = -(1/V)(\partial V/\partial P)_T$ might be a suitable candidate in the framework of the generalized *vdW* equation of state, further implicating dipole–dipole forces [77]

$$\beta_T \frac{RT}{V_l}(\eta Z_0'(\eta) - Z_0(\eta) - 2Z_\mu(\eta) + 2Z) = 1 \tag{14}$$

where Z, Z_0, and Z_μ, respectively, are the compressibility factors of the real liquid, of the HS liquid according to *Charnahan* and *Starling*, and of dipole–dipole interactions in the framework of the mean spherical approximation (MSA). The HS diameters so determined are found to be in excellent agreement with those derived from inert gas solubilities [76]. It may be noted that the method of *Ben-Amotz* and *Willis* [78], also based on β_T, uses the nonpolar HS liquid as the reference and, therefore, is applicable only to liquids of weak dipole–dipole forces. Of course, as the reference potential approaches that of the real liquid, the HS diameter of the reference liquid approximates more closely the actual hard-core length. Finally, because of its popularity, an older method should be mentioned that relies on the isobaric expansibility α_p as the probe, but this method is inadequate for polar liquids. It turns out that solvent expansibility is appreciably determined by attractive forces.

In Table 3, some representative values of η and σ based on Eqs. (14) and (9), including the two extreme cases, are given. Actually, water and n-hexadecane have the lowest and highest packing density, respectively, of all common solvents. Summed up, for the time being, the cavity formation energy may be calculated from Eq. (13) with values of σ taken from Refs. [76] and [74], *i.e.* for water, $\eta = 0.412$ and $\sigma = 2.87$ Å. These values are higher than those often used ($\eta = 0.363$ and $\sigma = 2.75$ Å) [68]. For Monte Carlo simulations, the effective HS diameter for water will critically depend on the water model used [79].

It should be noted that under isochoric conditions the free energy of cavity formation is a totally entropic quantity, that is

$$T\Delta S_{\text{cav},V} = -\Delta G_{\text{cav}}, \quad \Delta S_{\text{cav},V} = -(\partial \Delta G_{\text{cav}}/\partial T)_V$$

Table 3. Packing densities in some liquids; data are taken from Ref. [14]

Liquid	η	Free volume (%)	σ (Å)
H_2O	0.41	59	2.87
n-C_6H_{14}	0.50	50	5.94
Benzene	0.51	49	5.27
MeOH	0.41	59	3.77
Et_2O	0.47	53	5.38
n-$C_{16}H_{34}$	0.62	38	8.33

For the isobaric case which is the usual experimental situation, the chemical potential divides into entropy and enthalpy,

$$\Delta G = \Delta H - T\Delta S_p, \quad \Delta S_p = -(\partial \Delta G/\partial T)_p$$

and

$$\Delta S_{\text{cav},P} = \Delta S_{\text{cav},V} + \rho\alpha_p(\partial \Delta G_{\text{cav}}/\partial \rho)_T \tag{15}$$

The (positive) liquid expansibility term, however, is small for water due to its low expansibility ($\alpha_p = 0.26 \times 10^{-3}\,K^{-1}$), as compared to ordinary liquids ($\alpha_p \cong 1.1-1.3 \times 10^{-3}\,K^{-1}$). Along these lines, the large and negative entropy of cavity formation in water is traced to two particular properties of water: the small molecular size or, more specifically, the smaller size of the space between water molecules, and the low expansibility. The latter reflects the fact that chemical bonds cannot be stretched by temperature. There is also a recent perturbation approach showing that it is more costly to accommodate a cavity of molecular size in water than *e.g.* in hexane [80]. Considering the high fractional free volume for water (Table 3) it is concluded that the holes in water are distributed in smaller packets [81]. Compared to a H-bonding network, a hard-sphere liquid finds more ways to configure its free volume in order to make a cavity.

It may be mentioned that for the isobaric condition also cavity formation remains a quantity that is dominated by entropy. For example, ΔG_{cav} for methane in water is calculated to be 38.9 kJ/mol, which is made up of $\Delta H = 7.9$ kJ and $298\Delta S = 31.0$ kJ [82]. Thus the highly unfavourable cavity formation can be identified as a major contributor to the hydrophobic effect [83]. This can also be gauged from the solvation free energies of nitromethane as the solute in some aprotic solvents as shown in Table 4, where solvent ordering is, arbitrarily, according to the dielectric constant ε_s. The total solvation energy is a competition of the positive cavity formation energy and the negative solvation energy due to solute-solvent attractions (Eq. (10)). In the present case, only two kinds of attractions are relevant, that of dispersion and dipolar forces (neglecting quadrupolar forces, see below),

$$\Delta G_{\text{att}} = \Delta G_{\text{disp}} + \Delta G_{\text{dipolar}} \tag{16}$$

For the methods of calculation, the reader may consult the original paper [82]. Table 4 shows that the contributions of cavity formation and dispersion, both appreciable, substantially cancel each other. The trend found between the overall solvation free energies and the dielectric properties of solute and solvent is an

Table 4. Thermodynamic potentials (kJ/mol) of dissolution of nitromethane at 25°C; data are taken from Ref. [82]

Solvent	ε_s	ΔG_{cav}	ΔG_{disp}	$\Delta G_{dipolar}$	ΔG(calc)	ΔG(exp)
n-C_6H_{14}	1.9	23.0	−32.9	−2.2	−12.1	−12.1
cyclo-C_6H_{12}	2.0	28.1	−38.1	−2.8	−12.7	−12.0
Et_3N	2.4	24.2	−33.5	−2.9	−12.2	−15.2
Et_2O	4.2	22.6	−33.4	−5.8	−16.5	−17.5
EtOAc	6.0	28.0	−38.5	−9.7	−20.1	−21.2
THF	7.5	32.5	−41.5	−12.2	−21.1	−21.3
Cyclohexanone	15.5	35.1	−39.8	−17.9	−22.6	−21.8
2-Butanone	17.9	28.3	−33.9	−18.9	−24.5	−21.9
Acetone	20.7	27.8	−31.2	−22.1	−25.5	−22.5
DMF	36.7	38.1	−32.1	−28.5	−22.5	−23.7
DMSO	46.7	41.9	−31.1	−31.0	−20.2	−23.6

outcome of this competition. It is now easy to imagine the changes to be expected if we switch over to a nonpolar solute to be dissolved in water. From the above, the cavity formation energies are dramatically increased. For instance, ΔG_{cav} for the nitromethane solute in water is calculated to be as high as 59.0 kJ/mol. On the other hand, dipolar forces can be neglected, of course, whereas the contribution of dispersion (see below) may well remain virtually unchanged [82]. This renders nonpolar substances very poorly soluble in water.

Solute–water and water–water correlations

A major difficulty in studying the effect of nonpolar solutes on water has, of course, been their low solubility. In this case any experimental signal – IR or otherwise – from the perturbed hydrating waters is masked by the much larger contribution from bulk water. Due to this fact, computer simulations have got ahead of experiment in studying the solute-induced perturbations in water structure. The result has been a wide range of competing models. In recent years, fortunately, the situation has changed dramatically due to the considerable improvements made to various experimental techniques. In this way molecular models derived from computer simulations are put on trial. It is gratifying that very diverse and independent experimental results suggest a fairly consistent picture of the local molecular order present in aqueous solutions.

(1) There is a well defined nearest neighbour hydration shell.
(2) Water molecules in the first coordination shell of nonpolar molecules tend to be tangentially orientated with respect to the surface of the former.
(3) The OH bonds perpendicular to the first coordination shell have the preference of forming linear H-bonds to outer-sphere water.
(4) The H-bonds of the water molecules in the immediate hydration zone appear to be somewhat stronger than in the bulk.
(5) Structural properties of water beyond the second solvation shell are similar to those of bulk water, *i.e.* longer range correlations are only weakly apparent.

Experimental evidence comes from neutron diffraction and isotopic substitution studies on solute/water samples pressurized to increase the concentration. The solutes used were argon [84], krypton [85], and methane [86]. The secondary structures are relatively weak, suggesting that the solute-water interaction is limited to the first hydration zone. The tangential orientation of the first-sphere waters follows from the coincidence of the first-peak positions of the solute-oxygen and solute-hydrogen PDFs. Such an edge-on arrangement of water molecules has also been reported to occur around the tetramethylammonium ion [87] and the methyl groups of tertiary butanol [88]. Another experiment involves FT-IR spectroscopy of water in the presence of trimethylamine-N-oxide, which is, by its positive hydration heat capacity, judged to be hydrophobic-like [89]. In agreement with the recent view of the water structure described above, the authors interpret the IR spectra in terms of two H-bond angle populations, a more distorted and a less distorted (more linear, *i.e.* ice-like) one, with the presence of hydrophobic groups increasing the population of the less distorted H-bonds. This corresponds to the classical structure enhancement, as indicated by a significant ^{1}H NMR low-field shift of the protons of water surrounding the alkyl group of ethanol [90] (but contradicted by a recent simulation study [91]). However, since it is restricted largely to the first hydration zone, the term 'microscopic iceberg' is less appropriate as it suggests three-dimensional structures. Better terms might be 'frozen patch' [92] or 'elasticated net' [93] (see below). The sphere in the vicinity of the cavity can synonymously be characterized in terms of the coordination number, since less distorted H-bonds mean that the coordination number of water, or the local density, is lowered compared to bulk water [58]. In view of the ambiguity of the coordination number in the case that non-nearest neighbours become important, it may more precisely be argued that the portion of *vdW*-bonded water molecules is reduced. This, in combination with strong attractive solute–water interactions, is ultimately the reason for the slowing down of the translational and rotational mobilities of water near a nonpolar solute. Experimental evidence is now available from NMR techniques [94–96], corroborating simulation results [97, 98]. Reverting to the edge-on arrangement of first-neighbour water molecules it is worth mentioning that the same conclusion has been reached by a variety of computer simulations, despite the divergent potential functions as well as different simulation times and methods of calculation employed [99–107].

Another peculiar feature of hydrophobicity is the at first unexpected strength of the water-solute interaction. The aforementioned term 'elasticated net' was coined by *Feakins et al.* [93] in their illuminating papers on the thermodynamics of solutions in order to picture the water structure around nonpolar groups. These authors argued that the large positive activation parameters found for the viscous flow in aqueous solutions require strong interactions to occur between the alkyl group and the solvent. More direct evidence of the peculiar action of water in caging hydrophobic solutes is apparent in the sensitive solvent dependence of the NMR chemical shift of ^{129}Xe, which is an effective probe of the hydrophobic environment [108]. It is remarkable that the shift in water is similar to that in benzene or olive oil, and hence is much greater than would be expected from the water's refractive index (note the linear trend between the gas-to-solution shift of ^{129}Xe and the refractive index function of common solvents [108]). Moreover, the shift is much the same as the average of those found in the solid water clathrate. The relevance of this work is that it suggests not only that the xenon atom is caged in the water solution, but also that it is interacting in a noticeable fashion with the cage.

The occurrence of strong solute–water attractions follows, albeit indirectly, from Eq. (10), given that the favourable enthalpy in hydrophobic hydration is predominantly due to solute-water and not to water–water interactions and that the repulsive contributions can reliably be assessed from the HF formulas (Eqs. (11)–(13)). On the basis of the relationship $\Delta G_{att} = \Delta G_{sol} - \Delta G_{cav}$, *de Souza* and *Ben-Amotz* [109] put forward that experimental solvation free energies typically represent a near cancellation of large repulsive and attractive contributions. Thus, higher values of ΔG_{cav} in water (due to the small water hard-core diameter) would correspond to more favourable attractions. As an example, whereas the results in water and *n*-hexane are qualitatively similar, the magnitudes of both parts are larger by a factor of about 2 in water. It may also be noted that values of ΔG_{att} so obtained have been found to correlate with the solute polarizabilities, suggesting a dispersive mechanism for attractive solvation, and are in fair agreement with dispersion energy estimates. Along these lines, the at first unusual experimental finding of increasing solubility of the noble gases with increasing size (in contrast to aliphatic hydrocarbons whose solubility decreases with size) can be rationalized [82]. This differential behaviour is straightforwardly explained in terms of the high polarizability of the heavy noble gases, having a large number of weakly bound electrons strengthening the *vdW* interactions with water. It can be shown that for noble gases of increasing size the *vdW* interactions increase more rapidly than the work of cavity creation, thus enhancing solubility; the opposite holds for hydrocarbons, thus lowering the solubility with increasing size [67].

In view of the ever increasing awareness of the importance of the dispersion component of non-ionic solvation, an adequate assessment is urgently required. For this purpose, the *LJ* potential as given by the first right-hand term in Eq. (1), albeit a simplification, is employed. The *LJ* potential may be viewed as an effective potential representing a rotationally averaged soft repulsive core and dispersive attractions of the molecules at contact. A salient case in revealing the significance of the nonpolar solute-water interaction is the numerical value of the *LJ* energy of water either extracted from experimental data or else needed in simulation work to reproduce the thermodynamic, structural, and dynamic characteristics of water. In a recent paper [74], *LJ* energies ε_{LJ} of molecular fluids have been recalculated using some updated methods. Experimental sources of ε_{LJ} are typically gas solubilities (GS), second virial coefficients (SVC), fluid viscosity, or liquid expansibilities. Of these, *LJ* energies extracted from GS data would seem to be the most reliable. The traditional procedure is based on *Pierotti's* theory giving $\varepsilon_{LJ}(H_2O) = 79$ K and $\sigma = 2.77$ Å (where, as in the following, ε_{LJ} is given as ε_{LJ}/k_B) [110]. In this model, however, the effect of dispersions is treated in a continuum-like fashion. Another weak point is the manner of treating the repulsive contribution. A recalculation including cavity formation energy according to Eq. (12) and the *Barker-Henderson* theory for dispersion forces gives a much higher value of $\varepsilon_{LJ} = 332$ K ($\sigma = 2.87$ Å) [74].

Another method is based on the generalized *van der Waals* (G*vdW*) equation of state (Eq. (17)) where the liquid compressibility factor Z is composed of the compressibility factor Z_0 for intermolecular repulsions, modeled by the HS potential, and the attraction forces term. Since, however, the *vdW* constant *a* involves the gross effect of the liquid attractions, contributions from other sources must at first be separated out. At present, this is adequately feasible only for dipole–dipole forces. Values of ε_{LJ} so obtained range from 57 to 846 K [74]. Furthermore, calculations from viscosity data afforded $\varepsilon_{LJ} = 808$ K ($\sigma = 2.64$ Å) [111]. Other values, ranging from 167 to 775 K, have been discussed by *Pierotti* [112]. Unfortunately, all *LJ* energies found in the literature suffer more or less from the problem of containing contributions from more than dispersion forces alone.

$$Z = PV/RT = Z_0(\eta) - \beta a \eta \qquad (17)$$

Taken together, *LJ* energies extracted from GS data appear to be at present the least ambiguous choice for the following reasons: Firstly, a non-polar solute probes mainly the dispersion component of intermolecular attractions with dipolar and specific effects largely absent. Secondly, the free energies of dispersive forces, needed for the determination of liquid *LJ* energies from GS data, can adequately be calculated from *BH* and *WCA* theories. A disadvantage of the GS approach, however, is the necessity to employ arbitrary combining rules. It should be mentioned that upon using $\varepsilon_{LJ} = 332$ K in a thermodynamic analysis of hydration, and modifying the combining rules for larger solutes, the free energies of hydration for nonprotic solutes are well reproduced [82]. Beyond that, detailed comparisons with the *LJ* energy of water needed in simulation work to reproduce experimental data do not appear to be particularly valuable, since the numerical value to be chosen varies with the potential function employed [113]. Thus, for the TIP4P interaction potential of Eq. (1) the *LJ* parameters chosen were $\varepsilon_{LJ} = 88$ K and $\sigma = 3.15$ Å [114], and similarly for the TIP5P model, $\varepsilon_{LJ} = 80.6$ K and $\sigma = 3.12$ Å [41]. It should be remembered that the treatment of MD simulations is typically based on terms of the configuration of a pair of water molecules and does not (yet) take into consideration the aforementioned next-nearest neighbour aspect of liquid water. This raises serious questions about the realism of the results.

In conclusion, the relatively high ε_{LJ} value of liquid water as derived from the GS data (332 K) may be taken as another criterion for an enhanced nonpolar solute–water interaction. The physical basis for such strong interaction is intriguing. It has been suggested that the tangential ordering of O–H bonds of water molecules with respect to the surface of a non-polar solute might lead to enhanced proton tunneling (proton hopping), thereby creating a rapidly changing electric field that might strengthen the *London* dispersion interactions between water and the more polarizable solutes [67]. Thus, the interaction of the nonpolar group with its cage will not only be the sum of the *vdW* interactions of the group with the solvent molecules constituting the cage, but also be affected by the integrity of the cage.

In this context, it may be relevant to note that of the many crystalline clathrate hydrates known to date, guest molecules are always found to be separated by a single layer of shared solvent [115] as, for instance, in the helium hydrates [116]; a hydrophobic contact configuration of guest molecules has never been crystallized. This can well be indicative of the geometric constraints encountered by water molecules in an attempt to maintain a fully hydrogen bonded network surrounding the juxtaposed solutes. This is effected if the water molecules in the shell comprise four bonds to other water molecules, three bonds forming the surface of a cage and one pointing outward. Small non-polar solutes may tend to occupy the same position with respect to the *quasi*-tetrahedral lattice, so that the more weakly interacting water (cf. Fig. 1) would, in effect, be competing for this mismatch site. The types of cages to be envisaged in aqueous solution may involve forms like those observed in the dimethylether hydrate [117]. For large nonpolar solutes, however, the intimate hydration mechanism may change as recently suggested [118] and disputed [119].

For computational studies, it has been shown recently that the performance of polarizable water force fields in calculating phase equilibria is improved dramatically by introducing an additional coupling of the intramolecular structure to the electronic environment [120]. To analyze the involvement of cooperative effects, the through-space coupling (TSC) concept, which is the molecular orbital representation of *vdW* repulsive-attractive forces [121], may be mentioned. Non-bonded (*vdW*) repulsion is a manifestation of covalent antibonding and can be reduced upon charge redistribution, *i.e.* if electron density is released from high-lying antibonding TSC orbitals. This approach should open opportunities for the further study of water structure and hydrophobic (as well as ionic) hydration.

Nonpolar vs. polar (ionic) hydration

As noted in the Introduction, the study of the hydration heat capacity ΔC_p should provide a deeper insight into the role of changes in water structure upon hydration than analysis of entropy or enthalpy changes alone. There are at least two reasons for this. Firstly, whereas the hydration of both nonpolar and ionic solutes is accompanied by a decrease in entropy, values of ΔC_p have the opposite sign, a large positive ΔC_p for hydration of non-polar groups and a negative one in the case of ionic solutes. Thus, explanations for the positive ΔC_p of hydrophobic hydration in terms of the induction of more ordered water which 'melts' at increasing temperatures do not explain the negative ΔC_p for polar and ionic species. The latter phenomenon is perhaps the more puzzling one in view of the negative entropies of ionic hydration. Secondly, heat capacity relates the other three major thermodynamic quantities describing solvation (ΔG, ΔH, ΔS; Eq. (18)) where $\langle \Delta E^2 \rangle$ is the mean-squared fluctuation in total energy and k is *Boltzmann's* constant. These four equivalent definitions of C_p illustrate the diverse thermodynamic implications for processes involving changes in C_p.

$$\Delta C_p = \frac{\partial \Delta H}{\partial T} = T\frac{\partial \Delta S}{\partial T} = -T^2\frac{\partial^2 \Delta G}{\partial T^2} = \frac{\langle \Delta E^2 \rangle}{kT^2} \tag{18}$$

All recent explanations of the heat capacity puzzle are based on a two-state model of water structure. Before summarizing some relevant points, it is useful to compare the structural features of hydration of argon and the isoelectronic potassium cation, which differ dramatically with respect to the coordination number and the orientation of the hydrating water molecules. The argon atoms possess a nearest neighbour hydration shell composed of 16(2) water molecules in the range 2.8–5.4 Å, with the hydration shell molecules orientated with the dipole moment nearly tangentially to the Ar surface [84]. K^+, on the other hand, has 6–8 nearest neighbours at an internuclear K-O distance of 2.71–2.86 Å, with the hydrated water molecules, due to strong electrostatic solute-water interaction, directing their dipoles toward the central ion (with the M^+-O-H angle close to 127°) [122, 123].

Using a combination of Monte Carlo simulations and the random network model of water, *Sharp* and *Madan* [10, 124] have interpreted the IR spectra in terms of two major H-bond angle populations, a more distorted (more bent) and a less distorted (more linear, *i.e.* ice-like) one. The presence of hydrophobic groups increases the population of the less distorted H-bonds relative to the other, whereas polar and ionic solutes have the opposite effect. In addition, distortions in H-bond length and angle take place in a highly concerted fashion. Thus, the hydrogen bonds between the water molecules in the first hydration shell of a non-polar solute are shorter on average and less bent, giving rise to an increase in the water heat capacity whose sign is reversed for polar hydration. In terms of the last expression in Eq. (18), the water–water interaction contribution to ΔC_p is determined by the fluctuation in enthalpy given by the disparity in energy of different water microstates and the population of the energy levels. In other words, ΔC_p reflects the number of water molecules, *i.e.* the number of hydrogen bonds in the first shell, that are perturbed by the solute and further the extent of that perturbation. The positive sign of ΔC_p for non-polar hydration can be traced to (*i*) the fact that relatively many water molecules are engaged in the hydration shell and (*ii*) that the perturbation of the water structure is only weak so that the water molecules can fluctuate between these states. In the case of ion hydration, few water molecules are involved which, due to the strong electrostatic solute-water interaction, are so strongly associated with the solute that the water molecules cannot fluctuate easily between these states. Consequently, there is a net decrease in heat capacity.

The physical basis for the hydrophobic effect has also been studied using the statistical mechanical *MB* model of water in which the water molecules are represented as *Lennard-Jones* disks with hydrogen bonding arms [125] (the name of the *MB* model originates from the resemblance of each model water to the *Mercedes-Benz* logo). Similarly, the authors suggest a subdivision of the H-bonds in two classes which they call 'circumferential waters' and 'radial waters'. The former are 'cage waters' forming H-bonds to neighbouring first-shell water molecules but not to the solute. It is found that the insertion of a non-polar solute into cold water causes ordering and strengthening of the H-bonds in the first shell, but the reverse applies in hot water. Based on simple geometrical considerations of hydrophobic hydration, *Matubayasi* differentiates between matching and mismatching water regions [126], the former being characterized by stronger H-bonds than the latter. Insertion of a hydrophobic solute induces a local environment which is favourable for the strengthening of H-bonds between the neighbouring water molecules. In other words, H-bonding networks are stabilized by the presence of guest molecules.

It should also be mentioned in this context that pressure increases the solubility [127]. The effect of pressure on the entropy has been examined, and it was found that increasing pressure causes a reduction of orientational correlations, in agreement with the idea of pressure as a 'structure breaker' in water. Actually, frozen clathrate hydrates trapped beneath oceans and arctic permafrost may contain more than 50% of the world's organic carbon reserves [128, 129]. Likewise, the solubility of aromatics is increased at high pressure and temperature, with π-bond interactions involved [130].

Entropy-enthalpy compensation

Correlations between ΔH and ΔS known as the compensation effect are ubiquitous in physical chemistry [131–135], although the molecular origin remains obscure. In a simplified view, the origin of the compensation effect lies in the competing nature of maximum energy and minimum strain. Not unexpectedly, this phenomenon is also encountered in the thermodynamics of aqueous solutions. Thus, a linear trend between enthalpies and entropies of solution of quite diverse nonpolar gaseous solutes in water is apparent from Fig. 2. A closer view reveals that homologous series, or series of structurally similar compounds, separate into parallel straight lines (note the slightly deviating slope of the noble gases because they are monatomic and therefore lack degrees of freedom). This is particularly evident in Fig. 3, where also polar and protic solutes are included.

A very instructive example of enthalpy–entropy compensation is found in the thermodynamics of transfer of argon from cyclohexane as a reference solvent to water and to hydrazine (Table 5) [136]. Whereas the enthalpies and entropies are obviously very different, they almost completely compensate each other, rendering

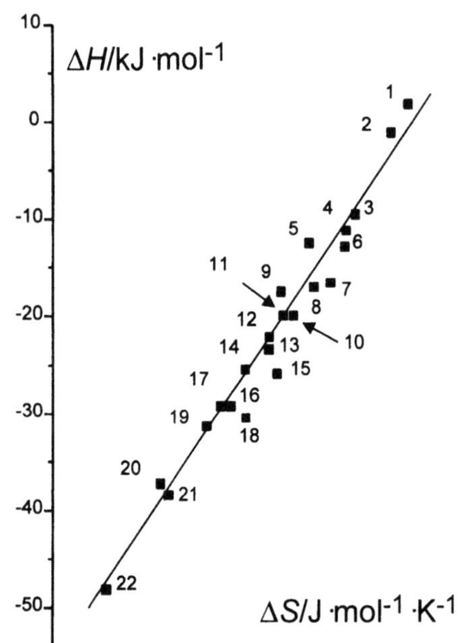

Fig. 2. Thermodynamics of solution of gaseous solutes in water at 298 K. The solutes are He (1), Ne (2), Ar (3), Kr (4), Xe (5), CF_4 (6), SF_6 (7), CH_4 (8), C_2H_6 (9), C_3H_8 (10), i-C_4H_{10} (11), n-C_4H_{10} (12), Me_4C (13), n-C_5H_{12} (14), c-C_5H_{10} (15), n-C_6H_{14} (16), Me_4Sn (17), c-C_6H_{12} (18), n-C_7H_{16} (19), n-C_8H_{18} (20), Et_4C (21), and Et_4Sn (22). Data are taken from *Abraham MH, Nasehzadeh A* (1981) J. Chem. Soc., Faraday Trans. 1 **77**: 321. Note that a correction for the enthalpy of sublimation or evaporation has been applied to non-gaseous solutes

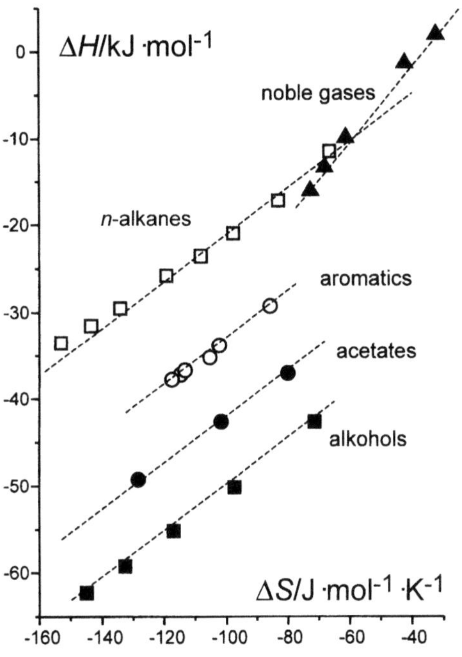

Fig. 3. Thermodynamics of solution of gaseous solutes in water at 298 K; data are taken from Ref. [82]; note that a correction for the enthalpy of sublimation/evaporation has been applied where necessary

Table 5. Thermodynamic data for the transfer of argon from cyclohexane to water and to hydrazine at 25°C; data are taken from Ref. [136]

Transfer of 1 mol Ar	$\frac{\Delta G_{tr}}{kJ/mol}$	$\frac{\Delta H_{tr}}{kJ/mol}$	$\frac{298\Delta S_{tr}}{kJ/mol}$
$C_6H_{12} \rightarrow N_2H_4$	11.88	9.50	−2.37
$C_6H_{12} \rightarrow H_2O$	10.38	−11.21	−21.58

the *Gibbs* free energies of transfer very similar. Since both solvents are hydrogen bonded, the peculiar nature of water cannot simply result from hydrogen bonding. Rather, water is unique by its molecular structure, having the same number of donors (protons) and acceptors (lone pairs) arranged tetrahedrally. This allows water to form a cage around a nonpolar solute without sacrificing much of the H-bonding. Moreover, the water structure can close on the solute like an elastic net, trapping the solvent and bringing the molecules close to it, enhancing their mutual *vdW* interactions. These could well be comparable in strength with the water–water interactions they replace [93]. Of course, the exothermic solute–solvent interaction occurs at the expense of orientational strain of the water cage. In contrast to water, hydrazine, due to its structure, is not a good cage former, rendering the transfer of argon from cyclohexane endothermic but, to compensate, entropically not so unfavourable. At higher temperatures water also gradually loses its tendency to form cages. On the other hand, cavity formation becomes less severe. Note that in going from 25 to 100°C the HS diameter σ of water changes from 2.87 to 2.51 Å, and the packing fraction η from 0.412 to 0.267 [137, 78]. Thus, on the basis of Eq. (13), ΔG_{rep} reduces from 38.9 to 23.5 kJ/mol, *i.e.* the solution of inert gases becomes less favourable by enthalpy but more favourable by entropy in a compensatory manner. This is reflected by the solubility data of methane in water as displayed in Table 2. It can be seen that the free energy is much less sensitive to temperature than are either enthalpy and entropy.

Another factor supporting enthalpy-entropy compensation is implicit in the curious agreement between the calculated and experimental ebullioscopic and cryoscopic constants. As is known from first-course physical chemistry, the theoretical derivation of these constants relies on the assumption that the free energy of the solvent varies with the number but not the nature of dissolved solute molecules [138].

Unfortunately, a detailed analysis of experimental enthalpy-entropy data such as those presented in Figs. 2 and 3 has not yet been set about; the problem may be approached in a way similar to that taken by *Linert* and *Jameson* in the treatment of isoparametric relationships [139]. What is eagerly being sought instead is to separate the hydrophobic process into individual steps by model calculations, looking for compensation behaviour of sub-systems, especially that of solvent reorganization, which may be defined as the difference of the average solvent–solvent interaction energy with and without the presence of the solute. On the basis of thermodynamic and statistical mechanical arguments it has been argued that the solvent reorganization energy and entropy exhibit complete cancellation [140, 48, 49].

Grundwald's arguments [48] proceed as follows: When solute B is dissolved in solvent A,

$$\mathrm{B}(g) \leftrightarrow \mathrm{B}(a) \tag{19}$$

the solvent is affected such that (s_{B} = mean coordination number of B)

$$\mathrm{B}(g) + s_{\mathrm{B}}\mathrm{A}(a) \leftrightarrow \mathrm{B}(a) + s_{\mathrm{B}}\mathrm{A}(b) \tag{20}$$

where A(a) denotes molecules A surrounded by A, and A(b) denotes molecules A in the neighbourhood of B. This equation is divisible into two parts:

$$\mathrm{B}(g) \leftrightarrow \mathrm{B}(a) \qquad \text{nominal equation}$$

$$s_{\mathrm{B}}\mathrm{A}(a) \leftrightarrow s_{\mathrm{B}}\mathrm{A}(b) \qquad \text{environmental equation}$$

In terms of mole numbers,

$$n_{\mathrm{B}(a)} = n_{\mathrm{B}} \quad \text{B in the environment of A} \tag{21a}$$

$$n_{\mathrm{A}(b)} = s_{\mathrm{B}} n_{\mathrm{B}} \quad \text{A in the environment of B} \tag{21b}$$

$$n_{\mathrm{A}(a)} = n_{\mathrm{A}} - s_{\mathrm{B}} n_{\mathrm{B}} \quad \text{A in the environment of A} \tag{21c}$$

Thus,

$$G = n_{\mathrm{A}(a)} G_{\mathrm{A}(a)} + n_{\mathrm{A}(b)} G_{\mathrm{A}(b)} + n_{\mathrm{B}(a)} G_{\mathrm{B}(a)} \tag{22a}$$

or

$$G = (n_{\mathrm{A}} - s_{\mathrm{B}} n_{\mathrm{B}}) G_{\mathrm{A}(a)} + s_{\mathrm{B}} n_{\mathrm{B}} G_{\mathrm{A}(b)} + n_{\mathrm{B}} G_{\mathrm{B}(a)} \tag{22b}$$

and

$$G = n_A G_{A(a)} + s_B n_B (G_{A(b)} - G_{A(a)}) + n_B G_{B(a)} \tag{22c}$$

Allowing s_B to relax to equilibrium (*le Châtelier's* principle) with n_A, n_B, T, p constant,

$$\left(\frac{\partial G}{\partial s_B}\right)_{n_A,n_B} = n_B(G_{A(b)} - G_{A(a)}) = 0$$

gives

$$\Delta G_{env} \equiv G_{A(x)} - G_{A(a)} = 0 \tag{23}$$

Hence,

$$\Delta H_{env} = T\Delta S_{env} \tag{24}$$

According to other treatments, however, the solvent reorganization energy and entropy certainly partly compensate but do not cancel each other [141, 142]. It is important to note, however, that discussion about cancellation between portions of the excess entropy and energy of solution is, to some extent, semantic, as pointed out by *Lee* [143]. Some of the reasons for the different results are a matter of physical interpretation, mathematical manipulation, and conditions chosen [144]. For instance, the analysis by *Yu* and *Karplus* [49] refers to solute insertion at constant volume (and not pressure). As another minor point, the thermal motion of the solute has not been considered (which is, however, unimportant in the case of water) [142b]. Anyway, the basic problem seems to rest on the artificial apportionment of both energy and entropy. Albeit the evaluation of solvent reorganization, with enthalpy-entropy compensation involved, is of fundamental interest by itself, it should be kept in mind that it ultimately features just a subsystem of the hydrophobic effect whose overall enthalpy and entropy arguably do vary in some compensatory manner.

As a first approximation, the net compensation behaviour can be described on the basis of the fundamental decoupling of the (total, *i.e.* experimental) solvation enthalpy into direct and response components [82]

$$\Delta H = \Delta H_{sw} + \Delta H_{ww} \tag{25a}$$

where the first term represents the solute–water interaction energy as the direct enthalpic perturbation and the second term is the enthalpy change due to the solvent reorganization that happens as a result of the perturbation. The situation with entropy is trickier. Basically, all entropy change upon hydration is due to solvent reorganization. If, however, one considers the excluded volume effect of cavity formation as independent of water structure, a dissection analogously to the above may be envisaged:

$$\Delta S = \Delta S_{sw} + \Delta S_{ww} \tag{25b}$$

Whereas the reorganization components are still unavailable theoretically for real systems, there are schemes for calculating both ΔH_{sw} and ΔS_{sw}. Most notable is the linear relationship found between ΔH_{sw}, calculated by direct temperature differentiation of the dispersive chemical potential, and ΔS_{sw}, assessed by means of Eq. (15), for nonpolar solutes. This result is at the origin of the hydrophobic effect and reflects the opposite effect exerted by solute size on cavity formation and on the attractive solute–water interaction. Whilst an increase in solute size enhances the cavity formation energy, polarizability is also increased, and this leads to a stronger solute–water attraction. Ultimately, compensation behaviour follows from the fact that cavity formation largely determines the excess entropy, whereas dispersive forces determine the excess enthalpy. These features appear to be the basis for the success of surface area correlations for hydrophobic phenomena [145, 146].

Another interesting result concerns the values of ΔH_{ww} and ΔS_{ww} as obtained by subtraction of the solute-solvent parts from experimental enthalpies and entropies, $\Delta H_{ww} = \Delta H^{exp} - \Delta H_{sw}$ and $\Delta S_{ww} = \Delta S^{exp} - \Delta S_{sw}$, in that there is an exact equality between ΔH_{ww} and $298\Delta S_{ww}$ in accordance with Eq. (24). This is all the more meaningful as both parts are derived from equations (and theories) that are fully independent of one another.

However, there is also a less pleasant feature in this analysis in that the calculated values of both ΔH_{ww} and ΔS_{ww} are positive and unexpectedly high (*e.g.* $\Delta H_{sw} = -21.55$ kJ/mol and $\Delta S_{sw} = -98.1$ J/K · mol for the methane/water system); this would point to H-bond weakening. Along with the experimental data of Table 1 one obtains $\Delta H_{ww} = +10$ kJ/mol and $\Delta S_{ww} = +31.4$ J/K · mol. On the other hand, the orientational rearrangement of vicinal water molecules in forming the water cage should be negative in entropy. Furthermore, according to the experimental and theoretical evidence described above, the water structure beyond the second solvation layer may be considered as rather unperturbed relative to the first hydration zone. This latter finding, by the way, might explain the relative success of continuum water models [52, 147–151]. It has therefore to be concluded that the calculation scheme of cavity formation for water should be modified so as to include the rearrangement of the vicinal water molecules [152]. Suggestions to modify SPT are not new [153]. Basically, it seems questionable whether positional and orientational entropies can really be decoupled.

The "Like-Dissolves-Like" Rule

In some respects, the phenomenon behind the like-dissolves-like rule is the converse of the hydrophobic effect. The buzzword 'polarity', derived from the dielectric approach to solvent effects, is certainly the most popular word dealing with solvent effects. However, the famous rule of thumb *similia similibus solvuntur* (like-dissolves-like), applied when discussing solubility and miscibility, has many exceptions. For instance, methanol and toluene, with ε_s of 32.6 and 2.4, respectively, are miscible, as are water (78.4) and isopropanol (18.3). The problem lies in exactly what is meant by a 'like' solvent. Originally, the term 'polarity' was meant to be an abbreviation of 'static dipolarity' and was thus associated with the dielectric properties of the solvent only.

Clearly, neither the dielectric constant nor the dipole moment is an adequate means to define polarity. The reason is that there are liquids whose constituent molecules have no net dipole moment, for symmetry reasons, but nevertheless have local polar bonds. This is quite obvious by a comparison of *E*-1,2-dichloroethene (formerly *trans*-1,2-dichloroethylene) with *Z*-1,2-dichloroethene (formerly *cis*-1,2-

dichloroethylene). The former has a molecular dipole moment of $\mu = 1.76\,\mathrm{D}$ and $\varepsilon_s = 9.20$, whereas the latter has $\mu = 0\,\mathrm{D}$ and $\varepsilon_s = 2.14$, but the polarity of the carbon-chlorine bonds are certainly the same in both cases. A polar molecule can be defined as having a strongly polar bond, but need not necessarily be a dipole. Recently, there has been growing awareness of the significant contributions of higher multipoles to intermolecular forces, of which the quadrupole moment is of primary importance. For the polarity of the C–H bonds it should be remembered that electronegativity is not an intrinsic property of an atom, but instead varies with hybridization. Only the $C(sp^3)$-H bond can be considered as truly nonpolar, but not the $C(sp^2)$–H bond [154]. Finally, ethine has hydrogen atoms which are definitely acidic.

It should be mentioned that higher moments or local polarities cannot produce a macroscopic polarization and thus be detected in infinite wavelength dielectric experiments yielding a static dielectric constant close to the squared refractive index. Because of their short range, quadrupolar interactions do not directly contribute to the dielectric constant, but are only reflected in the *Kirkwood* g_K factor which decreases due to breaking the angular dipole-dipole correlations with increasing quadrupolar strength.

As is well known, the dipole moment of an uncharged body can be thought of as being formed by separating positive and negative charges, the magnitude of the dipole being the product of charge and separation. Similarly, the quadrupole moment of a system with zero dipole moment can be thought of as arising from a separation of equal and opposite dipoles, the magnitude of a quadrupole being proportional to the product of dipole moment and separation. Thus, the linear carbon dioxide molecule possesses a quadrupole moment, the two C–O dipoles being opposed to but separated from one another. Similarly, any linear molecules, even dihydrogen or dioxygen, and planar ones, such as boron trifluoride and benzene, are quadrupolar. On the other hand, regular tetrahedral molecules like methane and carbon tetrachloride have octupoles as their leading multipoles, and octahedral ones like sulfur hexafluoride possess hexadecapoles [155].

Beyond this it should be remarked that many liquids have both a dipole and a quadrupole moment (water for example). In a recent perturbation theory study of solvation thermodynamics it has been shown that small solute dipoles are even more effectively solvated by solvent quadrupoles than by solvent dipoles [156]. Whereas for dipolar solvents such as acetonitrile, acetone, and dimethyl sulfoxide the dipolar solvation mechanism will be prevailing, for less dipolar solvents like tetrahydrofurane quadrupoles and dipoles might equally contribute to the solvation energetics. According to the *Matyushov-Voth* theory [156], the ratio of the quadrupolar solvation energy E_q to the dipolar solvation energy E_d is approximately equal to

$$\frac{E_q}{E_d} \approx 2.0 \frac{Q^2}{m^2 \sigma^2} r_{0s}^{-4/3} \tag{26}$$

where Q is the solvent quadrupole moment, m is the solvent dipole moment, $r_{0s} = r_0/\sigma + 0.5$ is the reduced distance of closest approach of the solute and solvent hard cores, r_0 is the solute radius, and σ is the HS diameter of the solvent. It has also been calculated that the effect of quadrupolar forces on various thermodynamic properties is much greater than dipolar forces at comparable parameter values [157, 158]. For the ion-hydration problem, it has been proposed recently to use a hybrid potential in which the ion-water interaction is modeled by ion-dipole plus ion-quadrupole terms [159]. In doing so, a bifurcated H-bond structure around the anion (two hydrogens are equally bonded), which is a consequence of the monopole-dipole potential, was avoided. Finally, the aforementioned problem with calculating the dispersion solvation component in the framework of *LJ* parameters, *i.e.* the necessity to employ arbitrary combining rules, should be recalled. It has long ago been suggested, but in the meantime forgotten, that the failure of the common combination rules $\varepsilon_{1,2} = (\varepsilon_1 \cdot \varepsilon_2)^{1/2}$ and $\sigma_{1,2} = (\sigma_1 + \sigma_2)/2$ for *LJ* parameters corresponding to the interaction of dissimilar pairs of molecules [160] is at least partly due to the neglection of quadrupolar effects [155].

The interplay (from classical theories unexpectedly strong) between the π-electron cloud of an aromatic system (or any other π system such as alkenes) with a cation is called cation-π interaction. This is an electrostatic binding largely determined by ion-quadrupole interactions. For instance, the gas-phase binding energy between K^+ and benzene of -77 kJ/mol is even slightly higher than that of K^+ and water at -75 kJ/mol [161]. In addition, this interaction is size-dependent: whereas in the case of K^+_{aq} benzene will displace some water molecules from direct contact with the ion, Na^+_{aq} is resistant towards dehydration in an aromatic environment, giving rise to selectivity in some K^+-channel proteins. Recently, this type of interaction has been hypothesized to play important roles in molecular recognition [162]. Similarly, the interactions of benzene with Li^+, BeH^+, and LiH were calculated by *ab initio* SCF and MO theory to be quite exothermic, *i.e.* -353, -193, and -572 kJ/mol, respectively [163]. Even between atomic lithium and benzene a significant electronic interaction takes place [164]. Consideration of quadrupole–quadrupole interactions also improves the description of solid–solid and solid–fluid equilibria in a molecular model of benzene [165]. Furthermore, the mutual solubilities of water and aromatic hydrocarbons, as measured by IR spectroscopy, are increased in high temperature/pressure mixtures [166], and this may be explained in these terms.

In the light of these findings it is strongly recommended to redefine polarity [14]. Instead of meaning solely dipolarity, it should also include higher multipolar properties, *i.e.* polarity = dipolarity + quadrupolarity + octopolarity.

Solvents which comprise molecules with permanent dipole moment of small or zero magnitude but with finite higher-order multipoles are called nondipolar in the literature [167, 168]. As to the relative importance of each contribution, it should be noted that the interaction energy decreases more rapidly the higher the order of the multipole. Thus, for the interaction of an n-pole with an m-pole, the potential energy varies with distance according to $E \propto 1/(r^{n+m+1})$. The reason for the faster decrease is that the array of charges seems to blend into neutrality more rapidly with distance the higher the number of individual charges contributing to the multipole. Therefore, quadrupolar forces die off faster than dipolar forces on the one hand, and on the other hand, are more important than the higher multipole forces.

The implication of quadrupolarity finally reconciles the notorious troublemakers in solvent reactivity correlations, the aromatic and chlorinated solvents (in

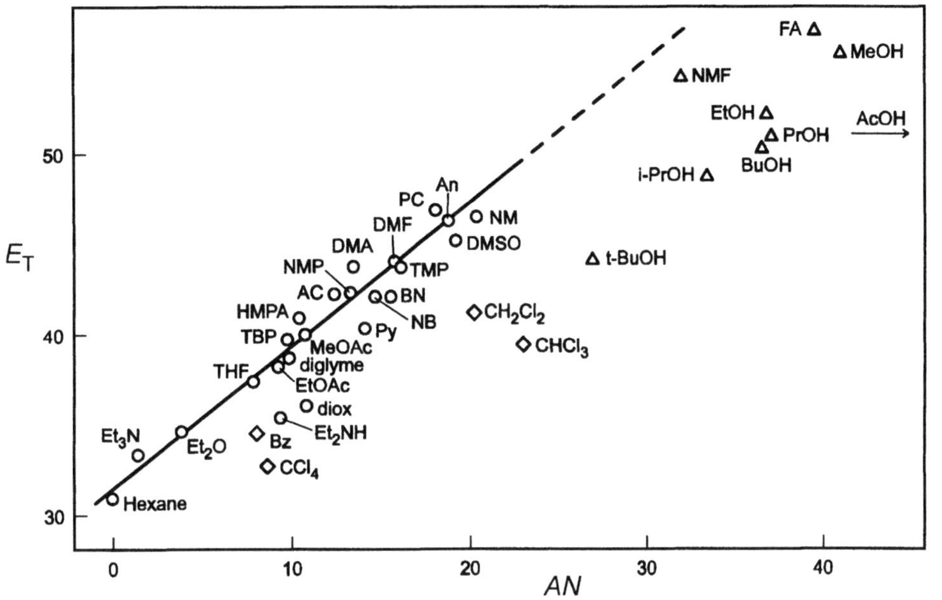

Fig. 4. Relationship between the $E_T(30)$ values and the acceptor number; triangles: protic solvents, squares: aromatic and chlorinated solvents; data are taken from Ref. [14]

addition to the protic solvents). A typical plot is shown in Fig. 4 relating two famous empirical solvent parameters ($E_T(30)$ and acceptor numbers). Originally, the deviations of the aromatic and chlorinated solvents have been interpreted in terms of polarizability effects. This gave rise to the addition of a polarizability correction term in the LSFERs (linear solvation free energy relationships) [14]. However, in a thermodynamic analysis of some empirical polarity scales in terms of long-range solute–solvent interactions due to induction, dispersion, and dipole–dipole forces, the deviations remained, although polarizability effects were properly taken into account [169]. Thus, two alternative explanations for the exceptional behaviour of the aromatic and chlorinated solvents are conceivable: charge transfer (solute–solvent π orbital overlap resulting in exciplex formation) or solute–quadrupole interactions. Although the authors of Ref. [71] were biased in favour of the former, the latter possibility appears to be preferable [168]. The quadrupole and charge transfer mechanisms, reflecting inertial and inertialess solvation pathways, could be distinguished by a comparative analysis of absorption and fluorescence shifts. In terms of quadrupolarity, the discrepant solvent classes of the aromatic and chlorinated solvents in Fig. 4 are due to a size effect. Thus, the betaine-30 molecule, which is the probe dye for the $E_T(30)$ scale, is much larger in size than triethyl phosphine oxide, which forms the basis of the acceptor number scale. Therefore, quadrupolar solvent-dipolar solute interactions are more important at the latter polarity scale. In other words, quadrupolar (and octupolar) solvents have a higher solvating power according to the *AN* relative to the $E_T(30)$ value.

Also, the so-called dioxane anomaly [170] (this solvent appears as a polar solvent despite of a very low dielectric constant) can be rationalized partly in terms of the high quadrupole moment of dioxane. In addition, it appears that dioxane can communicate its polar nature over a relatively long spatial extent, either because of the large charge separation in dioxane itself or because of strong intermolecular interactions between dioxane molecules [171].

Another phenomenon arguably related to quadrupolarity is the high solubility of O_2 in perfluorocarbons (*PFC*), of which the most investigated ones are perfluorodecaline, perfluorotripropylamine, and perfluorooctyl bromide. Intravenously administered *PFCs* may become an essential part in allogeneic transfusion sparing strategies, including preoperative autologous predonation, acute normovolemic hemodilution, and intraoperative blood recovery [172]. Though still awaiting confirmation, the most probable solution mechanism is the interaction between the negative charges on the fluorine centers and the dioxygen quadrupole moment, leading to an additional attractive stabilization. Thus, a variety of examples have been presented leading to the assumption that quadrupolar solvent effects on solvation and reactivity will receive increasing attention in the future [173, 174].

Notwithstanding the modified definition of polarity (*vide supra*), the problem remains that positive and negative charge solvation are not distinguished. Most importantly, there is no general relationship between polarity and cation solvation tendency. For example, although nitromethane and *DMF* have the same dielectric constant, the extent of ion pairing in $MeNO_2$ is much greater than that in *DMF*. This observation is attributed to the weak basicity of $MeNO_2$ which poorly solvates cations. As a result, ion pairing is stronger in $MeNO_2$ in spite of the fact that long-range ion-ion interactions in the two solvents are equal.

Finally, a potential problem with polarity rests in the fact that this term is typically associated with enthalpy. But caution is urged in interpreting the like-dissolves-like rule in terms of enthalpy. It has often been stated for example that non-polar liquids such as octane and carbon tetrachloride are miscible because the molecules are held together by weak dispersion forces. However, the importance of entropy may not be forgotten in the spontaneous mixing of the two phases.

References

[1] Ben-Naim A (1980) Hydrophobic Interactions. Plenum, New York
[2] Evans DF, Ninham BW (1986) J Phys Chem **90**: 226
[3] Tanford C (1973) The Hydrophobic Effect: Formation of Micelles and Biological Membranes. Wiley, New York
[4] Dill KA (1990) Biochemistry **29**: 7133
[5] Førrisdahl OK, Kvamme B, Haymet ADJ (1996) Mol Phys **89**: 819
[6] Wang H, Ben-Naim A (1996) J Med Chem **39**: 1531
[7] Ashbaugh HS, Kaler EW, Paulaitis ME (1999) J Am Chem Soc **121**: 9243
[8] Battino R, Clever HL (1966) Chem Rev **66**: 395
[9] Ben-Naim A, Marcus Y (1984) J Chem Phys **81**: 2016
[10] Madan B, Sharp K (1996) J Phys Chem **100**: 7713
[11] Lazaridis T (1998) J Phys Chem B **102**: 3531
[12] Marmur A (2000) J Am Chem Soc **122**: 2120
[13] Blokzijl W, Engberts JBFN (1993) Angew Chem Int Ed Engl **32**: 1545
[14] Schmid R (2001) Solvent Effects on Chemical Reactivity. In: Wypych G (ed) Handbook of Solvents. ChemTec Publishing, Toronto, p 737
[15] Giguère PA (1984) J Raman Spectrosc **15**: 354
[16] Wall TT, Hornig DF (1965) J Phys Chem **43**: 2079
[17] Walrafen GE (1968) J Chem Phys **48**: 244

[18] D'Arrigo G, Maisano G, Mallamace F, Migliardo P, Wanderlingh F (1981) J Chem Phys **75**: 4264
[19] Giguère PA (1987) J Chem Phys **87**: 4835
[20] Walrafen GE, Fisher MR, Hokmabady MS, Yang W-H (1986) J Chem Phys **85**: 6970
[21] a) Hare DE, Sorensen CM (1990) J Chem Phys **93**: 25; b) Hare DE, Sorensen CM (1990) J Chem Phys **93**: 6954; c) Hare DE, Sorensen CM (1992) J Chem Phys **96**: 13
[22] Maréchal Y (1991) J Chem Phys **95**: 5565
[23] Libnau FO, Toft J, Christy AA, Kvalheim OM (1994) J Am Chem Soc **116**: 8311
[24] Soper AK, Phillips MG (1986) Chem Phys **107**: 47
[25] Svishchev IM, Kusalik PG (1993) J Chem Phys **99**: 3049
[26] Svishchev IM, Zassetsky AY (2000) J Chem Phys **112**: 1367
[27] Soper AK (1994) J Chem Phys **101**: 6888
[28] De Santis A, Rocca D (1997) J Chem Phys **107**: 9559
[29] Giguère PA (1984) J Raman Spectrosc **15**: 354
[30] Némethy G, Scheraga HA (1962) J Chem Phys **36**: 3382
[31] Bosio L, Chen S-H, Teixeira J (1983) Phys Rev A **27**: 1468
[32] Sciortino F, Geiger A, Stanley HE (1990) Phys Rev Lett **65**: 3452
[33] Robinson GW, Cho CH, Urquidi J (1999) J Chem Phys **111**: 698
[34] Vedamuthu M, Singh S, Robinson GW (1994) J Phys Chem **98**: 2222 and references therein
[35] Kamb B (1964) Acta Cryst **17**: 1437
[36] Kamb B, Hamilton WC, LaPlaca SJ, Prakash A (1971) J Chem Phys **55**: 1934
[37] Benson SW, Siebert ED (1992) J Am Chem Soc **114**: 4269
[38] Falk M, Ford TA (1966) Can J Chem **44**: 1699
[39] Giguère PA, Pigeon-Gosselin M (1986) J Raman Spectrosc **17**: 341
[40] Cho CH, Singh S, Robinson GW (1997) J Chem Phys **107**: 7979
[41] Mahoney MW, Jorgensen WL (2000) J Chem Phys **112**: 8910
[42] Arthur JW, Haymet ADJ (1998) J Chem Phys **109**: 7991
[43] Frank HS, Evans MW (1945) J Chem Phys **13**: 507
[44] Mizuno K, Miyashita Y, Shindo Y, Ogawa H (1995) J Phys Chem **99**: 3225
[45] Goldammer EV, Hertz HG (1970) J Phys Chem **74**: 3734
[46] Pottel R, Kaatze V (1969) Ber Bunsenges Phys Chem **73**: 437
[47] Silveston R, Kronberg B (1989) J Phys Chem **93**: 6241
[48] Grunwald E, Steel C (1995) J Am Chem Soc **117**: 5687
[49] Yu HA, Karplus M (1988) J Chem Phys **89**: 2366
[50] Qian H, Hopfield JJ (1996) J Chem Phys **105**: 9292
[51] Lee B (1985) J Chem Phys **83**: 2421
[52] Vaisman II, Brown FK, Tropsha A (1994) J Phys Chem **98**: 5559
[53] Re M, Laria D, Fernández-Prini R (1996) Chem Phys Lett **250**: 25
[54] Soper AK, Finney JL (1993) Phys Rev Lett **71**: 4346
[55] Turner J, Soper AK (1994) J Chem Phys **101**: 6116
[56] Soper AK, Luzar A (1996) J Phys Chem **100**: 1357
[57] Meng EC, Kollman PA (1996) J Phys Chem **100**: 11460
[58] Ikeguchi M, Seishi S, Nakamura S, Shimizu K (1998) J Phys Chem B **102**: 5891
[59] Ben-Naim A (1987) Solvation Thermodynamics. Plenum, New York
[60] Abraham MH (1982) J Am Chem Soc **104**: 2085
[61] Ben-Naim A (1978) J Phys Chem **82**: 792
[62] Schmid R, Miah AM, Sapunov VN (2000) Phys Chem Chem Phys **2**: 97
[63] Owicki JC, Scheraga HA (1977) J Am Chem Soc **99**: 7403
[64] Bridgeman CH, Buckingham AD, Skipper NT (1996) Chem Phys Lett **253**: 209
[65] Reiss H (1966) Adv Chem Phys **9**: 1
[66] de Souza LES, Stamatopoulou A, Ben-Amotz D (1994) J Chem Phys **100**: 1456

[67] Graziano G (1998) J Chem Soc Faraday Trans **94**: 3345
[68] Graziano G (1999) Phys Chem Chem Phys **1**: 1877
[69] Boublik T (1970) J Chem Phys **53**: 471
[70] Mansoori GA, Carnahan NF, Starling KE, Leland Jr TW (1971) J Chem Phys **54**: 1523
[71] Matyushov DV, Ladanyi BM (1997) J Chem Phys **107**: 5815
[72] a) Pierotti RA (1963) J Phys Chem **67**: 1840; b) Pierotti RA (1976) **76**: 717
[73] a) Wilhelm E, Battino R (1971) J Chem Phys **55**: 4021; b) Wilhelm E, Battino RJ (1971) J Chem Thermodyn **3**: 761; Wilhelm E, Battino RJ (1973) J Chem Phys **58**: 3558
[74] Matyushov DV, Schmid R (1996) J Chem Phys **104**: 8627
[75] Ben-Amotz D, Herschbach DR (1990) J Phys Chem **94**: 1038
[76] Schmid R, Matyushov DV (1995) J Phys Chem **99**: 2393
[77] Matuyshov DV, Schmid R (1994) J Phys Chem **98**: 5152
[78] Ben-Amotz D, Willis KG (1993) J Phys Chem **97**: 7736
[79] Floris FM, Selmi M, Tani A, Tomasi J (1997) J Chem Phys **107**: 6353
[80] Prévost M, Oliveira IT, Kocher JP, Wodak SJ (1996) J Phys Chem **100**: 2738
[81] Pohorille A, Pratt LR (1990) J Am Chem Soc **112**: 5066
[82] Matyushov DV, Schmid R (1996) J Chem Phys **105**: 4729
[83] Stamatopoulou A, Ben-Amotz D (1998) J Chem Phys **108**: 7294
[84] Broadbent RD, Neilson GW (1994) J Chem Phys **100**: 7543
[85] Filipponi A, Bowron DT, Lobban C, Finney JL (1997) Phys Rev Lett **79**: 1293
[86] De Jong PHK, Wilson JE, Neilson GW, Buckingham AD (1997) Molec Phys **91**: 99
[87] Turner JZ, Soper AK, Finney JL (1995) J Chem Phys **102**: 5438
[88] Bowron DT, Finney JL, Soper AK (1998) J Phys Chem B **102**: 3551
[89] Sharp KA, Bhupinder M, Manas E, Vanderkooi JM (2001) J Chem Phys **114**: 1791
[90] Mizuno K, Miyashita Y, Shindo Y, Ogawa H (1995) J Phys Chem **99**: 3225
[91] Fidler J, Rodger PM (1999) J Phys Chem B **103**: 7695
[92] Guillot B, Guissani Y, Bratos S (1991) J Chem Phys **95**: 3634
[93] a) Feakins D, Canning FM, Mullally JJ, Waghorne WE (1989) Pure Appl Chem **61**: 133; b) Feakins D, Bates FM, Waghorne WE, Lawrence KG (1993) J Chem Soc Faraday Trans **89**: 3388
[94] Holz M, Haselmeier R, Mazitov RK, Weingärtner H (1994) J Am Chem Soc **116**: 801
[95] Haselmeier R, Holz M, Marbach W, Weingärtner H (1995) J Phys Chem **99**: 2243
[96] Nakahara M, Wakai C, Yoshimoto Y, Matabayasi N (1996) J Phys Chem **100**: 1345
[97] Sciortino F, Geiger A, Stanley HE (1992) J Chem Phys **96**: 3857
[98] García-Tarrés L, Guàrdia E (1998) J Phys Chem B **102**: 7448
[99] Owicki JC, Scheraga HA (1977) J Am Chem Soc **99**: 7413
[100] Swaminathan S, Harrison SW, Beveridge DL (1978) J Am Chem Soc **100**: 5705
[101] Geiger A, Rahman A, Stillinger FH (1979) J Chem Phys **70**: 263
[102] Pangali C, Rao M, Berne J (1979) J Chem Phys **71**: 2982
[103] Alagona G, Tani A (1980) J Chem Phys **72**: 580
[104] Rapaport DC, Schegara HA (1982) J Am Chem Soc **86**: 873
[105] Vaisman II, Brown FK, Tropsha A (1994) J Phys Chem **98**: 5559
[106] Hernández-Cobos J, Ortega-Blake I (1995) J Chem Phys **103**: 9261
[107] Fois E, Gamba A, Redaelli C (1999) J Chem Phys **110**: 1025
[108] a) Miller KW, Reo NV, Uiterkamp AJMS, Stengle DP, Stengle TR, Williamson KL (1981) Proc Natl Acad Sci USA **78**: 4946; b) Stengle TR, Hosseini SM, Basiri HG, Williamson KL (1984) J Solution Chem **13**: 779
[109] de Souza LES, Ben-Amotz D (1994) J Chem Phys **101**: 9858
[110] Pierotti RA (1976) Chem Rev **76**: 717
[111] Reid CR, Prauznitz JM, Poling BE (1987) The Properties of Gases and Liquids. McGraw-Hill, New York
[112] Pierotti RA (1965) J Phys Chem **69**: 281

[113] Jorgensen WL, Chandrasekhar J, Madura JD, Impey RW, Klein ML (1983) J Chem Phys **79**: 926
[114] Lazaridis T, Karplus M (1996) J Chem Phys **105**: 4294
[115] Zichi DA, Rossky PJ (1985) J Chem Phys **83**: 797
[116] a) Londono D, Kuhs WF, Finney JL (1988) Nature **332**: 141; b) Londono D, Finney JL, Kuhs WF (1992) J Chem Phys **97**: 547
[117] Udachin KA, Ratcliffe CI, Ripmeester JA (2001) Angew Chem **113**: 1343
[118] Southall NT, Dill KA (2000) J Phys Chem B **104**: 2391
[119] a) Graziano G (2001) J Phys Chem B **105**: 2079; b) Southall NT, Dill KA (2001) J Phys Chem B **105**: 2082
[120] Chen B, Xing J, Siepmann JI (2000) J Phys Chem B **104**: 2391
[121] Sapunov VN, Kirchner K, Schmid R (2001) Coord Chem Rev **214**: 143
[122] Marcus Y (1988) Chem Rev **88**: 1475
[123] Ohtaki H, Radnai T (1993) Chem Rev **93**: 1157
[124] a) Sharp K, Madan B (1997) J Phys Chem B **101**: 4343; b) Madan B, Sharp K (1997) J Phys Chem B **101**: 11237
[125] Silverstein KAT, Haymet ADJ, Dill KA (1999) J Chem Phys **111**: 8000
[126] Matubayasi N (1994) J Am Chem Soc **116**: 1450
[127] Rick SW (2000) J Phys Chem B **104**: 6884
[128] Collett TS (1997) Chem Eng News **75**: 60
[129] Silverstein TP (1998) J Chem Educ **75**: 116
[130] Furutaka S, Ikawa S (1998) J Chem Phys **108**: 5159
[131] a) Linert W (1989) Chem Phys **129**: 381; b) Linert W, Jameson RF (1989) Chem Soc Rev **18**: 477; c) Linert W (1994) Chem Soc Rev **23**: 429
[132] Boots HMJ, de Bokx PK (1989) J Phys Chem **93**: 8240
[133] Vailaya A, Horváth C (1996) J Phys Chem **100**: 2447
[134] Stolov AA, Herrebout WA, van der Veken BJ (1998) J Am Chem Soc **120**: 7310
[135] Abraham MH, Whiting GS, Fuchs R, Chambers EJ (1990) J Chem Soc Perkin Trans **2**: 291
[136] Ramadan MS, Evans DF, Lumry R (1983) J Phys Chem **87**: 4538
[137] Matyushov DV, Schmid R (1994) Ber Bunsenges Phys Chem **98**: 1590
[138] Atkins PW (1990) Physical Chemistry, 4th edn. Oxford University Press, Oxford, p 167ff
[139] Linert W, Jameson RF (1993) J Chem Soc Perkin Trans **2**, 1415
[140] Ben-Naim A (1978) J Phys Chem **82**: 874
[141] Matubayasi N, Reed LH, Levy RM (1994) J Phys Chem **98**: 10640
[142] a) Lazaridis T (1998) J Phys Chem B **102**: 3531; b) Lazaridis T (2000) J Phys Chem B **104**: 4964
[143] a) Lee B (1994) Biophys Chem **51**: 271; b) Lee B (1995) Methods in Enzymol **259**: 555
[144] a) Smith DE, Laird BB, Haymet ADJ (1993) J Phys Chem **97**: 5788; b) Lazaridis T, Paulaitis ME (1993) J Phys Chem **97**: 5789
[145] Ashbaugh HS, Kaler EW, Paulaitis ME (1999) J Am Chem Soc **121**: 9243
[146] Murray JS, Abu-Awwad F, Politzer P (1999) J Phys Chem A **103**: 1853
[147] Rashin AA, Bukatin MA (1991) J Phys Chem **95**: 2942
[148] Pitarch J, Moliner V, Pascual-Ahuir JL, Silla E, Tuñón I (1996) J Phys Chem **100**: 9955
[149] Sitkoff D, Ben-Tal N, Honig B (1966) J Phys Chem **100**: 2744
[150] Bader JS, Cortis CM, Berne BJ (1997) J Chem Phys **106**: 2372
[151] Luo R, Moult J, Gilson MK (1997) J Phys Chem B **101**: 11226
[152] Guillot B, Guissani Y (1993) J Chem Phys **99**: 8075
[153] Stillinger FH (1973) J Solution Chem **2**: 141
[154] Bratsch SG (1988) J Chem Edu **65**: 34
[155] Buckingham AD (1959) Quart Rev Chem Soc **13**: 183
[156] Matyushov DV, Voth GA (1999) J Chem Phys **111**: 3630
[157] Patey GN, Valleau JP (1976) J Chem Phys **64**: 170

[158] Jiang S, Pitzer KS (1995) J Chem Phys **102**: 7632
[159] Liu Y, Ichiye T (1996) J Phys Chem **100**: 2723
[160] Cottrell TL, Hamilton RA, Taubinger RP (1956) Trans Faraday Soc **52**: 1310
[161] a) Carbacos OM, Weinheimer CJ, Lisy JM (1998) J Chem Phys **108**: 5151; b) Carbacos OM, Weinheimer CJ, Lisy JM (1999) J Chem Phys **110**: 8429
[162] Siu FM, Ma NL, Tsang CW (2001) J Am Chem Soc **123**: 3397
[163] Jemmis ED, Schleyer PVR (1982) J Am Chem Soc **104**: 4781
[164] Manceron L, Andrews L (1988) J Am Chem Soc **110**: 3840
[165] Schroer JW, Monson PA (2001) J Chem Phys **114**: 4124
[166] Furutaka S, Ikawa S (1998) J Chem Phys **108**: 5159
[167] a) Perng BC, Newton MD, Raineri FO, Friedman HL (1996) J Chem Phys **104**: 7153; b) Perng BC, Newton MD, Raineri FO, Friedman HL (1996) J Chem Phys **104**: 7177
[168] Reynolds L, Gardecki JA, Frankland SJV, Horng ML, Maroncelli M (1996) J Phys Chem **100**: 10337
[169] Matyushov DV, Schmid R, Ladanyi BM (1997) J Phys Chem B **101**: 1035
[170] Geerlings JD, Varma CAGO, Hemmert MCv (2000) J Phys Chem B **104**: 56
[171] Kauffman JF (2001) J Phys Chem A **105**: 3433
[172] Zuck TF, Riess JG (1994) Crit Rev Clin Lab Sci **31**: 295
[173] Jeon J, Kim HJ (2000) J Phys Chem A **104**: 9812
[174] Read I, Napper A, Zimmt MB, Waldeck DH (2000) J Phys Chem A **104**: 9385

Received May 8, 2001. Accepted (revised) May 23, 2001

Invited Review

Thermodynamic Investigation of Phase Equilibria in Metal Carbonate–Water–Carbon Dioxide Systems

Wolfgang Preis* and **Heinz Gamsjäger**

Institut für Physikalische Chemie, Montanuniversität Leoben, Franz-Josef-Strasse 18, A-8700 Leoben, Austria

Summary. Solubility measurements as a function of temperature have been shown to be a powerful tool for the determination of thermodynamic properties of sparingly-soluble transition metal carbonates. In contrast to calorimetric methods, such as solution calorimetry or drop calorimetry, the evaluation of solubility data avoids many systematic errors, yielding the enthalpy of solution at 298.15 K with an estimated uncertainty of $\pm 3\,kJ \cdot mol^{-1}$. A comprehensive set of thermodynamic data for otavite ($CdCO_3$), smithsonite ($ZnCO_3$), hydrozincite ($Zn_5(OH)_6(CO_3)_2$), malachite ($Cu_2(OH)_2CO_3$), azurite ($Cu_3(OH)_2(CO_3)_2$), and siderite ($FeCO_3$) was derived. Literature values for the standard enthalpy of formation of malachite and azurite were disproved by these solubility experiments, and revised values are recommended. In the case of siderite, data for the standard enthalpy of formation given by various data bases deviate from each other by more than $10\,kJ \cdot mol^{-1}$ which can be attributed to a discrepancy in the auxiliary data for the Fe^{2+} ion. A critical evaluation of solubility data from various literature sources results in an optimized value for the standard enthalpy of formation for siderite. The *Davies* approximation, the specific ion-interaction theory, and the *Pitzer* concept are used for the extrapolation of the solubility constants to zero ionic strength in order to obtain standard thermodynamic properties valid at infinite dilution, $T = 298.15$ K, and $p = 10^5$ Pa. The application of these electrolyte models to both homogeneous and heterogeneous (solid-solute) equilibria in aqueous solution is reviewed.

Keywords. Thermodynamics; Solubility; Transition metals; Carbonate.

Introduction

The thermodynamic properties of transition metal carbonates play a major role for a better understanding of a wide variety of geochemical and industrial processes involving equilibria between solid carbonates and aqueous solutions. The total concentration of transition metals in carbonate-bearing natural waters is predominantly determined by the solubilities of the respective hydroxides, oxides,

* Corresponding author. E-mail: gamsjaeg@unileoben.ac.at

hydroxide carbonates, and neutral carbonates [1]. The thermodynamic modeling of dissolution or precipitation of sparingly-soluble metal carbonates may serve as an important tool for the prediction of the concentrations of transition metals in mine pit lakes [2]. A careful determination of thermodynamic data of metal carbonates is an essential prerequisite for the geochemical modeling of the release of trace elements from waste repositories [3].

In general, the standard *Gibbs* free energy of formation of a solid compound can be obtained from the determination of equilibrium constants or the measurement of the *emf* (electromotive force) of an appropriate electro-chemical cell. In the case of sparingly-soluble metal carbonates the careful determination of the solubility constant has led to reliable values for the standard *Gibbs* free energies of formation [4–11]. An alternative path for the determination of this quantity is the application of electrochemical cells using carbonate electrodes of the third kind without liquid junction [12–14].

The standard enthalpy of formation, $\Delta_f H^\ominus$, of metal carbonates can be measured by application of both calorimetric and equilibrium methods. Direct measurements of the enthalpy of solution by means of solution calorimeters operating at moderate temperatures (293–323 K) enable the determination of the enthalpy of formation of the mineral phase [15–18]. In addition, the application of drop calorimetry [19–21], drop solution calorimetry [22–24], and DSC (differential scanning calorimetry) [10, 11] has been reported. By using adequate thermodynamic cycles, $\Delta_f H^\ominus$ was calculated from the directly determined enthalpy change associated with (*i*) the drop of the specimen into the hot calorimeter and subsequent decomposition regarding drop calorimetry, or (*ii*) the drop of the sample and subsequent dissolution in molten lead borate glasses concerning drop solution calorimetry.

Alternatively, the enthalpy of solution and the enthalpy of decomposition can be obtained from the temperature dependence of the solubility constant [9–11] and the decomposition pressure [25, 26], respectively. It should be noted that the decarbonation measurements at elevated temperatures are difficult to evaluate, since the compressibility of solid carbonates as well as the fugacities of carbon dioxide at extremely high pressures are rather uncertain [25]. If $\Delta_f H^\ominus$ is derived from solubility measurements at various temperatures, a powerful electrolyte theory is necessary for the extrapolation of the solubility constants to infinite dilution.

The standard entropy of the solid carbonate can be calculated from the entropy of solution obtained from solubility measurements as a function of temperature. However, the most accurate and precise method for the determination of the standard entropy of a solid phase is the integration of the heat capacity function from 0 to 298.15 K. Low-temperature measurements of the heat capacity have been carried out for a number of transition metal carbonates, resulting in highly reliable values for the standard entropy [19, 27–29].

It is the aim of this work to derive an internally consistent set of thermodynamic properties, *i.e.* standard enthalpy of formation, standard entropy, and standard *Gibbs* free energy of formation, for the following transition metal carbonates: otavite ($CdCO_3$), smithsonite ($ZnCO_3$), hydrozincite ($Zn_5(OH)_6(CO_3)_2$), malachite ($Cu_2(OH)_2CO_3$), azurite ($Cu_3(OH)_2(CO_3)_2$), and siderite ($FeCO_3$). Basically, the thermodynamic quantities are derived from solubility measurements as a function of temperature and ionic strength. Details on experimental procedures have been given elsewhere [4, 7–10, 30–32] and need not to be discussed further in this contribution. The application of different models for the extrapolation of the solubility constants to zero ionic strength is reviewed. This extrapolation allows the derivation of standard thermodynamic data valid at infinite dilution, $T = 298.15$ K, and $p = 10^5$ Pa. The solubility data are evaluated with the help of the optimization routine of ChemSage [33]. Whereas the standard entropies of the solid carbonates are generally given by low-temperature measurements of the heat capacity, the standard *Gibbs* free energy of formation as well as the standard enthalpy of formation are obtained from the solubility measurements. In the case of hydrozincite, no direct determination of the standard entropy has been reported so far. Thus, this quantity is likewise derived from the temperature dependence of the solubility constant. Additionally, all results will be compared with a large number of thermodynamic investigations published in literature.

Theoretical Aspects

The dissolution of a sparingly-soluble metal(II) carbonate $M(OH)_x(CO_3)_y(s)$ in an aqueous medium of a certain ionic strength I can be expressed as

$$M(OH)_x(CO_3)_y(s) + 2H^+(I) \rightleftharpoons M^{2+}(I) + yCO_2(g) + (1 + x/2)H_2O(I). \quad (1)$$

Owing to the charge balance equation, the relation $x + 2y = 2$ must be fulfilled. The solubility constant valid at a given temperature T and ionic strength I can be written as

$$\log {}^*K^I_{ps0} = \log(([M^{2+}]/m^\ominus)(p(CO_2)/p^\ominus)^y([H^+]/m^\ominus)^{-2}) \quad (2)$$

where the square brackets refer to the molalities of the respective species in the aqueous phase and $p(CO_2)$ denotes the partial pressure of carbon dioxide. The standard molality and the standard pressure are defined as $m^\ominus = 1\,\text{mol}\cdot\text{kg}^{-1}$ and $p^\ominus = 10^5$ Pa, respectively. According to Eq. (2), the solubility of the solid carbonate is determined by three parameters: the partial pressure of carbon dioxide, the molality of the metal ions, and the $p[H]$. The square brackets of $p[H]$ indicate that with the calibration system used H^+ ion molalities were measured. Usually, the CO_2 pressure is maintained constant during the solubility experiments, the molality of the metal ions can be determined by complexometric titrations with *EDTA*, and the $p[H]$ of the aqueous solution in equilibrium with the solid carbonate and the gas phase is obtained from potentiometric measurements.

By fixing the ionic strength, I, of the solvent (usually aqueous $NaClO_4$ solutions) at a constant value, reliable equilibrium constants $\log {}^*K^I_{ps0}$ can be obtained. These solubility constants are extrapolated to zero ionic strength, $\log {}^*K^0_{ps0}$, by calculating the ionic strength dependence of the individual activity coefficients ($\gamma(M^{2+})$, $\gamma(H^+)$) as well as the activity of water, $a(H_2O)$:

$$\log {}^*K^0_{ps0} = \log {}^*K^I_{ps0} + \log\gamma(M^{2+}) - 2\log\gamma(H^+) + (1 + x/2)\log a(H_2O). \quad (3)$$

The activity of water is correlated to the osmotic coefficient ϕ by Eq. (4) [34], where M_{H_2O} and $\sum m_i$ are the molar mass of water and the sum over the molalities of all species in solution.

$$\phi = -\frac{\ln a(\mathrm{H_2O})}{M_{\mathrm{H_2O}}\sum m_i} \tag{4}$$

The individual activity coefficients of the solutes in a multicomponent aqueous phase are related to the osmotic coefficient (activity of water) by the *Gibbs-Duhem* equation [35]:

$$-\sum m_i \mathrm{d}\phi - (\phi - 1)\sum \mathrm{d}m_i + \sum(m_i \mathrm{d}\ln\gamma_i) = 0. \tag{5}$$

When solubility experiments are performed in perchlorate media of fairly high ionic strengths ($I \geq 0.20\,\mathrm{mol\cdot kg^{-1}}$ $NaClO_4$), precise *pH* values can be guaranteed. We have applied the *Davies* approximation [36], the specific ion-interaction theory [37], and the *Pitzer* equations [38] to take into account the solute–solvent interactions at such high ionic strengths. In the following discussion of electrolyte models, quantity algebra was sacrificed in order to improve readability.

Davies approximation

In general, the *Davies* approximation is useful for the prediction of the ionic strength dependence of the mean activity coefficient of strong electrolyte solutions up to molalities around 0.1–$0.2\,\mathrm{mol\cdot kg^{-1}}$. The individual activity coefficient γ_i of an ionic species in aqueous solution is given by

$$\log\gamma_i = -Az_i^2\left(\frac{\sqrt{I}}{1+\sqrt{I}} - 0.3I\right) \tag{6}$$

with A and z_i denoting the *Debye-Hückel* parameter and the ionic charge number, respectively. In accordance with the *Gibbs-Duhem* relation (Eq. (5)), the osmotic coefficient reads

$$\phi = 1 - \frac{\ln(10)A}{\sum m_i}\left(2 + 2\sqrt{I} - \frac{2}{1+\sqrt{I}} - 4\ln(1+\sqrt{I}) - 0.3I^2\right). \tag{7}$$

If a 1–1 electrolyte, such as NaCl or $NaClO_4$, is employed to fix the ionic strength, the extrapolation of the solubility constant to infinite dilution can be expressed as

$$\log{}^*K^0_{ps0} = \log{}^*K^I_{ps0} - 2A\left(\frac{\sqrt{I}}{1+\sqrt{I}} - 0.3I\right) - (1 + x/2)\frac{2I\phi M_{\mathrm{H_2O}}}{\ln(10)}, \tag{8}$$

where ϕ is given by

$$\phi = 1 - \frac{\ln(10)A}{I}\left(1 + \sqrt{I} - \frac{1}{1+\sqrt{I}} - 2\ln(1+\sqrt{I})\right) + \frac{A\ln(10)}{2}0.3I. \tag{9}$$

It should be mentioned that the Davies formalism is an extension of the *Debye-Hückel* limiting law without any individual interaction parameters.

Specific ion-interaction theory

The specific ion-interaction theory is likewise based on the *Debye-Hückel* limiting law. However, individual ion-interaction parameters are introduced which enable the calculation of activity coefficients up to ionic strengths of 3–4 $mol \cdot kg^{-1}$. This concept was originally introduced by *Brønsted* [39, 40], *Guggenheim* [41], and *Scatchard* [42]. *Grenthe et al.* [37] published many examples for the efficiency of this formalism with respect to homogeneous equilibria in media of constant ionic strength. The individual activity coefficient of an ionic species is defined as

$$\log\gamma_i = -\frac{Az_i^2\sqrt{I}}{1+1.5\sqrt{I}} + \sum \varepsilon(i,j)m_j \tag{10}$$

where only interaction coefficients $\varepsilon(i, j)$ between ions of opposite charges are considered. In accordance with the traditional *Setchenov* equation the activity coefficient of uncharged species γ_k reads [37]

$$\log\gamma_k = \sum \varepsilon(k,j)m_j \tag{11}$$

where only interactions between the neutral species and ions are taken into account. The osmotic coefficient can be written as [37]

$$\phi = 1 - \frac{2\ln(10)A}{1.5^3\sum m_i}\left(1+\sqrt{I} - \frac{1}{1+\sqrt{I}} - 2\ln(1+\sqrt{I})\right) + \frac{\ln(10)}{2\sum m_i}\sum_i\sum_j \varepsilon(i,j)m_i m_j \tag{12}$$

In the case of solutions of 1–1 electrolytes (*NX*) serving as the inert ionic medium, the extrapolation of $\log {}^*K^I_{ps0}$ to zero ionic strength is given by

$$\log {}^*K^0_{ps0} = \log {}^*K^I_{ps0} - \frac{2A\sqrt{I}}{1+1.5\sqrt{I}} + (\varepsilon(M^{2+},X^-) - 2\varepsilon(H^+,X^-))I - (1+x/2)\frac{2I\phi M_{H_2O}}{\ln(10)}, \tag{13}$$

with the osmotic coefficient ϕ being

$$\phi = 1 - \frac{A\ln(10)}{1.5^3 I}\left(1+1.5\sqrt{I} - \frac{1}{1+1.5\sqrt{I}} - 2\ln(1+1.5\sqrt{I})\right) + \frac{\ln(10)}{2}\varepsilon(N^+,X^-)I. \tag{14}$$

The temperature dependence of the interaction parameters is given by the relation

$$\varepsilon(i,j) = \varepsilon(i,j;298.15\,K) + \frac{\partial\varepsilon(i,j)}{\partial T}(T - 298.15\,K). \tag{15}$$

The values for $\partial\varepsilon(i,j)/\partial T$ are based on the ionic strength dependence of the relative apparent molar enthalpy ${}^\phi L$ of the pertinent strong electrolytes [37]

$${}^\phi L = \frac{3}{4}(\nu_+ + \nu_-)\frac{A_L|z_+z_-|}{1.5^3 I}(t^2 - 4t + 2\ln t + 3) - \frac{2\ln(10)\nu_+\nu_-}{\nu_+z_+^2 + \nu_-z_-^2}RT^2 I\frac{\partial\varepsilon(i,j)}{\partial T} \tag{16}$$

where $t = 1 + 1.5\sqrt{I}$ and A_L is the *Debye-Hückel* parameter for enthalpy [38]. The symbols ν_+, ν_-, z_+, and z_- denote the stoichiometric indices and charge numbers of the respective ions.

Pitzer model

The *Pitzer* concept is even valid in the case of saturated solutions of highly-soluble strong electrolytes. In contrast to the former electrolyte theories a large number of individual interaction parameters is necessary. Single electrolyte solutions require at least the knowledge of three binary parameters. In the case of multicomponent systems, such as solubility equilibria of metal carbonates in constant ionic media, only an extensive set of binary and ternary interaction parameters allows a correct prediction of the solubility constant at ionic strengths above $3\,\mathrm{mol \cdot kg^{-1}}$ NaCl or $NaClO_4$.

The electrolyte models mentioned above are incorporated into the general *Gibbs* energy minimization program ChemSage [43] which permits a wide variety of equilibrium calculations in aqueous media. Furthermore, the optimization routine of ChemSage [33] allows the simultaneous evaluation of experimental data from different sources in order to obtain an internally consistent set of optimized thermodynamic constants.

A comparison between the specific ion-interaction theory and the *Pitzer* formalism is illustrated in Figs. 1 and 2. The mean activity coefficient and the osmotic coefficient are plotted as a function of molality of the 2–1 electrolyte $Zn(ClO_4)_2$ at $T = 298.15\,\mathrm{K}$. The *Pitzer* equations provide a perfect fit to the experimental data taken from Ref. [44] over the whole range of molality (0–$4.3\,\mathrm{mol \cdot kg^{-1}}$). The specific ion-interaction parameter $\varepsilon(Zn^{2+}, ClO_4^-) = 0.36$ is obtained from a fit of Eqs. (10) and (12), respectively, to the experimental data from 0 to $2.5\,\mathrm{mol \cdot kg^{-1}}$.

The application of the *Davies* approximation, the specific ion-interaction theory, and the *Pitzer* concept to the theoretical prediction of the ionic product

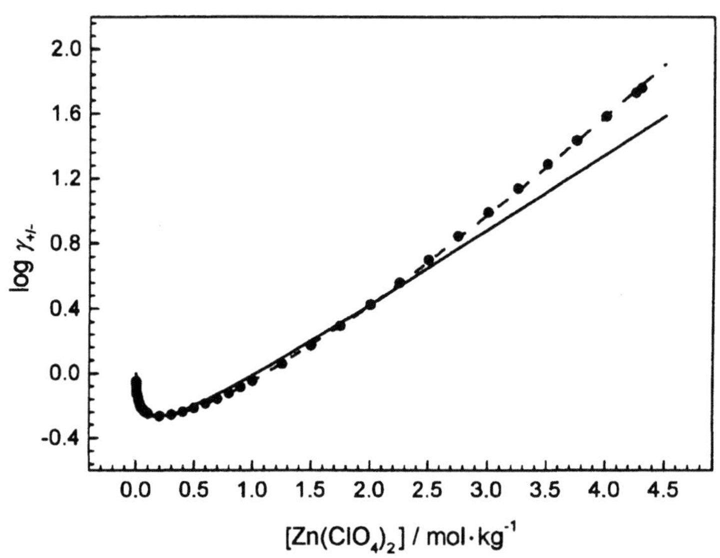

Fig. 1. Logarithm of the mean activity coefficient plotted *vs.* molality of $Zn(ClO_4)_2$ at 298.15 K; ●: *Goldberg* [44]; solid line: specific ion-interaction theory [37], $\varepsilon(Zn^{2+}, ClO_4^-) = 0.36$; dashed line: *Pitzer* model [38]

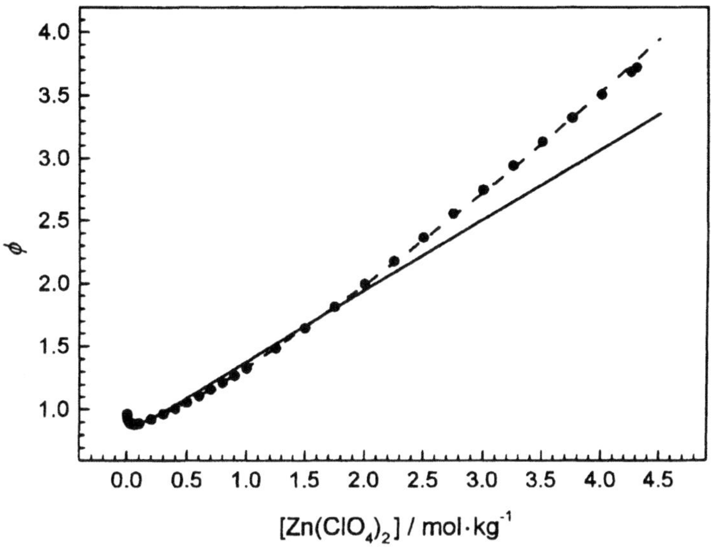

Fig. 2. Plot of the osmotic coefficient as a function of molality of $Zn(ClO_4)_2$ at 298.15 K; ●: *Goldberg* [44]; solid line: specific ion-interaction theory [37], $\varepsilon(Zn^{2+}, ClO_4^-) = 0.36$; dashed line: *Pitzer* model [38]

of water, pK_w^I,

$$pK_w^I = 14.00 + \log\gamma(H^+) + \log\gamma(OH^-) + \frac{2I\phi M_{H_2O}}{\ln(10)} \tag{17}$$

in solutions of sodium perchlorate and sodium chloride at 298.15 K is exemplified in Figs. 3 and 4, respectively. Again, the *Pitzer* equations allow a perfect description of the ionic strength dependence of pK_w^I by employing appropriate

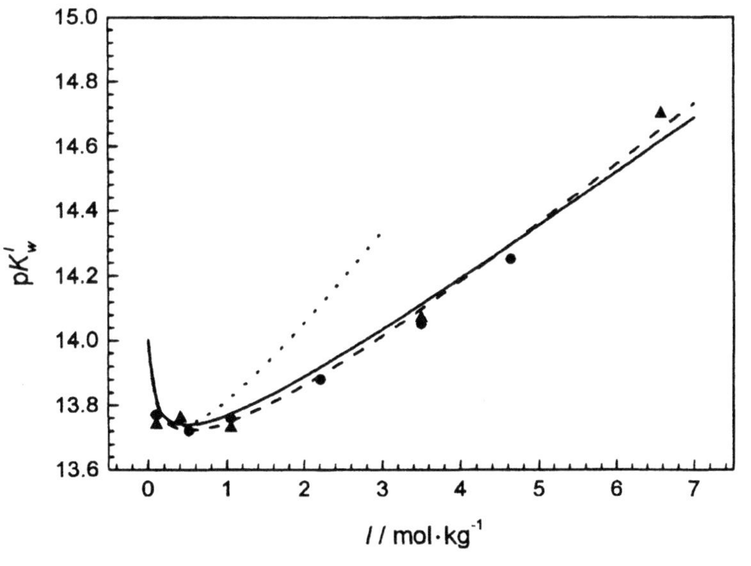

Fig. 3. Variation of pK_w^I with ionic strength I of $NaClO_4$ at 298.15 K; ●: *Fisher* and *Byé* [45]; ▲: *Carpéni et al.* [46]; solid line: specific ion interaction theory [37]; dashed line: *Pitzer* model [38]; dotted line: *Davies* approximation [36]

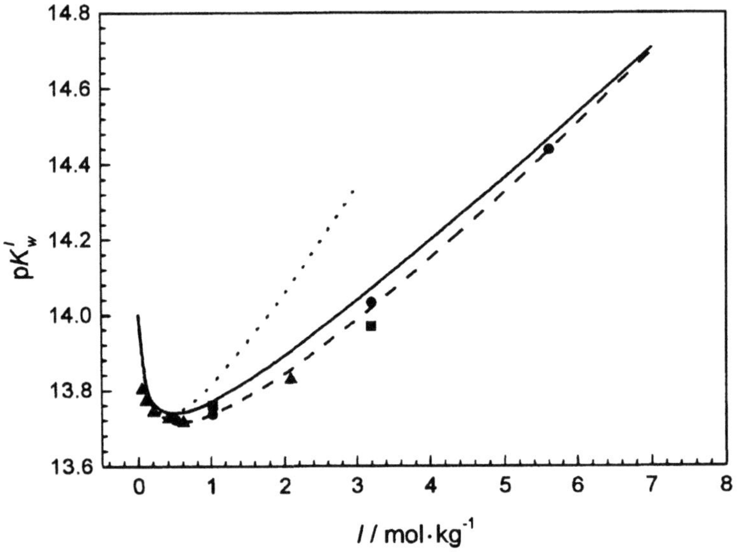

Fig. 4. Plot of pK_w^I *vs.* ionic strength I of NaCl at 298.15 K; ●: *Kron et al.* [47]; ▲: *Sjöberg et al.* [48]; ■: *Näsänen* and *Meriläinen* [49]; solid line: specific ion interaction theory [37]; dashed line: *Pitzer* model [38]; dotted line: *Davies* approximation [36]

ternary interaction parameters. Interestingly, the calculation of the ionic product of water in $NaClO_4$ solutions as a function of I, using the specific ion-interaction theory, is almost indistinguishable from the result of the *Pitzer* formalism (Fig. 3). In the case of NaCl the experimental data deviate from the calculated values employing the specific ion-interaction theory by less than 0.06 $\log K_w^I$ units (Fig. 4). The *Davies* formalism suffices to predict the ionic strength dependence of pK_w^I up to $1.0\,mol \cdot kg^{-1}$ $NaClO_4$ or NaCl with a maximum deviation around 0.1 $\log K_w^I$-units.

Apart from homogeneous equilibria (Figs. 3 and 4) the pertinent electrolyte theories are also applied to heterogeneous equilibria (Figs. 5 to 8) involving the dissolution of sparingly-soluble metal carbonates in constant ionic media consisting of $NaClO_4$ or NaCl.

The variations of the solubility constants of the transition metal carbonates $CdCO_3$ (Fig. 5) and $ZnCO_3$ (Fig. 6) with ionic strength I have been investigated at 298.15 K. The *Davies* formalism is valid up to an ionic strength of $1.0\,mol \cdot kg^{-1}$ $NaClO_4$. The specific ion-interaction theory provides a reasonable fit to the experimental data. Whereas the *Pitzer* model coincides perfectly with the experimental values in the case of cadmium carbonate (Fig. 5), a fairly large deviation occurs with respect to zinc carbonate especially at $I = 3.0\,mol \cdot kg^{-1}$ $NaClO_4$ (Fig. 6). This discrepancy may be attributed to the lack of ternary interaction parameters for the system $Zn(ClO_4)_2$–$NaClO_4$. It can be concluded that the specific ion-interaction theory leads to an even more accurate description of the ionic strength dependence of the solubility constant than the *Pitzer* model when essential ternary *Pitzer* parameters are unknown.

The application of the *Pitzer* equations and the specific ion-interaction theory to the modeling of the solubility of the alkaline earth metal carbonate calcite in

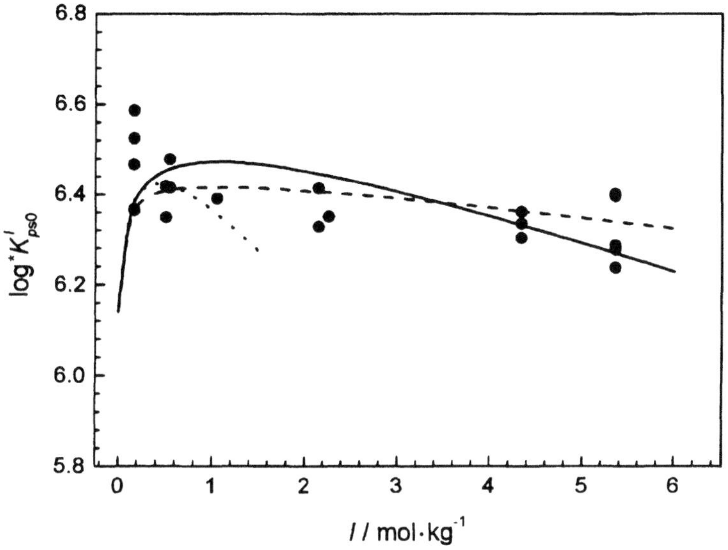

Fig. 5. Solubility constant of cadmium carbonate plotted *vs.* ionic strength at 298.15 K (inert electrolyte: $NaClO_4$); $\log {}^*K^0_{ps0} = 6.14 \pm 0.10$; ●: *Gamsjäger et al.* [9]; solid line: specific ion interaction theory [37]; dashed line: *Pitzer* model [38]; dotted line: *Davies* approximation [36]

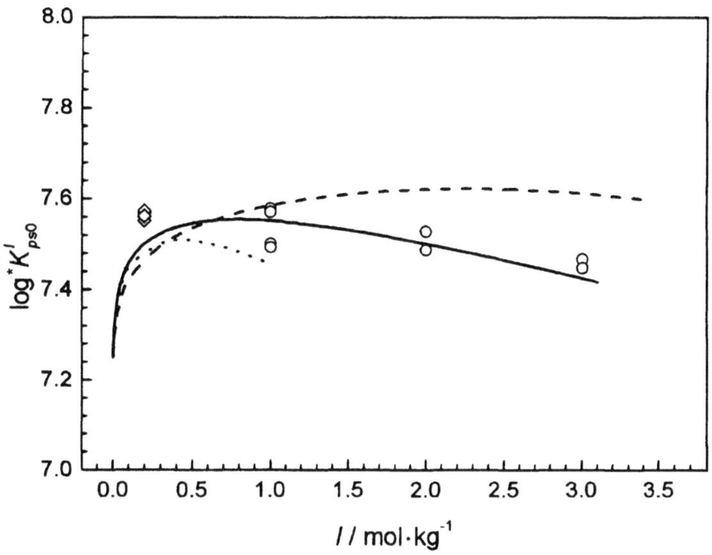

Fig. 6. Variation of $\log {}^*K^I_{ps0}$ of zinc carbonate with ionic strength at 298.15 K (inert electrolyte: $NaClO_4$); $\log {}^*K^0_{ps0} = 7.25 \pm 0.10$; ○: *Preis et al.* [10]; ◇: *Schindler et al.* [6]; solid line: specific ion interaction theory [37]; dashed line: *Pitzer* model [38]; dotted line: *Davies* approximation [36]

solutions of sodium chloride at three different temperatures (283.15, 298.15, and 333.15 K) is shown in Fig. 7. In addition, Fig. 8 represents the variation of the solubility constant of the rare earth metal carbonate lanthanite ($La_2(CO_3)_3 \cdot 8H_2O$) with ionic strength at 298.15 K. All relevant ternary *Pitzer* parameters are sufficiently known, and the application of the *Pitzer* concept leads to an excellent

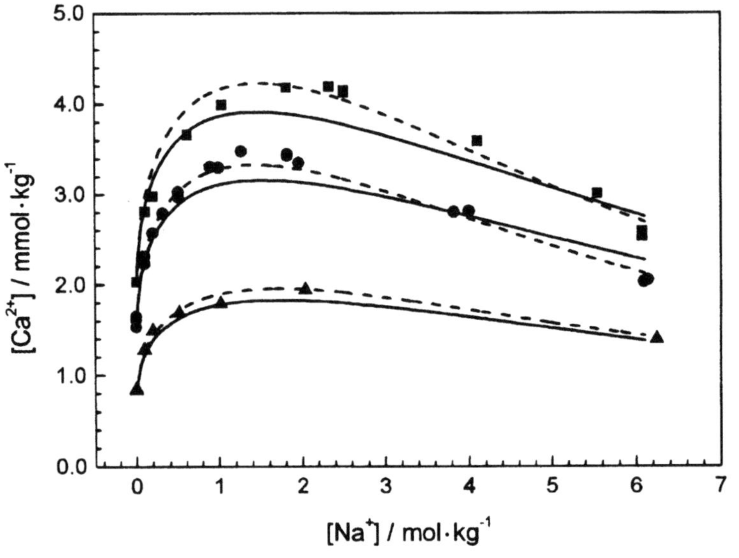

Fig. 7. Plot of solubility of calcite as a function of molality of sodium chloride; experimental data: *Wolf et al.* [50], ■: $T = 283.15\,K$, $p(CO_2) = 0.0094$ bar, ●: $T = 298.15\,K$, $p(CO_2) = 0.0092$ bar, ▲: $T = 333.15\,K$, $p(CO_2) = 0.0078$ bar; solid line: specific ion interaction theory [37]; dashed line: *Pitzer* model [38]

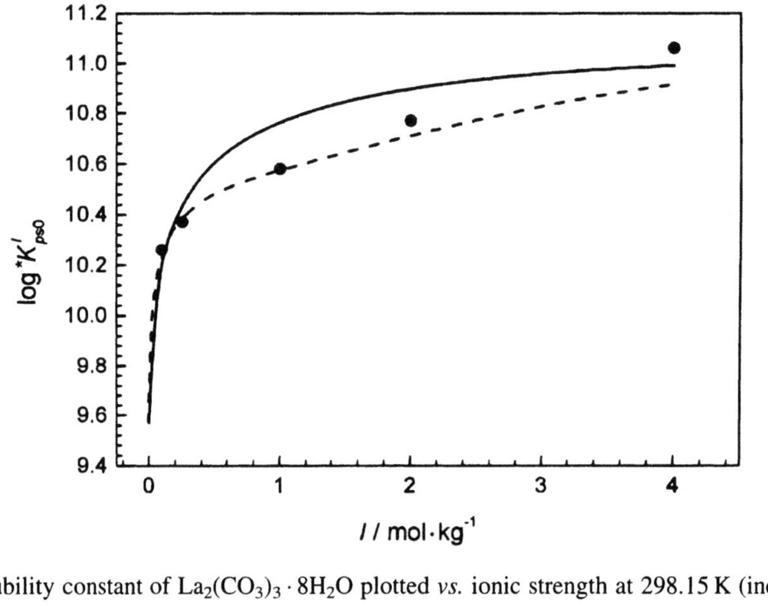

Fig. 8. Solubility constant of $La_2(CO_3)_3 \cdot 8H_2O$ plotted *vs.* ionic strength at 298.15 K (inert electrolyte: $NaClO_4$); $\log {}^*K^0_{ps0} = 9.57 \pm 0.10$; ●: *Nguyen et al.* [7]; solid line: specific ion interaction theory [37]; dashed line: *Pitzer* model [38]

agreement of the model calculations with the experimental results. However, it is worth emphasizing that the calculations by means of the specific ion-interaction theory concur reasonably with the experimental data, such that the maximum deviation is less than $0.2 \log {}^*K^0_{ps0}$ units in the range of 0–4.0 $mol \cdot kg^{-1}$ $NaClO_4$

Table 1. Specific ion-interaction parameters

$\varepsilon(i,j)$	$\varepsilon(i,j;\ 298.15\,K)$	$10^3\ \partial\varepsilon(i,j)/\partial T$	Ref.
$\varepsilon(H^+, ClO_4^-)$	0.12	0.58	[10, 37]
$\varepsilon(Na^+, ClO_4^-)$	0.01	1.3	[10, 37]
$\varepsilon(H^+, Cl^-)$	0.12	−0.45	[37], this study
$\varepsilon(Na^+, Cl^-)$	0.03	0.46	[37], this study
$\varepsilon(Na^+, OH^-)$	0.04	0.27	[37], this study
$\varepsilon(Na^+, HCO_3^-)$	0.00	–	[37]
$\varepsilon(Na^+, CO_3^{2-})$	−0.06	–	[37]
$\varepsilon(Ca^{2+}, ClO_4^-)$	0.27	0.40	[37], this study
$\varepsilon(Ca^{2+}, Cl^-)$	0.14	−0.24	[45], this study
$\varepsilon(Zn^{2+}, ClO_4^-)$	0.36	0.40	[10, 37]
$\varepsilon(Cu^{2+}, ClO_4^-)$	0.32	–	[37]
$\varepsilon(Cd^{2+}, ClO_4^-)$	0.33	–	this study
$\varepsilon(La^{3+}, ClO_4^-)$	0.47	–	[37]
$\varepsilon(CO_2(aq), ClO_4^-)$	−0.061	–	[a]
$\varepsilon(CO_2(aq), Cl^-)$	−0.04	–	[a]
$\varepsilon(CO_2(aq), Na^+)$	0.087	–	[a]

[a] $\varepsilon(k,j) = 2\lambda(k,j)/\ln(10)$, calculated from the corresponding *Pitzer* parameters taken from Ref. [51]

with regard to lanthanite (Fig. 8). The solubility of calcite in solutions containing up to $6.2\,mol \cdot kg^{-1}$ NaCl agrees even better with analogous predictions (Fig. 7).

All *Pitzer* parameters necessary for the calculations in this work can be found in Ref. [51]. The additional binary and ternary interaction parameters for the system $La(ClO_4)_3$–$NaClO_4$ are taken from *Kim* and *Frederick* [52, 53]. The binary *Pitzer* parameters for $Cd(ClO_4)_2$ can be found in Ref. [52], and the ternary parameters for the system $Cd(ClO_4)_2$–$NaClO_4$ have been given by *Gamsjäger et al.* [9]. Finally, the binary interaction coefficients for $Zn(ClO_4)_2$ are taken from Ref. [38].

The relevant parameters for the specific ion-interaction theory are listed in Table 1. Our coefficients are consistent within the limits of uncertainty with the data tabulated by *Grenthe et al.* [37]. The coefficient $\varepsilon(Cd^{2+}, ClO_4^-)$ has been obtained from a fit of Eq. (10) to experimental mean activity coefficients of aqueous solutions of $Cd(ClO_4)_2$ reported by *Kálmán et al.* [54] in the range from 0 to $2.0\,mol \cdot kg^{-1}$. The parameters $\partial\varepsilon(i,j)/\partial T$ have been derived by fitting Eq. (16) to the data for ${}^{\phi}L$ of the corresponding strong electrolytes taken from the NBS tables [55] up to a molality of $2.0\,mol \cdot kg^{-1}$.

Thermodynamic Properties of Transition Metal Carbonates

The temperature dependence of the solubility constant at infinite dilution can be expressed as

$$\log {}^*K^0_{ps0}(T) = -\frac{\Delta_{sol}H^{\ominus}(298.15)}{R\ln(10)} \cdot \frac{1}{T} + \frac{\Delta_{sol}S^{\ominus}(298.15)}{R\ln(10)} - \frac{1}{R\ln(10)}\left(\frac{1}{T}\int_{298.15}^{T} \Delta_{sol}C_p^{\ominus}\,dT - \int_{298.15}^{T} \frac{\Delta_{sol}C_p^{\ominus}}{T}\,dT\right) \quad (18)$$

where $\Delta_{sol}H^{\ominus}(298.15)$ and $\Delta_{sol}S^{\ominus}(298.15)$ refer to the standard enthalpy and the standard entropy of solution, respectively. The symbol $\Delta_{sol}C_p^{\ominus}$ denotes the change of the heat capacity for the reaction of Eq. (1). In general, $\Delta_{sol}H^{\ominus}(298.15)$ as well as $\Delta_{sol}S^{\ominus}(298.15)$ can be obtained from a fit of Eq. (18) to experimental data determined at various temperatures. If a suitable thermodynamic model, including the auxiliary data for all species involved in the solid-solute phase equilibrium, is available, the standard enthalpy of formation and the standard entropy of the solid compound can be derived from $\Delta_{sol}H^{\ominus}(298.15)$ and $\Delta_{sol}S^{\ominus}(298.15)$. When the standard entropy of the solid carbonate is well-known from low-temperature heat capacity measurements, solubility data are a reliable source for the standard enthalpy of formation of the solid phase. The thermodynamic properties of transition metal carbonates investigated so far are summarized in Table 2, and the auxiliary data for the respective metal ions are listed in Table 3 (all uncertainties in this work correspond to 2σ). The present set of thermodynamic constants can be regarded as an extension of the thermodynamic model published previously by

Table 2. Thermodynamic properties of relevant mineral phases, $T = 298.15$ K

Mineral	Formula	$\Delta_f H^{\ominus}$/kJ · mol^{-1}	$S^{\ominus}$/J · mol^{-1} · K^{-1}	$\Delta_f G^{\ominus}$/kJ · mol^{-1}
Otavite	$CdCO_3$	−752.1±0.8[a]	103.9±0.2[b]	−674.2±0.8[a]
Smithsonite	$ZnCO_3$	−818.9±0.6[c]	81.2±0.2[d]	−737.3±0.6[c]
Hydrozincite	$Zn_5(OH)_6(CO_3)_2$	−3584±15[e]	436±50[e]	−3164.6±3.0[e]
Zincite	ZnO	−350.46±0.27[f]	43.65±0.40[f]	−320.48±0.30[f]
Malachite	$Cu_2(OH)_2CO_3$	−1067.1±3.4[g]	166.3±2.5[h]	−903.3±1.2[g]
Azurite	$Cu_3(OH)_2(CO_3)_2$	−1675.1±5.1[g]	254.4±3.8[h]	−1434.2±1.8[g]
Tenorite	CuO	−156.1±2.1[i]	42.6±0.4[i]	−128.3±2.1[i]
Siderite	$FeCO_3$	−752.0±1.2[k]	95.47±0.15[l]	−678.9±1.2[k]

[a] *Gamsjäger et al.* [9]; [b] *Archer* [29]; [c] *Preis et al.* [10]; [d] *Robie et al.* [28]; [e] *Preis* and *Gamsjäger* [11]; [f] CODATA [56]; [g] *Preis* and *Gamsjäger* [57]; [h] *Kiseleva et al.* [19]; [i] JANAF tables [69]; [k] *Preis* and *Gamsjäger* [76]; [l] *Robie et al.* [27]

Table 3. Thermodynamic properties of aqueous species, $T = 298.15$ K

Species	$\Delta_f H^{\ominus}$/kJ · mol^{-1}	$S^{\ominus}$/J · mol^{-1} · K^{-1}	$\Delta_f G^{\ominus}$/kJ · mol^{-1}
Cd^{2+}	−75.92±0.60[a]	−72.8±1.5[a]	−77.73±0.60[a]
Zn^{2+}	−153.39±0.20[a]	−109.8±0.5[a]	−147.20±0.20[a]
$ZnHCO_3^+$	−839.8±5.3[b]	28±17[b]	−742.2±5.3[b]
Cu^{2+}	64.9±1.0[a]	−98±4[a]	65.0±0.1[a]
$CuHCO_3^+$	−598.8±5.3[b]	129±17[b]	−533.9±5.3[b]
$CuCO_3(aq)$	–	–	−501.3±0.6[c]
$Cu(CO_3)_2^{2-}$	–	–	−1046.9±0.6[c]
Fe^{2+}	−90.0±0.5[d]	−101.6±3.7[d]	−90.5±1.0[d]
$FeCO_3(aq)$	–	–	−648.7±1.6[e]
$Fe(CO_3)_2^{2-}$	–	–	−1186.9±1.6[e]

[a] CODATA [56]; [b] *Bauman* [58]; [c] *Baes* and *Mesmer* [59]; [d] *Parker* and *Khodakovskii* [72]; [e] *Bruno et al.* [75]

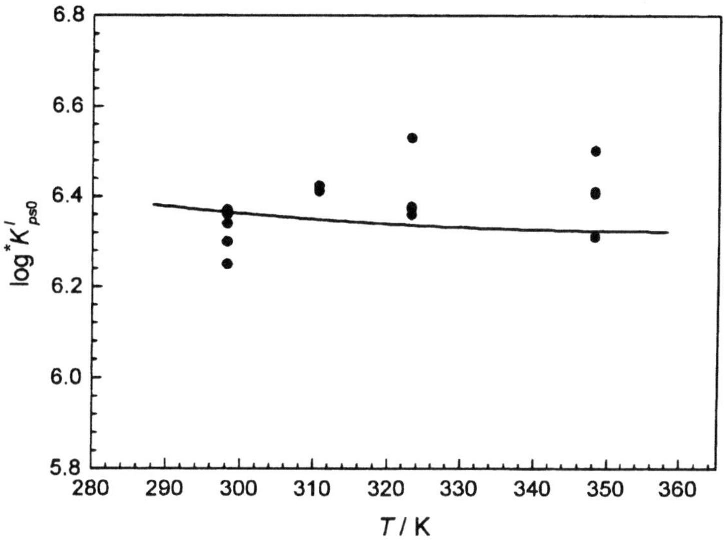

Fig. 9. Solubility constant of $CdCO_3$ plotted *vs.* temperature, $I = 1.00\,\text{mol}\cdot\text{kg}^{-1}$ $NaClO_4$; ●: experimental data [9]; solid line: *Davies* approximation [36]

Königsberger et al. [51]. It is worth mentioning that all auxiliary data are consistent with the CODATA key values for thermodynamics [56].

Cadmium carbonate

According to the dissolution reaction

$$CdCO_3 + 2H^+(I) \rightleftharpoons Cd^{2+}(I) + CO_2(g) + H_2O(I), \quad (19)$$

the solubility constant of otavite reads

$$\log {}^*K^I_{ps0} = \log([Cd^{2+}]p(CO_2)) + 2p[H]. \quad (20)$$

As can be seen from Fig. 9, the solubility constant is independent of temperature within the limits of experimental error, leading to a vanishing enthalpy of solution $\Delta_{sol}H^{\ominus}(298.15) = (-3.2 \pm 3.4) \approx 0\,\text{kJ}\cdot\text{mol}^{-1}$. The solubility constant of otavite at 298.15 K concurs well with the results reported by *Gamsjäger et al.* [4] and *Stipp et al.* [60]. The standard entropy of otavite listed in Table 2 is based on low-temperature measurements of the heat capacity from 4.5 to 350 K [29]. A direct determination of the enthalpy of formation by using a calorimetric method has been reported in the literature [15]. These original data were evaluated by *Wagman et al.* [55], leading to $\Delta_f H^{\ominus}(CdCO_3) = -750.6\,\text{kJ}\cdot\text{mol}^{-1}$. However, this result has been criticized recently by *Archer* [29, 61]. Thus, measuring the temperature dependence of the solubility seems to yield the most reliable value for $\Delta_f H^{\ominus}(CdCO_3)$.

Zinc carbonates

The variations of the solubility constants of smithsonite ($ZnCO_3$) and hydrozincite ($Zn_5(OH)_6(CO_3)_2$) with temperature at $I = 1.00\,\text{mol}\cdot\text{kg}^{-1}$ $NaClO_4$ are depicted

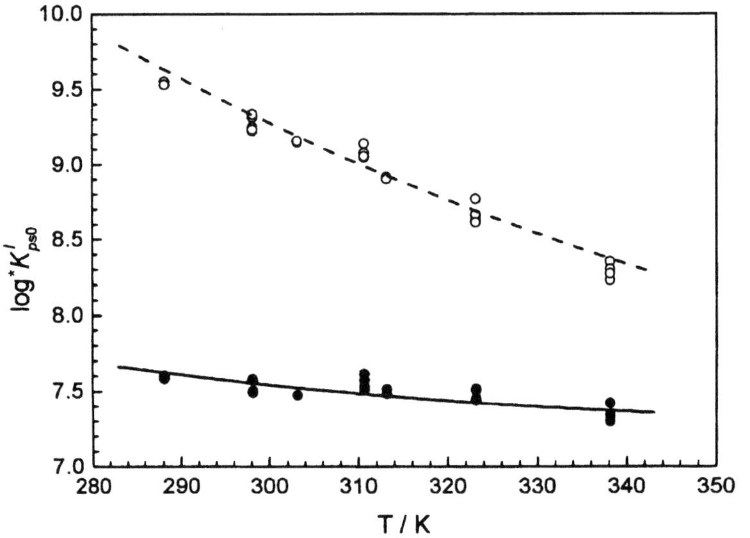

Fig. 10. Solubility constants of zinc carbonates plotted *vs.* temperature, $I = 1.00\,\mathrm{mol \cdot kg^{-1}}$ $NaClO_4$; experimental data: ●: $ZnCO_3$ [10], ○: $Zn_5(OH)_6(CO_3)_2$ [11]; model calculations using the specific ion-interaction theory [37]: solid line: $ZnCO_3$, dashed line: $Zn_5(OH)_6(CO_3)_2$

in Fig. 10. The corresponding dissolution reactions can be written as

$$ZnCO_3(s) + 2H^+(l) \rightleftharpoons Zn^{2+}(l) + CO_2(g) + H_2O(l) \qquad (21)$$

and

$$0.2Zn_5(OH)_6(CO_3)_2(s) + 2H^+(l) \rightleftharpoons Zn^{2+}(l) + 0.4\,CO_2(g) + 1.6H_2O(l). \qquad (22)$$

Figure 10 reveals that hydrozincite shows a pronounced dependence of the solubility constant on temperature in contrast to the neutral transition metal carbonate $ZnCO_3$. In the case of smithsonite, our solubility constants coincide satisfactorily with those given by *Smith* [62], *Schindler et al.* [6], and *Reiterer* [63]. The standard entropy was derived from low-temperature measurements of the heat capacity by *Robie et al.* [28]. The enthalpy of solution was found to be consistent with the heat of decomposition obtained from DSC experiments [10]. Our enthalpy of formation (Table 2) agrees well with $\Delta_f H^\ominus(ZnCO_3) = -817\,\mathrm{kJ \cdot mol^{-1}}$ estimated from the decarbonation pressure at elevated temperatures [25]. The standard enthalpy of formation $\Delta_f H^\ominus(ZnCO_3) = -814.9\,\mathrm{kJ \cdot mol^{-1}}$, based on new auxiliary data and reported by *Roth* and *Chall* [16], deviates significantly from the value given in Table 2. The solubility constant of hydrozincite at 298.15 K is in close agreement with that published by *Schindler et al.* [6] and *Alwan* and *Williams* [64]. The standard *Gibbs* free energy of formation of hydrozincite coincides remarkably well with the results of *emf* measurements carried out by *Mercy et al.* [14]. The standard entropy of hydrozincite has not yet been determined directly by low-temperature measurements of the heat capacity. Thus, both the standard enthalpy of formation and the standard entropy listed in Table 2 are based on solubility measurements as a function of temperature [11].

Copper carbonates

The dissolution reactions for the copper carbonates malachite ($Cu_2(OH)_2CO_3$) and azurite ($Cu_3(OH)_2(CO_3)_2$) read

$$0.5\,Cu_2(OH)_2CO_3(s) + 2\,H^+(I) \rightleftharpoons Cu^{2+}(I) + 0.5CO_2(g) + 1.5H_2O(I) \tag{23}$$

and

$$1/3\,Cu_3(OH)_2(CO_3)_2(s) + 2H^+(I) \rightleftharpoons Cu^{2+}(I) + 2/3CO_2(g) + 4/3H_2O(I). \tag{24}$$

The respective solubility constants, plotted as a function of temperature in Fig. 11, were found to be of the same order of magnitude. The temperature dependence of the solubility of $Cu_2(OH)_2CO_3$ seems to be somewhat more pronounced than that of $Cu_3(OH)_2(CO_3)_2$.

The solubility constants agree satisfactorily with those obtained by *Free* [65], *Silman* [66], *Schindler et al.* [5], and *Symes* and *Kester* [67]. The standard entropies of malachite and azurite selected by *Robie* and *Hemingway* [68] are based on low-temperature measurements of the heat capacity [19]. Both solution and drop calorimetry were applied to determine the standard enthalpies of formation of malachite and azurite, leading to $\Delta_f H^\ominus$ $(Cu_2(OH)_2CO_3) = -1054.0\,kJ \cdot mol^{-1}$ [18] and $\Delta_f H^\ominus(Cu_3(OH)_2(CO_3)_2) = -1632.2\,kJ \cdot mol^{-1}$ [17, 19]. These values were finally adopted by *Robie* and *Hemingway* [68]. According to this set of thermodynamic data, the equilibrium partial pressure of carbon dioxide for the coexistence of malachite and tenorite (CuO) in aqueous media at 323.15 K should amount to 3.1 bar. However, by bubbling CO_2 gas through an aqueous suspension of copper oxide it was possible to transform tenorite into pure malachite at

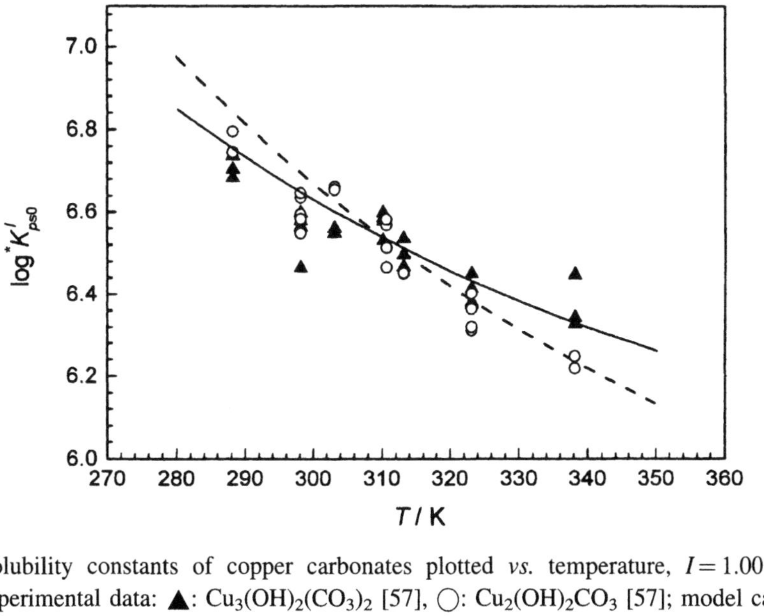

Fig. 11. Solubility constants of copper carbonates plotted *vs.* temperature, $I = 1.00\,mol \cdot kg^{-1}$ $NaClO_4$; experimental data: ▲: $Cu_3(OH)_2(CO_3)_2$ [57], ○: $Cu_2(OH)_2CO_3$ [57]; model calculations using the specific ion-interaction theory [37]: solid line: $Cu_3(OH)_2(CO_3)_2$, dashed line: $Cu_2(OH)_2CO_3$

$p(CO_2) = 0.84$ bar and $T = 323.15$ K. This observation clearly disproves the prediction based on the values listed in Ref. [68]. The thermodynamic data for CuO are consistent with the values of the JANAF tables [69] and are regarded to be fairly reliable. Thus, the thermodynamic properties of malachite and azurite were reinvestigated by performing solubility measurements as a function of temperature between 288.15 and 338.15 K and constant ionic strength ($I = 1.00\,\mathrm{mol \cdot kg^{-1}}$ $NaClO_4$) [57]. The results of these solubility measurements confirm the values for the standard entropies determined by *Kiseleva et al.* [19], whereas revised values for $\Delta_f H^\ominus$ of both malachite and azurite are presented in Table 2.

Iron(II) carbonate

Whereas solubility data for siderite ($FeCO_3$) coincide remarkably well [70], a tremendous controversy occurs in literature with respect to the enthalpy of formation. *Wagman et al.* [55] as well as *Knacke et al.* [71] have selected -740.5 $\mathrm{kJ \cdot mol^{-1}}$ for the enthalpy of formation, whereas $\Delta_f H^\ominus(FeCO_3) = -755.9$ $\mathrm{kJ \cdot mol^{-1}}$ has been tabulated by *Robie* and *Hemingway* [68]. *Stubina* and *Toguri* [26] derived $\Delta_f H^\ominus(FeCO_3) = -749\,\mathrm{kJ \cdot mol^{-1}}$ from measurements of the decomposition pressure of siderite at elevated temperatures. *Chai* and *Navrotsky* [20] used a calorimetric method for the determination of this quantity and obtained $\Delta_f H^\ominus(FeCO_3) = -750.6\,\mathrm{kJ \cdot mol^{-1}}$. The large difference in the values for the enthalpy of formation listed in data compilations seems to be caused by a discrepancy in the standard *Gibbs* free energy of formation of the iron(II) ion (NBS-tables [55]: $\Delta_f G^\ominus(Fe^{2+}) = -78.9\,\mathrm{kJ \cdot mol^{-1}}$; *Parker* and *Khodakovskii* [72]: $\Delta_f G^\ominus(Fe^{2+}) = -90.53\,\mathrm{kJ \cdot mol^{-1}}$). The thermodynamic properties of Fe^{2+}

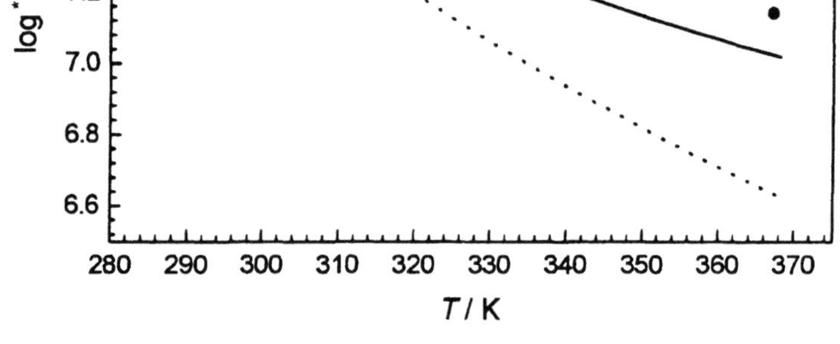

Fig. 12. Solubility constant of iron(II) carbonate plotted *vs.* temperature, $I = 0$; ▲: *Smith* [73]; ■: *Reiterer et al.* [63, 74]; ●: *Greenberg* and *Tomson* [77]; ▼: *Bruno et al.* [75]; solid line: *Preis* and *Gamsjäger* [76], $\Delta_{sol}H^\ominus = (-17.3 \pm 1.2)\,\mathrm{kJ \cdot mol^{-1}}$; dashed line: *Wagman et al.* [55], $\Delta_{sol}H^\ominus = -27.9\,\mathrm{kJ \cdot mol^{-1}}$

recommended recently by *Parker* and *Khodakovskii* [72] were incorporated into our thermodynamic model in order to evaluate solubility data of siderite given by several literature sources. This critical evaluation led to an optimized value for the standard enthalpy of formation of siderite of $\Delta_f H^\ominus(FeCO_3) = -752.0\,kJ\cdot mol^{-1}$. The standard entropy of $FeCO_3$ is taken from *Robie et al.* [27]. In contrast to the set of thermodynamic quantities listed in the NBS tables [55], the application of our thermodynamic model leads to a close agreement between the calculation of the temperature dependence of the solubility constant and various experimental data [73–75, 77]. Hence, Fig. 12 provides evidence for the reliability of both our value for $\Delta_f H^\ominus(FeCO_3)$ and the Fe/Fe^{2+} standard potential $\left(E^\ominus_{Fe/Fe^{2+}} = (-469.0 \pm 2.6)\,mV\right)$ recommended by *Parker* and *Khodakovskii* [72].

Discussion and Conclusions

The most accurate technique for the determination of the standard entropy of a solid compound is the integration of low-temperature heat capacities. When such low-temperature measurements have not yet been published, the standard entropy is given by the entropy of solution obtained from solubility measurements at various temperatures.

Calorimetric methods as well as the evaluation of solubility measurements as a function of temperature yield the standard enthalpy of formation of metal carbonates. The enthalpies of solution of smithsonite [16], malachite [17, 18], and azurite [17] were studied by solution calorimetry where the respective mineral phase was dissolved in solutions of hydrochloric acid. Although the results show a high precision (low random errors), high systematic errors (low accuracy) might occur during the calorimetric measurements due to the evolution of unknown quantities of $CO_2(g)$. The enthalpies of formation of malachite [19], azurite [19], and siderite [20] were determined by drop calorimetry. Especially, the decomposition experiments on malachite and azurite may suffer from systematic deviations, because the thermodynamic state of the microcrystalline metal oxide formed rapidly during the decarbonation might be poorly defined. In contrast to these direct methods it was shown in this work that a proper analysis of the temperature dependence of solubility constants yields the enthalpy of solution of sparingly-soluble carbonates. Obviously, the precision (absence of random error) of our equilibrium approach is expected to be much lower compared to the calorimetric methods. The evaluation of the temperature dependence of solubility data, however, avoids systematic errors, leading to a fairly high accuracy. Therefore, the enthalpy of solution was obtained with an estimated uncertainty of $\pm 3\,kJ\cdot mol^{-1}$ per mole dissolved metal cation as summarized in Table 4. Using appropriate auxiliary data, the standard enthalpy of formation can be easily derived from the enthalpy of solution. All thermodynamic data derived in this work are consistent with the CODATA key values of thermodynamics [56]. The *Davies* approximation, the specific ion-interaction theory, and the *Pitzer* concept were used for the extrapolation of the solubility constants to zero ionic strength in order to obtain standard thermodynamic properties valid for infinite dilution. This procedure enables the simultaneous evaluation of solubility experiments performed in different ionic media (solvents) at various temperatures. Finally, from Table 4 it

Table 4. Solubility constants and enthalpies of solution at $T = 298.15\,\mathrm{K}$ and $I = 0$ for relevant transition metal carbonates

Mineral	$\log {}^*K^0_{ps0}$	Dissolution reaction	$\Delta_{sol}H^{\ominus}/\mathrm{kJ \cdot mol^{-1}}$
Otavite	6.14	$CdCO_3(s) + 2H^+(aq) \rightarrow Cd^{2+}(aq) + CO_2(g) + H_2O(l)$	−3.2
Smithsonite	7.25	$ZnCO_3(s) + 2H^+(aq) \rightarrow Zn^{2+}(aq) + CO_2(g) + H_2O(l)$	−13.8
Hydrozincite	9.00	$0.2Zn_5(OH)_6(CO_3)_2(s) + 2H^+(aq) \rightarrow Zn^{2+}(aq) + 0.4\ CO_2(g) + 1.6\ H_2O(l)$	−51.3
Malachite	6.34	$0.5Cu_2(OH)_2CO_3(s) + 2H^+(aq) \rightarrow Cu^{2+}(aq) + 0.5\ CO_2(g) + 1.5\ H_2O(l)$	−27.1
Azurite	6.30	$2/3Cu_3(OH)_2(CO_3)_2(s) + 2H^+(aq) \rightarrow Cu^{2+}(aq) + 2/3CO_2(g) + 4/3H_2O(l)$	−20.2
Siderite	7.56	$FeCO_3(s) + 2H^+(aq) \rightarrow Fe^{2+}(aq) + CO_2(g) + H_2O(l)$	−17.3

can be concluded that the temperature dependence of the solubility of basic transition metal carbonates is more pronounced than that of neutral carbonates, *i.e.* the enthalpy of solution becomes more negative when the ratio $n(OH)/n(CO_3)$ in the solid mineral increases.

References

[1] Stumm W, Morgan JJ (1996) Aquatic Chemistry 3rd edn. Wiley, New York
[2] Eary LE (1999) Appl Geochem **14**: 963
[3] Grauer R (1997) Solubility Limitations: An "Old Timer's" view. In: Grenthe I, Puigdomenech I (eds) Modelling in Aquatic Chemistry. OECD NEA, Paris, pp 131–152
[4] Gamsjäger H, Stuber HU, Schindler P (1965) Helv Chim Acta **48**: 723
[5] Schindler P, Reinert M, Gamsjäger H (1968) Helv Chim Acta **51**: 1845
[6] Schindler P, Reinert M, Gamsjäger H (1969) Helv Chim Acta **52**: 2327
[7] Nguyen AM, Königsberger E, Marhold H, Gamsjäger H (1993) Monatsh Chem **124**: 1011
[8] Gamsjäger H, Marhold H, Königsberger E, Tsai YJ, Kolmer H (1995) Z Naturforsch **50 a**: 59
[9] Gamsjäger H, Preis W, Königsberger E, Magalhães MC, Brandão P (1999) J Solution Chem **28**: 717
[10] Preis W, Königsberger E, Gamsjäger H (2000) J Solution Chem **29**: 605
[11] Preis W, Gamsjäger H (2001) J Chem Thermodynamics (in press)
[12] Rock PA, Casey WH, McBeath MK, Wallin EM (1994) Geochim Cosmochim Acta **58**: 4281
[13] Casey WH, Rock PA, Chung JB, Walling EM, McBeath MK (1996) Amer J Sci **296**: 1
[14] Mercy MA, Rock PA, Casey WH, Mokarram MM (1998) Amer Mineral **83**: 739
[15] Thomsen J (1883) Thermochemische Untersuchungen, vol 3, Metals. Johann Ambrosius, Leipzig
[16] Roth WA, Chall P (1928) Z Elektrochem **34**: 185
[17] Roth WA, Berendt H, Wirths G (1941) Z Elektrochem **47**: 185
[18] Richardson DW, Brown RR (1974) U S Bur Mines Rep Inv **7851**: 1–5
[19] Kiseleva IA, Ogorodova LP, Melchakova LV, Bisengalieva MR, Becturganov NS (1992) Phys Chem Minerals **19**: 322
[20] Chai L, Navrotsky A (1994) Amer Mineral **79**: 921
[21] Chai L, Navrotsky A (1996) Geochim Cosmochim Acta **60**: 4377
[22] Navrotsky A, Rapp RP, Smelik E, Burnley P, Circone S, Chai L, Bose K, Westrich HR (1994) Amer Mineral **79**: 1099

[23] Kiseleva IA, Kotelnikov AR, Martynov KV, Ogorodova LP, Kabalov JK (1994) Phys Chem Minerals **21**: 392
[24] Casey WH, Chai L, Navrotsky A, Rock PA (1996) Geochim Cosmochim Acta **60**: 933
[25] Haselton HT, Goldsmith JR (1987) Geochim Cosmochim Acta **51**: 261
[26] Stubina NM, Toguri JM (1989) ISS Transactions **10**: 87
[27] Robie RA, Haselton HT, Hemingway BS (1984) Amer Mineral **69**: 349
[28] Robie RA, Haselton HT, Hemingway BS (1989) J Chem Thermodynamics **21**: 743
[29] Archer DG (1996) J Chem Eng Data **41**: 852
[30] Schindler P (1963) Chimia **17**: 313
[31] Heindl R, Gamsjäger H (1977) Monatsh Chem **108**: 1365
[32] Gamsjäger H, Reiterer F (1979) Environment International **2**: 419
[33] Königsberger E, Eriksson G (1995) CALPHAD **19**: 207
[34] Robinson RA, Stokes RH (1959) Electrolyte Solutions, 2nd edn. Butterworths, London
[35] Lewis GN, Randall M (1961) Thermodynamics. McGraw-Hill, New York, NY (revised by Pitzer KS, Brewer L)
[36] Davies CW (1962) Ion Association. Butterworths, London
[37] Grenthe I, Plyasunov AV, Spahiu K (1997) Estimations of medium effects on thermodynamic data. In: Grenthe I, Puigdomenech I (eds) Modelling in Aquatic Chemistry. OECD NEA, Paris, pp 325–426
[38] Pitzer KS (1991) Ion interaction approach: Theory and data correlation. In: Pitzer KS (ed) Activity Coefficients in Electrolyte Solutions, 2nd edn. CRC Press, Boca Raton, Florida, pp 76–153
[39] Brønsted JN (1922) J Am Chem Soc **44**: 877
[40] Brønsted JN (1922) J Am Chem Soc **44**: 938
[41] Guggenheim EA (1935) Phil Mag **19**: 588
[42] Scatchard G (1936) Chem Rev **19**: 309
[43] Eriksson G, Hack K (1990) Metall Trans **21 B**: 1013
[44] Goldberg RN (1981) J Phys Chem Ref Data **10**: 1
[45] Fischer R, Byé J (1964) Bull Soc Chim, pp 2920–2929
[46] Carpéni G, Boitard E, Pilard R, Poize S, Sabiani N (1973) J Chim Phys **69**: 1445
[47] Kron I, Marshall SL, May PM, Hefter G, Königsberger E (1995) Montsh Chem **126**: 819
[48] Sjöberg S, Högglund Y, Nordin A, Ingri N (1983) Mar Chem **13**: 35
[49] Näsänen R, Meriläinen P (1960) Suomen Kem **B33**: 149
[50] Wolf M, Breitkopf O, Puk R (1989) Chem Geol **76**: 291
[51] Königsberger E, Königsberger L-C, Gamsjäger H (1999) Geochim Cosmochim Acta **63**: 3105
[52] Kim H-T, Frederick Jr WJ (1988) J Chem Eng Data **33**: 177
[53] Kim H-T, Frederick Jr WJ (1988) J Chem Eng Data **33**: 278
[54] Kálmán E, Horn G, Schwabe K (1970) Z Phys Chem **244**: 106
[55] Wagman DD, Evans WH, Parker VB, Schumm RH, Halow I, Bailey SM, Churney KL, Nuttall RL (1982) J Phys Chem Ref Data **11** (Suppl 2)
[56] Cox JD, Wagman DD, Medvedev VA (1989) CODATA Key Values for Thermodynamics. Hemisphere Publishing Corporation, Washington, New York, London
[57] Preis W, Gamsjäger H (2001) J Chem Thermodynamics (submitted)
[58] Bauman Jr JE (1981) U S Bur Mines Inf Circ **8852**: 268
[59] Baes CF, Mesmer RE (1976) The Hydrolysis of Cations. Wiley, New York
[60] Stipp SL, Parks GA, Nordstrom DK, Leckie JO (1993) Geochim Cosmochim Acta **57**: 2699
[61] Archer DG (1998) J Phys Chem Ref Data **27**: 915
[62] Smith HJ (1918) J Amer Chem Soc **40**: 883
[63] Reiterer F (1980) Löslichkeitskonstanten und Freie Bildungsenthalpien neutraler Übergangsmetallcarbonate. PhD Thesis, Montanuniversität Leoben, Austria
[64] Alwan AK, Williams PA (1979) Trans Met Chem **4**: 128
[65] Free RE (1908) J Amer Chem Soc **30**: 1366

[66] Silman JFB (1958) PhD Thesis, Harvard University
[67] Symes JL, Kester DR (1984) Geochim Cosmochim Acta **48**: 2219
[68] Robie RA, Hemingway BS (1995) Thermodynamic properties of minerals and related substances at 298.15 K and 1 bar (10^5 Pascals) pressure and at higher temperatures. US Geological Survey Bulletin 2131, Washington
[69] Chase MW (1998) NIST-JANAF Thermochemical Tables, Part II, Cr-Zr 4th edn. American Chemical Society, Gaithersburg, Maryland
[70] Grauer R (1994) Bereinigte Löslichkeitsprodukte von M(II)-Schwermetallcarbonaten. Paul Scherrer Institut, Internal Technical Report, TM-44-94-05, pp 1–35
[71] Knacke O, Kubaschewski O, Hesselmann K (1991) Thermochemical Properties of Inorganic Substances, 2nd edn. Springer-Verlag, Verlag Stahleisen G.m.b.H., Berlin, Düsseldorf
[72] Parker VB, Khodakovskii IL (1995) J Phys Chem Ref Data **24**: 1699
[73] Smith HJ (1918) J Amer Chem Soc **40**: 879
[74] Reiterer F, Johannes W, Gamsjäger H (1981) Mikrochim Acta **I**: 63
[75] Bruno J, Wersin P, Stumm W (1992) Geochim Cosmochim Acta **56**: 1149
[76] Preis W, Gamsjäger H (2001) Critical evaluation of solubility data: Enthalpy of formation of siderite. Phys Chem Chem Phys (in preparation)
[77] Greenberg J, Tomson M (1992) Appl Geochem **7**: 185

Received June 26, 2001. Accepted July 2, 2001

Invited Review

The Solvent-Like Nature of Silica Particles in Organic Solvents

Yvonne Zimmermann, Mohamed El-Sayed, Silvio Prause, and **Stefan Spange***

Department of Polymer Chemistry, Institute of Chemistry, University of Technology Chemnitz, D-09107 Chemnitz, Germany

Summary. *Kamlet-Taft*'s α (hydrogen bond donor acidity) and π^* (dipolarity/polarizability) values of various silica batches measured in various solvents are presented. The α and π^* parameters for the various solid acids are analyzed by means of Fe(*phen*)$_2$(CN)$_2$ (*cis*-dicyano-*bis*-(1,10)-phenanthroline-iron(II), **1**), *Michler*'s ketone (4,4′-*bis*-(dimethylamino)-benzophenone, **2**), and two hydrophilic derivatives of **2**, (4-(dimethylamino)-4′-(di-2-hydroxyethyl)-amino-benzophenone (**3a**) and 4,4′-*bis*-(di-(2-hydroxyethyl)-amino)-benzophenone (**3b**) as well as coumarin 153 (**4**) as solvatochromic surface polarity indicators. Apparent β (hydrogen bond acceptor basicity) parameters for bare silica have been evaluated by means of an aminobenzodifuranone dye (**5**) as solvatochromic probe.

The chemical interpretation of the α and π^* parameters and the nature of the solvent/surface interaction which they reflect are discussed. It can be shown that an increase of the HBA (hydrogen bond accepting) capacity of the solvent significantly decreases the HBD (hydrogen bond donating) capacity of the surface environment, whereas the dipolarity/polarizability value of the silica/solvent interface is a composite of many effects. The classification of the polarity of silica particles in organic solvents compared to pure liquids is outlined.

Theoretical $E_T(30)$ values of the solid/solvent interfaces are calculated by applying linear solvation energy (LSE) relationships using the independently measured α and π^* values of the solid acids according to $E_T(30) = (E_T(30))_o + a\alpha + s\pi^*$.

Keywords. Donor-acceptor effects; Silica; Solvent effect; Surface polarity; UV/Vis spectroscopy.

Introduction

Solvent polarity is among the most widely used concepts in chemistry [1–9] (see Refs. [2–4], [7], and [9] for reviews). Most solvent polarity scales are empirical and based on kinetic, thermodynamic, or spectroscopic data relating to certain reference reactions [2–9]. Different empirical solvent polarity scales have been shown to correlate well with each other, pointing to the existence of an underlying common feature [3, 7–9]

As a consequence of this established concept in solution chemistry, there have been recent considerations of the changes in surface polarity associated with

* Corresponding author. E-mail: stefan.spange@chemie.tu-chemnitz.de

structural variations in organically functionalized silica particles including stationary chromatographic phases, various kinds of bare silica particles, and other solid acids [10–25].

However, the interaction of a surface environment with a solvent or an adsorbed molecule is a composite of many effects [11, 13, 19, 23–25]. Not only acid-base, but also dipole-dipole, induced dipole-dipole, and dispersion forces contribute to the overall adsorption energy of a solvent with an inorganic surface. To date, no general agreement upon a definition of the term surface polarity has emerged [21]. In the broadest and most general sense, the surface polarity can be viewed as the sum of all interactive forces between an adsorbed molecule and the occupied surface site or sites. This definition is based on the related interpretation of the term solvent polarity [1].

Multiple intermolecular solute/solvent interactions can be described by the LSE (linear solvation energy) relationship of *Kamlet* and *Taft* [7, 26] (Eq. 1).

$$XYZ = (XYZ)_o + m\delta_H^2 + a\alpha + b\beta + s(\pi^* + d\delta) \quad (1)$$

$(XYZ)_o$ is the solute property of a reference system, *e.g.* a nonpolar medium, δ_H^2 is a cavity term which relates to the *Hildebrandt* solubility parameter, α describes the HBD (hydrogen bond donating) acidity, β the HBA (hydrogen bond accepting) ability, and π^* the dipolarity/polarizability of the solvents. δ is a polarizability correction term which is 1.0 for aromatic, 0.5 for polyhalogenated, and zero for aliphatic solvents; *m*, *a*, *b*, *s*, and *d* are solvent-independent correlation coefficients [7, 26].

Empirical solvent polarity scales based on spectroscopic measurements usually employ changes in the UV/Vis absorption maximum of an indicator in different solvents (solvatochromism) [2, 6, 7]. Visible probing with suitable indicator dyes has also been recommended by several authors for investigating the surface polarities of solids [10–25].

Among others, *Marcus* has shown that the well established empirical $E_T(30)$ solvent polarity parameter and other polarity scales can be expressed by LSE relationships using the *Kamlet-Taft* solvent parameters α and π^* [2, 3].

Presuming a solvent-like behaviour for a surface environment, it is expected that an adsorbed solvatochromic probe reflects contributions of each of these interactions relating to a certain set of α, β, and π^* values, respectively, of the occupied surface site by shifts of its UV/Vis absorption maximum (ν_{max}). The simplified *Kamlet-Taft* LSE equation applied to single solvatochromic shifts, $XYZ = \nu_{max}(\text{probe})$ [2], is given in Eq. (2). Because volume changes do not contribute to $\nu_{max}(\text{probe})$ during electronic transition, the $m\delta_H^2$ term can be neglected.

$$\nu_{max}(\text{probe}) = (\nu_{max}(\text{probe}))_0 + a\alpha + b\beta + s(\pi^* + d\delta) \quad (2)$$

The position of the UV/Vis absorption maximum of the adsorbed dye ($\nu_{max}(\text{probe})$) has been used as a unit for surface polarity [11, 19]. As a consequence of this relation, the knowledge of two precise LSE equations (Eqs. (3a) and (3b)) for two different probes, where the solvent-independent coefficients *a* and *s* are known, is theoretically required in order to determine the pair α and π^* of an unknown material. Usually, for the determination of the underlying LSE equations, the solvatochromic probes are measured in more than 20 solvents which serve as the reference system [3, 27–30].

Then, the molecular probes Fe(*phen*)$_2$(CN)$_2$ (*cis*-dicyano-*bis*-(1,10)-phenanthroline-iron(II), **1**) and *Michler*'s ketone (4,4′-*bis*-(dimethylamino)-benzophenone, **2**) can be adsorbed on the solid material (or environment) of interest in order to determine $\nu_{max}(\mathbf{1})$ and $\nu_{max}(\mathbf{2})$, respectively. For example, $(\nu_{max}(\mathbf{1}))_0$ denotes the UV/Vis absorption maximum of probe **1** in a nonpolar reference environment.

$$\nu_{max}(\mathbf{1}) = f_1(\alpha, \pi^*) = (\nu_{max}(\mathbf{1}))_0 + a_1'\alpha + s_1'\pi^* \quad (3a)$$

$$\nu_{max}(\mathbf{2}) = f_2(\alpha, \pi^*) = (\nu_{max}(\mathbf{2}))_0 + a_2''\alpha + s_2''\pi^* \quad (3b)$$

In earlier reports we were able to show that a correlation analysis of the desired polarity parameter with the solvatochromic UV/Vis absorption maxima of two structural different polarity indicators according to Eq. 4 seems better suited for determining polarity parameters than are the functions according to Eqs. (3a) and (3b) [19, 28, 30].

$$\alpha = f(\nu_{max}(\mathbf{1}), \nu_{max}(\mathbf{2})) = a_0 + a_1\nu_{max}(\mathbf{1}) + a_2\nu_{max}(\mathbf{2}) \quad (4)$$

This is true when additional influences of the solvent, *i.e.* derived from the hydrophobicity or size (cavity effect) of the probes, on each ν_{max} are similar [30]. Thus, systematic solvation effects on $\nu_{max}(\mathbf{1})$ and $\nu_{max}(\mathbf{2})$, which are not included in the original α or π^* solvent parameters, then can be better excluded.

It is not easy to decide an adequate probe pair which is suited for an accurate determination of surface polarity parameters of solid materials [19, 30–32]. The following solvatochromic probe molecules have been well established as surface polarity indicators for several applications during the recent years: 2,6-diphenyl-4-(2,4,6-triphenyl-1-pyridinio)-phenolate (*Reichardt*'s standard dye) and various substituted derivatives [10–12, 18, 23], Fe(*phen*)$_2$(CN)$_2$ (**1**) [11, 23, 31, 32], and *Michler*'s ketone (**2**) [11, 23, 31, 32]. Indicator formulas are shown in Scheme 1.

Fe(*phen*)$_2$(CN)$_2$ has been well established as HBD indicator for solvents as well as for solid materials. For instance, its UV/Vis absorption can be used for determining *AN* (acceptor number) of a variety of materials [27]. For separation of the contribution of the α term, we used in particular the ν_{max} values from Fe(*phen*)$_2$(CN)$_2$ and *Michler*'s ketone measured in the same solvents (taken from Refs. [27] and [29]). Multiple square analyses were performed for several dual combinations of $\nu_{max}(\mathbf{1})$ and ν_{max} of other indicators including coumarin 153 (**4**) [30]. However, two solvatochromic probe pairs have been found suitable as α indicators. For determining α values of polysaccharides and silicas, the use of $\nu_{max}(\mathbf{1})$ and $\nu_{max}(\mathbf{2})$ or $\nu_{max}(\mathbf{4})$ has been recommended [19]. A suitable probe pair for the determination of independent π^* parameters for moderately strong solid acids has not yet been established [11, 29, 30]. However, in Ref. [30] we have recommended the probe pair **1** and **4** for determining reasonable π^* parameters of polysaccharides.

There exists still a problem in determining reasonable β values for bare silica and other moderately strong solid acids [13, 14, 31, 32]. For this purpose, different UV/Vis probes are suitable as long as their UV/Vis absorption maxima adequately reflect the β value of a surface environment [31]. However, this is not always accomplished for UV/Vis probes bearing primary amino groups such as

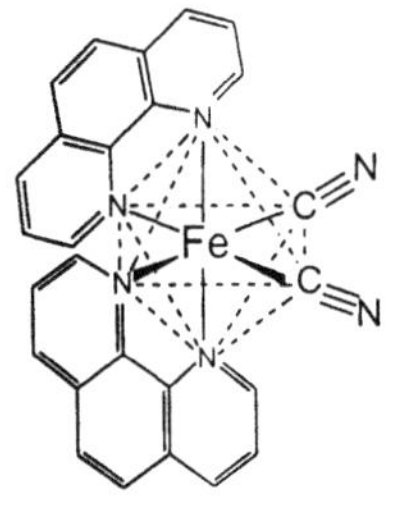

1

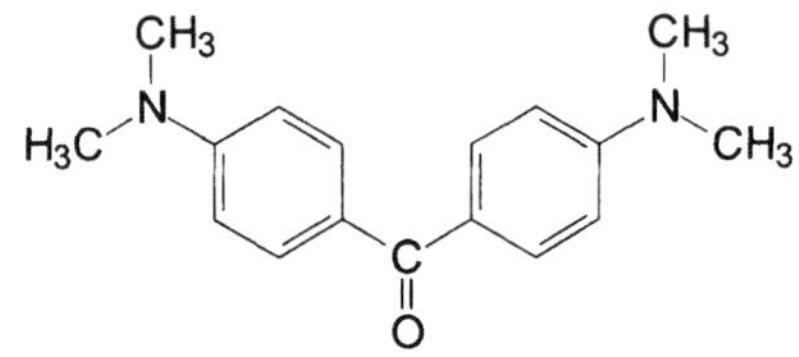

2

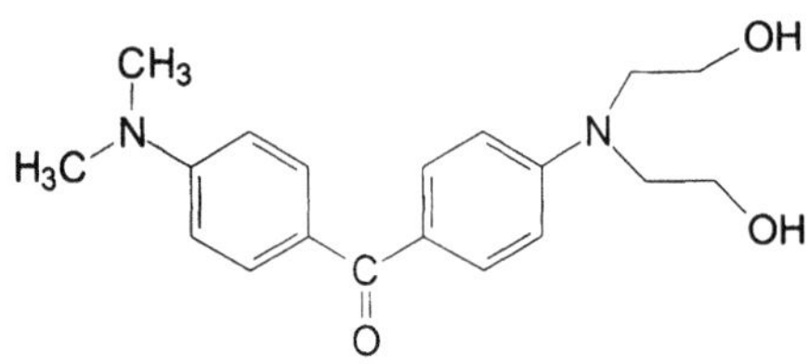

3a

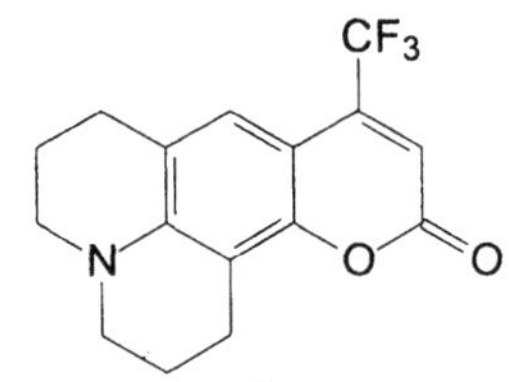

4

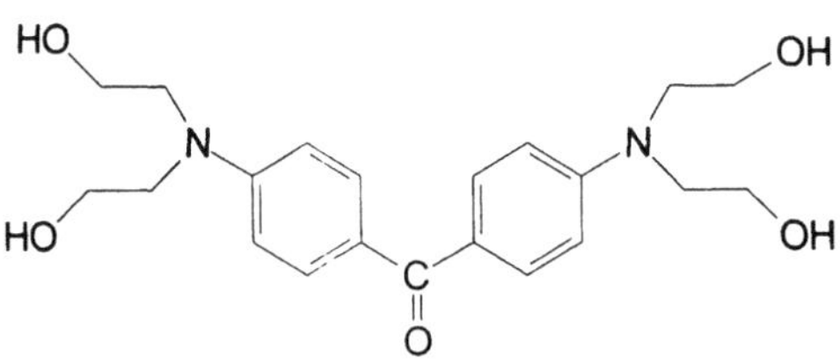

3b

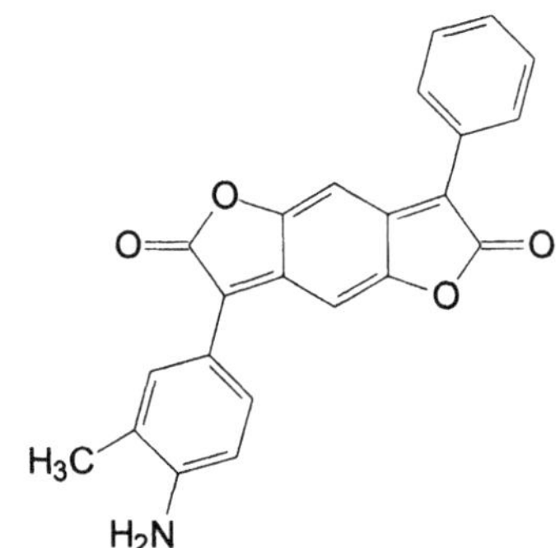

5

Scheme 1. Solvatochromic probe dyes used

4-aminobenzophenone [31, 32] and the aminobenzodifuranone dye (**5**) [14, 31, 33], because they do also interact with the lone electron pair of the nitrogen atom and/or the carbonyl oxygen and acidic sites of the surface. Because **5** has been found suitable as a HBA probe for synthetic polymers and solvents with HBA properties [19], its applicability as β indicator for some silica materials will be considered in this work. Nevertheless, the *Kamlet-Taft* parameters must be handled with caution in order to be mathematically orthogonal, and they neglect any other higher order term.

For the solvatochromic measurements, very low concentrations of the indicators are used to avoid multilayer adsorption on the surface. In the case of a mobile probe, the probe migrates to this site corresponding to a minimum of free energy (see Scheme 2). Using silica and weak basic probe molecules in organic liquids, it is assumed that the strongest acidic sites of the surface preferentially interact at low dye concentration, but the energies are average energies of all the sites to which the base coordinates [20, 34]. This average energy is a measure of the acid strength or

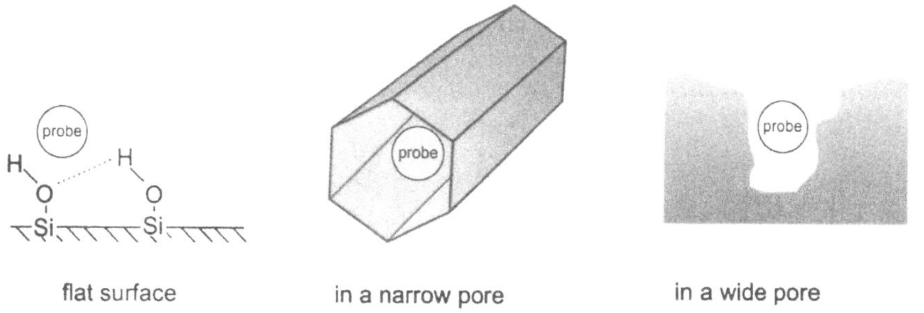

Scheme 2. Possible localizations of a solvatochromic probe on surface environments with different morphologies

polarity strength of the strongest site only if just the strongest site interacts. However, it is assumed that a freely accessible surface (particle or flat surface) does not restrict the mobility of a probe since the probe is well soluble in the liquid. The situation can change dramatically when the probe is adsorbed in a confined space (inside MCM-41) or encapsulated in a hydrogel or xerogel [34]. In this paper we will focus our interest to silica particles with different morphologies and pore size distribution where no confinement effects are expected, because a really solvent-like behaviour is only expected for freely accessible silica particle surfaces in a slurry where the solvents and the probe have enough available space and are not restricted in their thermal motion.

Therefore, it is of great interest to evaluate the relevance of the calculated polarity parameters α and π^* with regard to independent physicochemical parameters such as the catalytic activity of the solid acid catalyst, adsorption, and chromatographic properties [35–38]. *Park* and *Carr* have shown for the first time that the retention behaviour of solvents on a stationary silica phase is precisely determined by the HBA property (β term from Eq. (1)) and the dipolarity/polarizability (π^* term from Eq. (1)) of the solvents [35, 36]. It could also be shown that the rate of a polar reaction to be catalyzed by a moderately strong solid acid catalyst increases by increasing the α value of the solid acid [38]. These results clearly support the importance of the determined surface polarity parameters for practical applications. Furthermore, the linear correlation of the zeta-potential value measured in water with the β value determined in dichloromethane of organically functionalized silica particles shows the widespread conceptional background of the solvatochromic method for surface environments [14, 39].

The objective of the present review is the reinterpretation and critical reflection of some of our former results using an extended set of surface polarity indicators. Especially the influence of the structure of the probe (basicity and hydrophilicity) [29] and the suspending liquid on the α and π^* values for different silica samples will be discussed, because also weak HBA solvents, *i.e.* aromatics, halogenated solvents, or alkanes, can significantly change the surface reactivity of a solid acid [23, 38]. For this purpose, the established polarity indicators **1–5** have been selected (see Scheme 1). As suspending liquids we used weak HBA and moderately strong dipolar solvents. Their polarity parameters are given in Table 2.

Results

Influence of the structure of the probe on the surface polarity determined for bare silica

In several papers, various solvatochromic probes adsorbed on silica have been reported. Table 1 presents UV/Vis spectroscopic data for **1–4** and **5** when adsorbed on various silica samples.

Despite the silica samples altogether show quite the same overall surface polarity as expressed by their $E_T(30)$ value, they behave differently with respect to the probes **1**, **2**, **4**, and **5**. The corresponding polarity parameters α, β, and π^* calculated from Eqs. 5–8 for different probe pair combinations applied are compiled in Table 2.

It is obvious that the established probe pair **1** and **2** gives reasonable α and π^* values, whereas the probe pair **1** and **4** gives too large π^* and too low α values. We think that the bathochromic shift of the indicator **4** when adsorbed on silica contains

Table 1. UV/Vis absorption maxima of the indicator dyes **1**, **2**, **4**, and **5** when adsorbed on different silica samples from a 1,2-dichloroethane solution and related data for strong HBD solvents

Environment	$\nu_{max} \cdot 10^3/cm^{-1}$				$E_T(30)/kcal \cdot mol^{-1}$ (measured)
	1	**2**	**4**	**5**	
1,2-Dichloroethane (*DCE*)	16.39	28.25	23.75	17.42	41.3
Aerosil® 300/*DCE*	18.48	25.51	21.83	17.45	57.8
KG® 60/*DCE*	18.73	25.25	21.19	17.99	58.1
LC® 1500/*DCE*	18.98	25.13	21.51	17.48	58.5
Wacker silica[a]/*DCE*	18.69	25.25	21.93	19.42	–
2,2,2-trifluoroethanol	19.31	25.77	22.68	19.38	59.8
1,1,1,3,3,3-hexafluoro-2-propanol	20.04	24.94	21.98	20.04	65.3

[a] Nondried silica sample

Table 2. Calculated polarity parameters of four different silica samples utilizing ν_{max}(probe) of the probe pairs **1** and **2**, **1** and **4**, and **1** and **5**

Silica/solvent system	$\alpha(\mathbf{1},\mathbf{2})$[a]	$\alpha(\mathbf{1},\mathbf{4})$[b]	$\pi^*(\mathbf{1},\mathbf{2})$[c]	$\pi^*(\mathbf{1},\mathbf{4})$[d]	$\beta(\mathbf{1},\mathbf{5})$[e]
Aerosil 300®/*DCE*	1.01	−0.06	1.09	2.62	0.50
KG 60/*DCE*	1.11	−0.24	1.11	3.13	0.36
LC 1500/*DCE*	1.22	0.03	1.09	2.81	0.57
Wacker silica/*DCE*	1.10	0.08	1.12	2.50	−0.12

[a] Calculated by means of Eq. (5) ($\alpha = -7.900 + 0.453\nu_{max}(\mathbf{1}) \cdot 10^{-3} + 0.021\nu_{max}(\mathbf{2}) \cdot 10^{-3}$);
[b] calculated by means of Eq. (7) ($\alpha = -19.146 + 0.489\nu_{max}(\mathbf{1}) \cdot 10^{-3} + 0.460\nu_{max}(\mathbf{4}) \cdot 10^{-3}$);
[c] calculated by means of Eq. (6) ($\pi^* = 13.889 - 0.251\nu_{max}(\mathbf{1}) \cdot 10^{-3} - 0.320\nu_{max}(\mathbf{2}) \cdot 10^{-3}$);
[d] calculated by means of Eq. (8) ($\pi^* = 24.632 - 0.181\nu_{max}(\mathbf{1}) \cdot 10^{-3} - 0.855\nu_{max}(\mathbf{4}) \cdot 10^{-3}$);
[e] calculated by $\beta = 3.059 + 0.174\nu_{max}(\mathbf{1}) \cdot 10^{-3} - 0.331\nu_{max}(\mathbf{5}) \cdot 10^{-3}$

too much contributions of the π^* term, because it shows an unprecedented large bathochromic shift which is similar to the effect observed in *HFI* (1,1,1,3,3,3-hexafluor-2-propanol). May be probe **4** responses to a higher order term. However, the dubious results for the pair **1** and **4** show a general problem: is it really justified to seperate the α and π^* term from two units of measurement? As a consequence of this conclusion, the concept for determining rigid parameters for surface environments has to be abandoned. In a following paper we will discuss this important question again.

Using probe **5** as β indicator, the silica surface would exhibit a moderately strong HBA property with similar strength compared to tetrahydrofuran or acetonitrile. This is not reasonable, because silica becomes not protonated even by strong acids like CF_3SO_3H [40]. Furthermore, a β value of 0.4 for silica does not correspond to its isoelectric point from zeta-potential measurements [14, 39]. Therefore, we assume that **5** is adsorbed *via* the carbonyl oxygen atom on silica and not *via* the hydrogen atom of the amino group. In a following paper we will make use of **5** as indicator for several solids and discuss this point again, because one silica batch shows a reasonable β value with **5** as indicator.

Influence of the solvent property on the surface polarity of bare silicas

For investigating the influence of the solvent on the polarity of the silica/solvent interface, we have chosen Aerosil® 300 and the porous silica KG® 60 which show very similar catalytic activity and surface polarity parameters in 1,2-dichloroethane as solvent [23, 38]. Unfortunately, the iron complex **1** is insoluble in weak dipolar solvents such as hexane, cyclohexane, toluene, benzene, tetrahydrofuran, dioxane, or anisole [27]. The problem can be circumvented when **1** is adsorbed on the silica sample from a dichloromethane solution and the coloured silica is filtered and dried [41]. The silica sample with the adsorbed dye **1** is suspended in this solvent where **1** is insoluble. This is indicated in Table 3 for **1** on silica with an asterisk. However, probe **1** is then restricted in its naturally occuring motion which has to be taken into account when interpreting the results. In moderately and stronger polar solvents, in particular the indicators **2**, **3a**, and **4** adsorb weakly on silica. Therefore, sometimes the difference between the UV/Vis spectrum in the slurry from and that in the supernatant solution has been utilized. This is specifically indicated in the Table 3. When **1** adsorbs from 1,2-dichlorethane on silica, a characteristic change of colour from blue (supernatant solution) to red (silica surface) is observed. The supernatant solution remains colourless. The adsorption process of **2** or **3a** can be easily followed by the eye of the investigator. In toluene or 1,2-dichlorethane, the solution of **2** or **3a** is colourless; after adsorption on silica, a yellow colour appears on the surface. Figures 1a,b show UV/Vis spectra of the adsorbed polarity indicators **1** and **3b**, respectively, on Aerosil® 300 measured in various solvents.

1 does not adsorb on silica from strong HBD solvents such as *DMSO*, *DMF*, or N-methylpyrrolidone; similar results have been found for derivatives of *Michler*'s ketone. It should be emphasized, however, that the acid-sensitive indicator **1** does well adsorb from alcoholic solvents on the porous silica KG® 60, whereas the other indicators do not. On Aerosil® 300, no adsorption from alcoholic solvents has been observed.

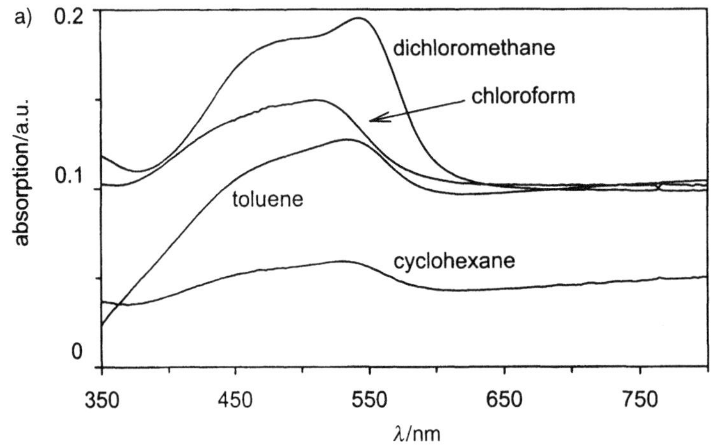

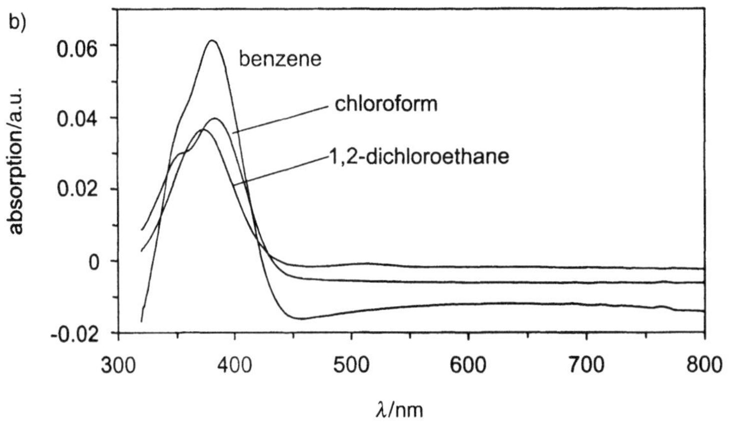

Fig. 1. a) UV/Vis absorption spectra of **1** adsorbed on Aerosil® 300 in dichloromethane, chloroform, toluene, and cyclohexane, respectively; the corresponding ν_{max} values are given in Table 3; b) UV/Vis absorption spectra of **3b** adsorbed on Aerosil® 300 in *DCE*, chloroform, and benzene, respectively; the corresponding ν_{max} values are given in Table 3

Surface polarity parameters of functionalized silicas

$E_T(30)$ values as well as UV/Vis absorption maxima of the indicators **1** and **2** when adsorbed on various organofunctionalized silica particles and solid acid catalysts have been reported [11, 23]. These data were taken in particular from Refs. [11] and [23] for the correlation analyses $E_T(30) = f(\alpha, \pi^*)$.

Discussion

The UV/Vis absorption of the acid sensitive surface polarity indicator **1** ($\nu_{max}(\mathbf{1})$) when adsorbed on Aerosil® 300 shows a significant bathochromic shift in increasing the basicity of the suspension solvent used. This is shown by Eq. (9a). Here and in the following, ν_{max} is expressed in cm^{-1}; *r*: correlation coefficient, *sd*: standard deviation, *n*: number of solvents, *F*: significance.

$$\nu_{max}(\mathbf{1}) \cdot 10^{-3} = 19.09 - 3.42\beta \quad (r = 0.91, sd = 0.39, n = 10, F < 0.0001) \quad (9a)$$

Table 3. UV/Vis absorption maxima ν_{max} of the probes **1**, **2**, **3a**, and **3b** when adsorbed on two different silica samples, KG® 60 (**S1**) and Aerosil® 300 (**S2**), respectively, in various solvents

Solvent	α	β	π^*	$\nu_{max} \cdot 10^{-3}/cm^{-1}$					
				1		**2**		**3a**	**3b**
				S1	**S2**	**S1**	**S2**	**S2**	**S2**
CH_2Cl_2	0.13	0.10	0.82	18.55	18.42	25.45	25.38	25.77	26.32
$CHCl_3$	0.20	0.10	0.58	18.52	18.42	25.58	25.58	25.84	26.04
Toluene	0	0.11	0.54	18.73*	18.76*	25.51	25.64	25.91	26.46
CCl_4	0	0.10	0.28	18.76*	b	26.53	26.53	26.11	d
Cyclohexane	0	0	0	18.52*	18.87*	26.32	26.18	d	d
DMF	0	0.69	0.88	b	16.20[e]	a/b	a/b	a/b	a/b
THF	0	0.55	0.58	17.57	17.70[e]	a	a/b	a	b
CH_3NO_2	(0.22)	0.06	0.85	18.73	18.80[e]	c	c	b/c	b/c
CH_3CN	0.19	0.40	0.75	b	18.00[e]	b	b	a	a/b
$ClCH_2CH_2Cl$	0	0.10	0.81	18.69	18.80	25.32	25.51	25.77	26.74
C_6H_6	0	0.10	0.59	b	19.44[e]	b	b	26.18	26.18
MeOH	0.93	0.62	0.60	18.42	a	a	a	a/b	a/b
EtOH	0.83	0.77	0.54	18.21	a	a	a	a/b	a/b
TFE	1.51	–	0.73	19.16	b	a	b	a/b	a/b
HFI	1.96	–	0.65	18.87	b	a	b	a/b	a/b
Acetone	0.08	0.43	0.71	17.86	b	a	a	a/b	a/b
Diethyl ether	0	0.47	0.27	17.70*	17.45*	26.04	a/c	29.15	29.15
n-Hexane	0	0	−0.04	18.42*	b	26.32	b	d	d
p-Xylene	0	0.12	0.43	b	b	b	b	25.91	26.25
$Cl_2CHCHCl_2$	0	0	0.95	b	b	b	b	26.32	26.60

[a] No adsorption; [b] not measured; [c] detection not possible; [d] probe is insoluble in the pure solvent; [e] taken from Ref. [42]; for *DMF* and CH_3CN, the UV/Vis spectrum of **1** on the surface interferes with that of the surrounding solution; * asterisks denote that **1** is adsorbed on silica from a *DCM* solution; then the silica sample has been dried and measured in the solvent indicated

This result is reasonable, because the acid-base interactions of the silanol groups with the solvent play an important role [42–45]. The nucleophilic attack of the lone electron pair of the solvent on the silanol groups decreases the hydrogen-bond donating capacity of the silica surface and thus the interacting strength with the probe molecule. This leads to a bathochromic shift of the UV/Vis absorption band of the adsorbed indicator **1**. The solvatochromic UV/Vis absorption band of **1** ($\nu_{max}(\mathbf{1})$) in well-behaving regular solvents shows a hypsochromic shift upon increasing the α and π^* value ($a = 2.24$ and $s = 1.63$ according to Eq. (2)) of the solvent [27, 28]. Using multiple correlations of $\nu_{max}(\mathbf{1})$ (on silica) with α, β, and π^*, the quality of the correlation is not improved, but the significance decreases.

$$\nu_{max}(\mathbf{1}) \cdot 10^{-3} = 19.14 - 3.40\beta - 0.08\pi^* \quad (r = 0.91, sd = 0.41, n = 10, F = 0.0008) \qquad (9b)$$

$$\nu_{max}(\mathbf{1}) \cdot 10^{-3} = 19.14 - 1.05\alpha - 3.53\beta + 0.07\pi^* \quad (r = 0.91, sd = 0.46, n = 9, F = 0.01) \qquad (11c)$$

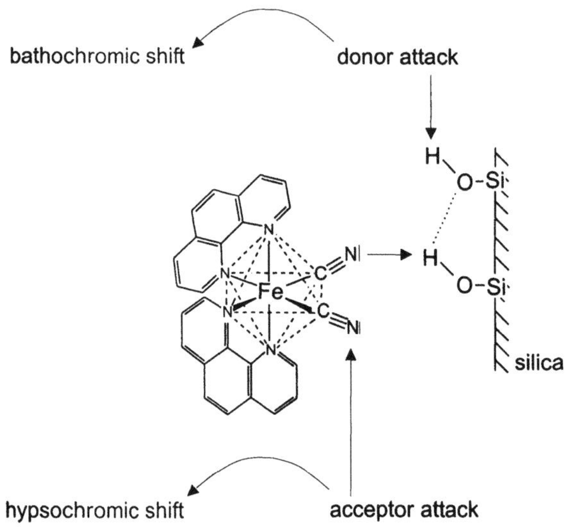

Scheme 3. Multiple interactions of $Fe(phen)_2(CN)_2$ with silanol groups and solvent molecules

The result of Eq. (9a) is in accordance with values from the literature, *i.e.* the silanol valency vibration of silica decreases with increasing *Gutmann* donor number (*DN*) of the solvent used [46]; *DN* and β correlate well with each other [3]. Because the correlation of $\nu_{max}(\mathbf{1})$ with the acceptor number (*AN*) of the solvent is also well established [27, 28, 47, 48], it seems obvious that the acid-base interaction between the silanol groups and the lone electron pair of the solvents plays the major role. Possible interactions of **1** with silica and the solvents are suggested in Scheme 3.

For *Michler*'s ketone as polarity indicator when completely adsorbed on silica in a suitable solvent, the best fit has been found for the influence of the π^* term on $\nu_{max}(\mathbf{2})$ (Eq. (10)).

$$\nu_{max}(\mathbf{2}) \cdot 10^{-3} = 26.40 - 1.19\pi^* \quad (r = 0.84, sd = 0.27, n = 5, F = 0.04) \qquad (10)$$

However, for **2** adsorbed on Aerosil® 300, only five solvents could be cleanly measured, and the significance of this relation is insufficient for a profound interpretation. A multiple correlation analyses also makes no sense for this series.

For the adsorbed hydrophilically substituted *Michler*'s ketone derivative **3a**, a significant influence of the β term of the solvent upon the UV/Vis absortion maximum has been found.

$$\nu_{max}(\mathbf{3a}) \cdot 10^{-3} = 25.33 + 7.51\beta \quad (r = 0.92, sd = 0.46, n = 8, F = 0.0005) \qquad (11a)$$

$$\nu_{max}(\mathbf{3a}) \cdot 10^{-3} = 24.85 + 8.25\beta + 0.65\pi^* \quad (r = 0.92, sd = 0.48, n = 8, F = 0.003) \qquad (11b)$$

$$\nu_{max}(\mathbf{3b}) \cdot 10^{-3} = 25.81 + 6.61\beta \quad (r = 0.91, sd = 0.44, n = 7, F = 0.001) \qquad (12a)$$

$$\nu_{max}(\mathbf{3b}) \cdot 10^{-3} = 24.17 + 9.17\beta + 2.07\pi^* \quad (r = 0.96, sd = 0.33, n = 7, F = 0.002) \qquad (12b)$$

$$\nu_{max}(\mathbf{3b}) \cdot 10^{-3} = 24.28 - 2.57\alpha + 8.96\beta + 2.10\pi^* \quad (r = 0.98, sd = 0.26, n = 7, F = 0.003) \quad (12c)$$

$\nu_{max}(\mathbf{3a})$ as well as $\nu_{max}(\mathbf{3b})$ reflect mainly the dipolarity/polarizability and the HBD capacity of well-behaving regular solvents [29]. However, it is shown by Eq. (9a) that the HBA property of the solvent has a significant influence upon the surface acidity. Thus, since the α value of silica decreases, $\nu_{max}(\mathbf{3a})$ and $\nu_{max}(\mathbf{3b})$ shift hypsochromically which is indicated by the positive sign of the coefficient for the β term in Eqs. (11a), (11b), and (12a–c). It seems that the additional polar alcoholic groups at the dimethylamino substituent are responsible for significant acid-base interactions with the silica surface compared to the nonsubstituted **2**. The shift of $\nu_{max}(\mathbf{3a})$ as well as $\nu_{max}(\mathbf{3b})$ for adsorbed **3a** and **3b**, respectively, is also in accordance with a strong acid-base interaction for these two adsorbed hydrophilic probes with silica. The influence of the terms α and π^* of the solvents on $\nu_{max}(\mathbf{3a})$ and $\nu_{max}(\mathbf{3b})$ play a less important role. This shows that the solvatochromic response of **3a** and **3b**, respectively, preferentially reflects the solvent influences on the acid-base interaction with silica and not the (intrinsic) solvatochromic property of the probe, whereas for **2** – which seems to be weaker adsorbed – the direct solvent influence on the probe itself is reflected. This result is very important for a correct interpretation of sorptio-solvatochromatic effects of adsorbed probes on surfaces.

For KG® 60, the following correlations have been found for the adsorbed polarity indicators **1** and **2** (the probe dyes **3a** and **3b** have not been considered for this material so far).

$$\nu_{max}(\mathbf{1}) \cdot 10^{-3} = 18.32 + 0.31\alpha \quad (r = 0.44, sd = 0.41, n = 14, F = 0.10) \quad (13a)$$

$$\nu_{max}(\mathbf{1}) \cdot 10^{-3} = 18.63 - 1.02\beta \quad (r = 0.67, sd = 0.31, n = 12, F = 0.01) \quad (13b)$$

$$\nu_{max}(\mathbf{1}) \cdot 10^{-3} = 18.29 + 0.30\pi^* \quad (r = 0.19, sd = 0.45, n = 14, F = 0.50) \quad (13c)$$

$$\nu_{max}(\mathbf{1}) \cdot 10^{-3} = 18.50 - 1.09\beta + 0.29\pi^* \quad (r = 0.70, sd = 0.32, n = 12, F = 0.33) \quad (13d)$$

The correlation equations (13a–d) gave no specific hint at a prefered silica/solvent interaction. Hence, for the porous silica a rather complex influence of the surrounding solvent upon the surface acidity seems to be operative. This is shown by the multiple correlation of $\nu_{max}(\mathbf{1})$ with α, β, and π^* (Eq. (14)) which shows sufficient significance and a satisfying correlation coefficient.

$$\nu_{max}(\mathbf{1}) \cdot 10^{-3} = 18.58 + 1.00\alpha - 1.85\beta + 0.17\pi^*$$
$$(r = 0.92, sd = 0.19, n = 12, F = 0.0007) \quad (14)$$

However, the HBA property of the solvents (β term or *DN*) mainly also decreases the HBD capacity (or *AN*) of the silica surface compared to Aerosil® 300 (Eq. (9a)). The influence of the π^* term is of minor importance but cannot be neglected. The influence of the α term is probably caused by the adsorbed HBD solvent molecules according to Scheme 3.

For the adsorption of **2** on KG® 60, a similar influence of the π^* term upon $\nu_{max}(\mathbf{2})$ is detectable as found for Aerosil® 300 (Eq. (15)).

$$\nu_{max}(\mathbf{2}) \cdot 10^{-3} = 26.40 - 1.26\pi^* \quad (r = 0.89, sd = 0.23, n = 7, F = 0.003) \quad (15)$$

It seems that the strength of adsorption of the probe molecule (P) on the surface determines the influence of the property of the solvent on the direction of the sorptio-solvatochromic UV/Vis shift of $\nu_{max}(P)$ according to Eq. (2). If the probe is more hydrophobic, the interaction with the evident hydrophilic bare silica surface environment is lowered. A similar result has been found for hydrophobically substituted nitroaniline derivatives as probes for micelle polarity [49].

Functionalized silica particles and other solid acids

For probing the well-established relationship $E_T(30) = f(\alpha, \pi^*)$ [2, 3] including an extended data set of organofunctionalized silicas and various solid acids [11, 23], we used α and π^* values determined by the probe pair **1** and **2** as well as directly measured $E_T(30)$ values by means of *Reichardt*'s betaine dyes. The result of the least squares analysis is given in Eq. (16) and Fig. 2.

$$E_T(30) = 40.68 + 11.45\alpha + 3.86\pi^* \quad (r = 0.86, sd = 1.98, n = 45, F < 0.0001) \qquad (16)$$

The correlation is only satisfactory which shows that the respective probe reflects the hydrogen bond acidity (α term) and dipolarity/polarizability (π^* term) of each surface environment differently in a certain section of the polarity scale. As a whole, the correlation is significant and shows, as expected from our former interpretations [11, 27], that the α term contributes stronger to $E_T(30)$ than does the π^* term [11, 19, 23]. However, we think that a systematic study using well defined solvatochromic probes concerning their basicity and hydrophobicity [49, 50] is necessary in order to evaluate the reliability of the parameters. At this time, the determined parameters are of practical relevance for specific applications.

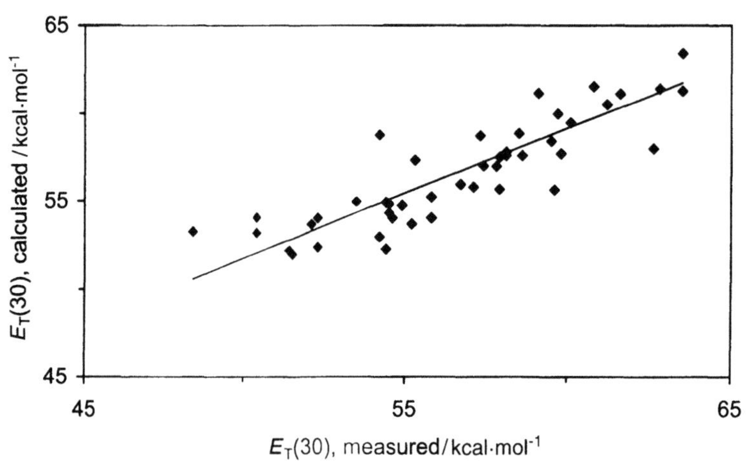

Fig. 2. Calculated *vs.* measured $E_T(30)$ values for organofunctionalized silicas and various solid acids including silicas, aluminas, titanium dioxides, and aluminosilicates; all data were taken from Refs. [11] and [23]

Conclusions

The interaction of a solvatochromic probe with a surface environment in a solvent can be well measured by the UV/Vis shift of the adsorbed polarity indicator when adsorbed on silica in transparent slurries. The quantitative adsorption of the probe dye requires specific conditions because the solvent competes with the probe. Therefore, unambiguously measurable UV/Vis absorption maxima of the adsorbed probe dye cannot be measured for each solvent used. It seems that a basic requirement is that the probe should be less basic than the solvent used for the silica slurry. Acid-base interactions between the probe and the silica surface predominate as shown by the correlation analyses of ν_{max}(probe) – measured on the surface in the slurry – with the *Kamlet-Taft* solvent parameters of the solvent. It can be shown that not only the chromophoric system of the dye is responsible for the UV/Vis response of the respective interaction, but that the hydrophilicity or rigidity of the probe is also of importance. Therefore, the silica surface can be treated like a smooth solvent shell which mediates preferential HBD solvation. As a consequence of this research, polarity parameters for silica and solid acids are dynamic quantities.

At this moment, an interpretation of the results in terms of morphological reasons is too speculative. We think this is an important topic for future work, because the morphology of silica seems of importance for the interaction with both probe and solvent.

Acknowledgements

Financial support for this research by the DFG and the *Fonds der Chemischen Industrie* is gratefully acknowledged.

Experimental

Chemicals

Fe(*phen*)$_2$(CN)$_2$ was prepared according to *Schilt* [51]. *Michler*'s ketone was purchased from Merck (Darmstadt), recrystallized twice from EtOH, and carefully dried before use. The hydrophilically substituted derivatives of **2**, **3a** and **3b**, were synthesized as previously published [29]. Coumarin dye 153 (**4**) was purchased from Radiant Dyes Laser Accessories GmbH. The aminobenzodifuranone dye (**5**) was kindly provided by BASF, Manchester. 1,1,1,3,3,3-hexafluor-2-propanol (*HFI*) and 2,2,2-trifluoroethanol (*TFE*) were used without further purification. All other solvents were dried and freshly distilled before use. The physical properties of the silica batches used in this study are given in Table 4.

Table 4. Physical properties and sources of the solid acids used in this work

Solid acid sample	BET-surface area (m^2g^{-1})	Specific pore volume (cm^3g^{-1})	Average pore diameter (nm)	Source
SiO_2, Aerosil 300®	240	–	–	Degussa
SiO_2, KG 60®	423	0.63	9.0	Merck
SiO_2, KG LC 1500®	30	–	5.5	Grace
SiO_2	200	–	–	Wacker

UV/Vis measurements

The UV/Vis absorption maxima of the dyes **1–5** adsorbed on the solid acid catalyst were recorded using a diode array spectrometer with glass fiber optics. The solid acids were heated at 400°C for 12 h. After cooling to room temperature under dried argon, a solution of the probe dye in the corresponding solvent (about $10^{-5}\,\mathrm{mol/dm^3}$) was added to the solid material. Care must be taken to avoid overloading the surface with the indicator used, as multilayer adsorption is expected in solution at higher concentrations as well as interfering absorptions from the nonadsorbed dye from the solution.

For the UV/Vis measurement of transparent slurries, the amount of the solid acid silica must be limited in order to achieve a sufficient particle concentration in the continuously stirred slurry. 0.1 to 0.2 g of silica in 15 cm^3 of the solvent are very suitable, but a lower amount is also possible. The amount of the silica portion has no influence on the position of the UV/Vis absorption maximum of the adsorbed probe dye.

The equipment employed was a UV/Vis spectrometer MCS 400 connected to an immersion cuvette TSM 5A (Zeiss). The measurement of the UV/Vis spectrum of the supernatant solution is possible in the same cell after the solid has deposited.

The measurements of silica particle suspensions are advantageous because they can be performed under inert conditions and reflect better the conditions suitable for liquid chromatography and catalysis in organic solvents. The reproducibility of the UV/Vis spectra of the adsorbed dyes is very good. For silicas, transparent suspensions are obtained in most liquids. This allows to record very good quality transmission spectra with an excellent reproducibility of less than $\lambda_{max} \pm 1$ nm. The position of the UV/Vis absorption maximum of the adsorbed dye on the silica particles remains constant during 1–2 h (*vide infra*).

Determination of the polarity parameters

The following dual correlations according to Eq. (4) have been used for the determination of the surface polarity parameters [19, 30]. $\alpha(\mathbf{1},\mathbf{2})$ denotes that the solvatochromism of the indicators **1** and **2** has been considered in the multiple square analyses for this equation and that the influence of $\nu_{max}(\mathbf{1})$ on α dominates, because the coefficient a_1 for $\nu_{max}(\mathbf{1})$ is significantly larger than a_2 for $\nu_{max}(\mathbf{2})$ according to Eq. (4). Equations (5) and (6) were taken from Ref. [19], Eqs. (7) and (8) from Ref. [30].

$$\alpha(\mathbf{1},\mathbf{2}) = -7.90 + 0.45 \cdot \nu_{max}(\mathbf{1}) \cdot 10^{-3} + 0.02 \cdot \nu_{max}(\mathbf{2}) \cdot 10^{-3} \quad (r = 0.95, sd = 0.17, n = 34, F = 0.00) \qquad (5)$$

$$\pi^*(\mathbf{2},\mathbf{1}) = 13.89 + 0.251 \cdot \nu_{max}(\mathbf{1}) \cdot 10^{-3} - 0.32 \cdot \nu_{max}(\mathbf{2}) \cdot 10^{-3} \quad (r = 0.57, sd = 0.15, n = 36, F = 0.00) \qquad (6)$$

$$\alpha(\mathbf{1},\mathbf{4}) = -19.15 + 0.46 \cdot \nu_{max}(\mathbf{4}) \cdot 10^{-3} + 0.49 \cdot \nu_{max}(\mathbf{1}) \cdot 10^{-3} \quad (r = 0.98, sd = 0.09, n = 18) \qquad (7)$$

$$\pi^*(\mathbf{4},\mathbf{1}) = 24.63 - 0.85 \cdot \nu_{max}(\mathbf{4}) \cdot 10^{-3} - 0.18 \cdot \nu_{max}(\mathbf{1}) \cdot 10^{-3} \quad (r = 0.94, sd = 0.08, n = 18) \qquad (8)$$

References

[1] Müller P (1994) Pure Appl Chem **66**: 1077
[2] Reichardt C (1994) Chem Rev **94**: 2319
[3] Marcus Y (1993) Chem Soc Rev 409
[4] Gutmann V (1976) Coord Chem Rev **18**: 225
[5] Liptay W (1965) Naturforsch **20a**: 1441
[6] Suppan P (1990) J Photochem Photobiol A **50**: 293
[7] Reichardt C (1988) Solvents and Solvent Effects in Organic Chemistry, 2nd edn. VCH, Weinheim, and references therein
[8] Novaki LP, El Seoud OA (1996) Ber Bunsenges Phys Chem **100**: 648
[9] Palm N, Palm V (1997) Organic Reactivity (Tartu) **104**: 141; Taverner SJ, Clark JH, Gray GW, Heath PA, Macquarrie D (1997) J Chem Soc Chem Commun 1147

[10] Michels JJ, Dorsey JG (1990) Langmuir **6**: 414
[11] Spange S, Reuter A (1999) Langmuir **15**: 141
[12] Spange S, Reuter A, Lubda D (1999) Langmuir **15**: 2103
[13] Spange S, Reuter A, Linert W (1998) Langmuir **14**: 3479
[14] Spange S, Reuter A, Prause S, Bellmann C (2000) J Adhesion Sci Technol **14**: 399
[15] Lindley SM, Flowers GC, Leffler JE (1985) J Org Chem **50**: 607
[16] Handreck GP, Smith TD (1988) J Chem Soc Faraday Trans I **84**: 1847
[17] Helburn RS, Rutan SC, Pompano J, Mitchern D, Patterson WT (1994) Anal Chem **66**: 610
[18] Macquarrie DJ, Taverner SJ, Gray GW, Heath PA, Rafelt JS, Saulzet SI, Hardy JJE, Clark JH, Sutra P, Brunel D, di Renzo F, Fajula F (1999) New J Chem **23**: 725
[19] Spange S, Vilsmeier E, Fischer K, Prause S, Reuter A (2000) Macromol Rapid Commun (Feature) **21**: 643
[20] Chronister C, Drago RS (1993) J Am Chem Soc **115**: 4793
[21] Jensen WB (1991) In: Mittal KL, Anderson HR (eds) Acid-Base-Interactions. VSP, Utrecht, p 3–23
[22] Dutta PK, Turbeville W (1991) J Phys Chem **95**: 4087
[23] Spange S, Vilsmeier E, Zimmermann Y (2000) J Phys Chem B **104**: 6417
[24] Rutan SC, Harris JM (1993) J Chromatogr A **656**: 197
[25] Helburn RS, Rutan SC, Pompano J, Mitchern D, Patterson WT (1994) Anal Chem **66**: 610
[26] a) Kamlet MJ, Abboud J-LM, Abraham MH, Taft RW (1983) J Org Chem **48**: 2877; b) Taft RW, Kamlet MJ (1979) J Chem Soc Perkin Trans 2, 1723
[27] Spange S, Keutel D (1992) Justus Liebigs Ann Chem 423
[28] Spange S, Keutel D, Simon F (1992) J Chim Phys **89**: 1615
[29] El-Sayed M, Müller H, Rheinwald G, Lang H, Spange S (2001) J Phys Org Chem **14**: 247
[30] Fischer K, Prause S, Spange S, Cichos F, Borzyskowski C (2001) J Phys Chem B (submitted)
[31] Spange S, Schmidt C, Kricheldorf HR (2001) Langmuir **17**: 856
[32] Fischer K, Spange S (2000) Macromol Chem Phys **201**: 1922
[33] Gorman AA, Hutchings MG, Wood PD (1996) J Am Chem Soc **118**: 8497
[34] Spange S, Zimmermann Y, Gräser A (1999) Chem Mat **11**: 3245
[35] Park JH, Carr PW (1989) J Chromatogr **465**: 123
[36] Brune BJ, Payne GF, Chaubal MV (1997) Langmuir **13**: 5766
[37] Rutan SC, Carr PW, Taft RW (1989) J Phys Chem **93**: 4292
[38] Adolph S, Spange S, Zimmermann Y (2000) J Phys Chem B **104**: 6429
[39] Voigt I, Esthel K, Simon F, Spange S (2001) Langmuir **17** (in press)
[40] Arnett EM, Ahsan T (1991) J Am Chem Soc **113**: 6861
[41] Spange S, Simon F, Heublein G, Jacobasch HJ, Börner M (1991) Coll Polym Sci **269**: 173
[42] Arnet EM, Cassidy KF (1988) Rev Chem Intermed **9**: 27
[43] Kittelmann U, Unger K, Kreis W (1980) Progr Coll Polym Sci **67**: 19
[44] Winde H, Fink P, Köhler A (1977) Z Chem **17**: 41
[45] Pohle W (1982) J Chem Soc Faraday Trans 1 **78**: 2101
[46] Gutmann V (1978) The Donor-Acceptor Approach to Molecular Interactions. Plenum Press, New York
[47] Soukup RW, Schmid W (1985) J Chem Edu **62**: 459
[48] Linert W, Jameson RF (1993) J Chem Soc Perkin Trans 2, 1415
[49] Helbrun R, Dijiba Y, Mansour G, Maxka J (1998) Langmuir **14**: 7147
[50] Gameiro P, Maia A, Pereira E, de Castro B (2000) Transition Metal Chemistry **25**: 283
[51] Schilt AA (1960) J Am Chem Soc **82**: 3000

Received February 2, 2001. Accepted (revised) March 3, 2001

Invited Review

Prediction of Electrolyte Solubilities from Minimal Thermodynamic Information

Erich Königsberger*

Department of Chemistry, School of Mathematical and Physical Sciences, Murdoch University, Murdoch WA 6150, Australia

Summary. A method is proposed that permits the proper extrapolation of thermodynamic quantities represented by the *Pitzer* equations from 25°C to other temperatures. The new method, which assumes temperature independent heat capacities, was tested for the $NaCl–H_2O$ system and found very satisfactory. A new evaluation of the $Na_2CO_3–H_2O$ system according to the CALPHAD method is presented, and solubilities for the quaternary $Na_2CO_3–NaCl–NaOH–H_2O$ and its ternary subsystems are predicted. Limitations of (*i*) these predictions and of (*ii*) the *Pitzer* model regarding extrapolations to high molalities are discussed.

Keywords. Solubility; Phase diagram; *Pitzer* model, SIT; Enthalpy; Heat capacity; Sodium chloride; Sodium carbonate.

Introduction

It is often believed that macroscopically observable properties of electrolyte systems, like thermodynamic quantities or solubilities, can be derived from solid-solute-solvent interactions once computers have become powerful enough to solve complicated equations in order to calculate 'everything' from first principles. In contrast, holism claims that completely new qualities emerge in macroscopic systems, and simple reductionism is a fundamentally wrong concept because microscopic and macroscopic properties are complementary to each other in the sense of *Niels Bohr*'s famous concept [1]. In fact, all molecular-level theories, but also empirical models, have failed so far to predict, for instance, solubilities of ionic solids over wide ranges of temperature, pressure, and composition in sufficient accuracy for industrial, geochemical, or environmental applications.

Nevertheless, in recent years the *semi*-empirical ion-interaction model developed by *Pitzer* [2] has been successfully applied to solubility calculations in multicomponent electrolyte solutions at ambient [3,4] and other temperatures

* On leave from Institut für Physikalische Chemie, Montanuniversität Leoben, A-8700 Leoben, Austria. E-mail: koenigsb@chem.murdoch.edu.au

ranging from $T = -60°C$ to $T = 250°C$ [5–13]. Whereas database development by geochemical groups [5, 8–13] usually involves fitting of solubility products to experimental solubility data as a function of temperature, *Pabalan* and *Pitzer* [6, 7] employed a procedure similar to what has been known for a long time as the CALPHAD method.

Kaufman and others originally developed the CALPHAD (CALculation of PHAse Diagrams) method [14] to perform equilibrium calculations on heterogeneous, high-temperature systems involving phases like alloys, slags, oxides, or silicates. Using so-called optimization programs, various kinds of experimental information (phase equilibria, activities, enthalpies, heat capacities, densities, *etc.*) are employed to evaluate *Gibbs* energies of the individual phases as functions of temperature, pressure, and composition. Thereby, parameters of appropriate *semi*-empirical models are adjusted that describe the excess properties of the various non-ideal phases as closely as possible. Since quaternary interactions are normally negligible, the optimizations are performed on binary and ternary systems in order to correlate their thermodynamic properties, and the model equations are then used to predict the thermodynamic properties of multicomponent systems.

The present author has employed the CALPHAD method for the assessment of electrolyte systems involving highly or sparingly soluble solid phases [15–20]. Carefully evaluated thermodynamic data have been used for the calculation of solubility equilibria with applications to geochemistry [16–26] and to the simulation of industrial processes [27]. In this communication, an attempt is made to predict solubilities of highly soluble ionic solids in the temperature range from 0 to 100°C using only limited experimental information valid for 25°C as model input. Therefore, proper extrapolation of solution thermodynamic properties to other temperatures is crucial and will be examined in detail. The *Pitzer* model [2] will be mainly used, but some remarks on the *Brønsted-Guggenheim-Scatchard* specific ion-interaction theory (SIT) proposed by *Grenthe et al.* [28] will be given as well. Solubility predictions for the $NaCl–NaOH–Na_2CO_3–H_2O$ system and its subsystems will be presented and their limitations will be discussed.

The *Pitzer* Model: A Story of Success with Limitations

In 1973, *Pitzer* [29] developed a set of equations that can accurately represent the excess properties of electrolyte solutions as a function of composition, temperature, and pressure. The total excess *Gibbs* energy, G^E, of a binary solution is given by

$$G^E = w_w \nu m RT(1 - \phi + \ln\gamma_\pm), \tag{1}$$

where R and T have their usual meaning, w_w is the mass of the solvent, $\nu = \nu_M + \nu_X$ (ν_M, ν_X: stoichiometric coefficients of cations and anions, respectively), m is the molality, ϕ is the osmotic coefficient of the solvent, and $\gamma_\pm$ is the mean activity coefficient of the solute. In the *Pitzer* model, G^E contains a *Debye-Hückel* term, a representation of the ionic strength dependence of the second virial coefficient B_{MX}, and a third viral coefficient C_{MX} that is independent of ionic strength:

$$G^E/(w_w RT) = -A_\phi(4Ib^{-1})\ln(1 + bI^{1/2}) + 2\nu_M\nu_X(m^2 B_{MX} + m^3 \nu_M z_M C_{MX}) \tag{2}$$

$$B_{\mathrm{MX}} = \beta_{\mathrm{MX}}^{(0)} + 2\beta_{\mathrm{MX}}^{(1)}(1 - (1 + \alpha_1 I^{1/2})\exp(-\alpha_1 I^{1/2}))(\alpha_1^2 I)^{-1} + 2\beta_{\mathrm{MX}}^{(2)}(1 - (1 + \alpha_2 I^{1/2})\exp(-\alpha_2 I^{1/2}))(\alpha_2^2 I)^{-1} \quad (3)$$

In Eqs. (2) and (3), I is the ionic strength on molality basis, A_ϕ is the *Debye-Hückel* coefficient for the osmotic function (at 25°C, $A_\phi = 0.3915\,\mathrm{kg}^{1/2} \cdot \mathrm{mol}^{-1/2}$), and z_i are the charges of the ions i. The constant b equals 1.2 for all solutes, $\alpha_1 = 2$ and $\alpha_2 = 0$ (*i.e.* $\beta_{\mathrm{MX}}^{(2)}$ is not required) unless both ions have a charge greater than or equal to 2, in which case $\alpha_1 = 1.4$ and $\alpha_2 = 12$. For 3–2 and 4–2 electrolytes, $\alpha_1 = 2$ and $\alpha_2 = 50$ have been used [2]. It should be noted that b, α_1, and α_2 are assumed to be temperature independent. Thus, at fixed pressure and temperature, G^{E} is expressed in terms of up to four adjustable parameters $\beta_{\mathrm{MX}}^{(0)}$, $\beta_{\mathrm{MX}}^{(1)}$, $\beta_{\mathrm{MX}}^{(2)}$, and C_{MX} per electrolyte, which are generally fitted to osmotic and/or activity coefficient data. Values of $\beta_{\mathrm{MX}}^{(2)}$ are relatively large and negative for solutes which are significantly ion-paired. However, if the value of the ion-association constant exceeds some critical value, the resulting complexes are usually taken into account as individual species [4]. Instead of C_{MX}, the quantity that is often tabulated is the parameter occurring in the expression for the osmotic coefficient, C_{MX}^{ϕ}, which is defined as

$$C_{\mathrm{MX}} = C_{\mathrm{MX}}^{\phi}(2|z_{\mathrm{M}} z_{\mathrm{X}}|^{1/2})^{-1} \quad (4)$$

Equations for ϕ and $\ln\gamma_\pm$, which are related to the partial molar excess *Gibbs* energies of solvent and solute, are obtained by appropriate differentiation [2]. *Pitzer* [2] also presents general equations for G^{E}, ϕ, and $\ln\gamma_\pm$ that are valid for multicomponent solutions. For the latter, higher-order electrostatic terms, $^{\mathrm{E}}\theta(I)$, which arise from asymmetric mixing, *e.g.* between HCO_3^- and CO_3^{2-}, and are functions of the ionic strength, are usually taken into account. In addition, the parameters $^{\mathrm{S}}\theta_{\mathrm{M,M'}}$, $^{\mathrm{S}}\theta_{\mathrm{X,X'}}$, $\psi_{\mathrm{M,X,X'}}$, and $\psi_{\mathrm{M,M',X}}$ are used to describe specific interactions in ternary systems. *Pitzer* [2] presents a convenient method to calculate $^{\mathrm{E}}\theta(I)$, as well as extensive tables of interaction parameters.

Equations for apparent molar volumes, enthalpies, and heat capacities can be derived from the excess *Gibbs* energy by appropriate differentiation with respect to pressure or temperature. In this communication, the effect of temperature on G^{E} is of particular interest. Thus, the relative enthalpy, L, and the relative heat capacity, J, of the solution must be known, which are defined according to Eqs. (5) and (6).

$$L = H - H^\circ = -T^2(\partial(G^{\mathrm{E}}/T)/\partial T)_{P,m} \quad (5)$$

$$J = C_p - C_p^\circ = (\partial L/\partial T)_{P,m} = -T^2(\partial^2(G^{\mathrm{E}}/T)/\partial T^2 + (2/T)(\partial(G^{\mathrm{E}}/T)/\partial T))_{P,m} \quad (6)$$

In Eqs. (5) and (6), the superscript ° denotes the total quantity of the components in their standard states. The apparent molar relative enthalpy, $^{\phi}L$, and the apparent molar heat capacity, $^{\phi}C_p$, are defined as

$$^{\phi}L = L/n_2 \quad (7)$$

and

$$^{\phi}C_p = (C_p - n_1 C_{p1}^\circ)/n_2 = C_{p2}^\circ + (\partial^{\phi}L/\partial T)_{P,m}, \quad (8)$$

where C_{p1}° and C_{p2}° are the molar heat capacity of pure water and the partial molar heat capacity of the electrolyte at infinite dilution, respectively, and $n_2 = w_{\mathrm{w}}m$ is the amount of solute.

Taking the appropriate derivative of Eq. (2) as given by Eqs. (5) and (7) results in the expression for the relative apparent molar enthalpy.

$$^{\phi}L = \nu|z_{\mathrm{M}}z_{\mathrm{X}}|A_L(2b)^{-1}\ln(1+bI^{1/2}) - 2\nu_{\mathrm{M}}\nu_{\mathrm{X}}RT^2(mB_{\mathrm{MX}}^{L} + m^2\nu_{\mathrm{M}}z_{\mathrm{M}}C_{\mathrm{MX}}^{L}) \quad (9)$$

$$B_{\mathrm{MX}}^{L} = (\partial B_{\mathrm{MX}}/\partial T)_{P,m} \quad (10)$$

$$\beta_{\mathrm{MX}}^{(\mathrm{i})L} = (\partial \beta_{\mathrm{MX}}^{(\mathrm{i})}/\partial T)_P \quad \text{for i} = 0,1,2 \quad (11)$$

$$C_{\mathrm{MX}}^{L} = (\partial C_{\mathrm{MX}}/\partial T)_P = (\partial C_{\mathrm{MX}}^{\phi}/\partial T)_P\,(2|z_{\mathrm{M}}z_{\mathrm{X}}|^{1/2})^{-1} \quad (12)$$

In Eq. (9), the *Debye-Hückel* coefficient for enthalpy is defined by

$$A_L = 4RT^2(\partial A_{\phi}/\partial T)_P \quad (13)$$

and has the value $A_L/RT = 0.801\,\mathrm{kg}^{1/2}\cdot\mathrm{mol}^{-1/2}$ at 25°C. The apparent molar heat capacity is given by

$$^{\phi}C_p = C_{p2}^{\circ} + \nu|z_{\mathrm{M}}z_{\mathrm{X}}|A_J(2b)^{-1}\ln(1+bI^{1/2})$$
$$- 2\nu_{\mathrm{M}}\nu_{\mathrm{X}}RT^2(mB_{\mathrm{MX}}^{J} + m^2\nu_{\mathrm{M}}z_{\mathrm{M}}C_{\mathrm{MX}}^{J}) \quad (14)$$

$$B_{\mathrm{MX}}^{J} = (\partial^2 B_{\mathrm{MX}}/\partial T^2)_{P,m} + (2/T)(\partial B_{\mathrm{MX}}/\partial T)_{P,m} \quad (15)$$

$$\beta_{\mathrm{MX}}^{(\mathrm{i})J} = (\partial^2 \beta_{\mathrm{MX}}^{(\mathrm{i})}/\partial T^2)_P + (2/T)(\partial \beta_{\mathrm{MX}}^{(\mathrm{i})}/\partial T)_P \quad \text{for i} = 0,1,2 \quad (16)$$

$$C_{\mathrm{MX}}^{J} = (\partial^2 C_{\mathrm{MX}}/\partial T^2)_P + (2/T)(\partial C_{\mathrm{MX}}/\partial T)_P$$
$$= ((\partial^2 C_{\mathrm{MX}}^{\phi}/\partial T^2)_P + (2/T)(\partial C_{\mathrm{MX}}^{\phi}/\partial T)_P)(2|z_{\mathrm{M}}z_{\mathrm{X}}|^{1/2})^{-1} \quad (17)$$

In Eq. (14), the *Debye-Hückel* coefficient for heat capacity is defined by

$$A_J = (\partial A_L/\partial T)_P \quad (18)$$

and has the value $A_J/R = 3.94\,\mathrm{kg}^{1/2}\cdot\mathrm{mol}^{-1/2}$ at 25°C.

The unchallenged strength of the *Pitzer* model rests in its capability to correlate and reproduce, within the experimental uncertainty of high-precision measurements, the thermodynamic properties of electrolyte solutions over wide ranges of temperatures, pressures, and concentrations up to saturation. This has been convincingly demonstrated for osmotic coefficients, enthalpies of dilution, heat capacities, and molar volumes measured with unsurpassed precision and accuracy at Oak Ridge National Laboratory (ORNL) [30]. However, it is a major limitation of the *Pitzer* model that extrapolation to molalities exceeding the range covered during parametrization is generally poor. The $ZnSO_4$–H_2O system [15] provides a striking example. When the relative apparent molar enthalpy of $ZnSO_4(aq)$ is modelled with parameters reported by *Silvester* and *Pitzer* [31] to be valid up to $1\,\mathrm{mol}\cdot\mathrm{kg}^{-1}$, wrong signs of the temperature coefficients of zinc sulfate heptahydrate and hexahydrate solubilities are predicted. This results from a completely wrong extrapolation of $^{\phi}L$ to saturation ($3.6\,\mathrm{mol}\cdot\mathrm{kg}^{-1}$). After adjusting the enthalpy parameters with respect to experimental data measured up to saturation, very good solubility predictions could be made for temperatures ranging from

10 to 70°C [15]. Other examples of wrong extrapolations will be discussed below.

To correlate high-precision experimental data over wide ranges of temperature, pressure, and composition, various modifications or extensions of the classical *Pitzer* equations have been proposed:

(*i*) To improve the fit, researchers have frequently changed the numerical value of α_1 (see *e.g.* Refs. [32–34]) or even used different α_1 values for the various components of a mixture [34]. Adjustable α_2 values and $\beta_{MX}^{(2)}$ parameters have also been employed for 1–1 electrolytes to account for ion association at the highest temperatures [32, 35]. Moreover, there is some confusion regarding the definition of the C^L parameters (sometimes denoted as C') owing to different forms of the equation used for $^{\phi}L$ [32, 35, 36].

(*ii*) To correlate electrolyte properties to very high ionic strengths (*e.g.* $33\,mol\cdot kg^{-1}$ for $CaCl_2$), the *Pitzer* equations have been extended using virial coefficients up to the sixth order [37]. *Archer* introduced an ionic-strength term at the third level for NaBr [38] and NaCl [39], and his equations have been generalized by *Clegg*, *Rard*, and *Pitzer* [40] to electrolyte mixtures. *Pitzer et al.* [41] recently presented a general form of equations that retains the expressions of the second virial level for all higher coefficients. Although similar to one another, these approaches are not completely compatible. Thus, either different equations have to be used or the parameters have to be refitted to the most general form given in Ref. [41].

(*iii*) The use of different equations for the dielectric constant of water and different equations of state of water resulted in various compilations of the *Debye-Hückel* limiting law slopes A_ϕ, A_L, and A_J as well as those for apparent molar volumes and compressibilities [42–44]. Nowadays, the *Bradley-Pitzer* equations [43] are most frequently used for the calculation of the *Debye-Hückel* slopes. Researchers of ORNL found in their recent study [45] that at temperatures from $T = 25$°C to 250°C and pressures $P \leq 400$ bar the values taken from Ref. [43] are essentially identical to those given more recently by *Archer* and *Wang* [44].

Therefore, care must be taken when combining *Pitzer* coefficients from different sources, as the parameters are generally not compatible with any single form of the *Pitzer* equations.

Making Predictions Using Limited Information

For a small number of about two dozen electrolytes, *Pitzer* parameters are known over a wide range of temperatures and are, for instance, fitted to equations of the form [46]

$$X(T) = w_1 + w_2(T/\mathrm{K})^{-1} + w_3 \ln(T/\mathrm{K}) + w_4(T/\mathrm{K}) + w_5(T/\mathrm{K})^2 + w_6(680 - T/\mathrm{K})^{-1} + w_7(T/\mathrm{K} - 227)^{-1} \tag{19}$$

For the vast majority of electrolytes, however, *Pitzer* parameters for osmotic and activity coefficients [2, 47, 48], apparent molar enthalpies [2, 31], and heat capacities [49, 50] are only available at 25°C. Nonetheless, there is an urgent need in many geochemical, environmental, and industrial applications to estimate activity coefficients in mixed electrolyte solutions in the temperature range from 0 to 100°C.

For carbonate systems, *Monnin* and *Schott* [5] as well as *Pitzer* and associates [51, 52] have used the following power series expression

$$X(T) = X_0 + a(T - T_0) + (1/2)b(T - T_0)^2 \qquad (20)$$

to estimate activity and osmotic coefficients from 0 to 50°C. In Eq. (20), $X(T)$ denotes the interaction parameter at temperature T, X_0 is its value at $T_0 = 298.15$ K, $a = (\partial X/\partial T)_P$, and $b = (\partial^2 X/\partial T^2)_P$ at $T = T_0$. The constants a and b can be expressed in terms of enthalpy and heat capacity parameter values valid at T_0, which are denoted as X^L and X^J, respectively. Using the definitions given in Eqs. (11), (12), (16), and (17), $a = X^L$ and $b = X^J - 2X^L/T_0$ are found.

Criss and *Millero* [49] suggested equations of the form

$$X(T) = X_0 + a(1/T - 1/T_0) + b(T^2 - T_0^2) \qquad (21)$$

and claimed that reasonable estimates of activity coefficients are provided from 0 to 75°C and from 0 to 2 $\text{mol} \cdot \text{kg}^{-1}$ in the ionic strength. In Eq. (21), $a = X^J T_0^3/3 - X^L T_0^2$ and $b = X^J/6$.

The new method proposed in this work (hereafter referred to as the present model) is based on temperature independent heat capacities.

$$X(T) = X_0 + a(1/T - 1/T_0) + b \ln(T/T_0). \qquad (22)$$

In Eq. (22), $a = X^J T_0^3 - X^L T_0^2$ and $b = X^J T_0^2$. If there are no heat capacity parameters available, $X^J = 0$ and thus $a = -X^L T_0^2$ and $b = 0$. Quite frequently, only X_0 parameters have been reported and neither X^L nor X^J are available. Then, Eq. (22) would result in $X(T) = X_0$, *i.e.* the contribution to the excess *Gibbs* energy would be purely entropic. It is better, however, to assume that the contribution to the excess *Gibbs* energy is solely enthalpic. In this case, the convention used in the F*A*C*T database [53, 54] (see below) is employed, and $a = X_0 T_0$ is defined in a way that

$$X(T) = X_0 T_0/T \quad \text{only if } X^L = X^J = 0 \qquad (23)$$

The applicability and limitations of Eqs. (20)–(23) will be investigated in the subsequent sections.

Solubility Calculations *via Gibbs* Energy Minimization

The widespread use and popularity of the CALPHAD method is based on the crucial role that the *Gibbs* energy plays in thermodynamic calculations. Not only

(*i*) all the other thermodynamic quantities can be derived by appropriate differentiation as indicated above, but also

(*ii*) the minimum of the *Gibbs* energy at constant temperature and pressure with respect to the composition of the various phases defines the thermodynamic equilibrium state of the heterogeneous system.

So-called *Gibbs* energy minimizers have been developed to perform these calculations reliably and, on modern computers, with sufficient speed even for multicomponent, multiphase systems. For each phase, the input data comprise parameters describing the temperature and pressure dependencies of (*i*) the standard *Gibbs* energies of all phase constituents and (*ii*) the excess *Gibbs* energies described by appropriate models. Among others, ChemSage [55] with its optimizer [15] and F*A*C*T (Facility for the Analysis of Chemical Thermodynamics) [53] are particularly useful for systems involving aqueous solutions. Both ChemSage and F*A*C*T feature the classical *Pitzer* model, in which the *Debye-Hückel* parameters are calculated according to *Bradley* and *Pitzer* [43]. There are two modifications of the model in which unsymmetrical, higher-order electrostatic mixing terms can be either taken into account or not. In ChemSage, the *Brønsted-Guggenheim-Scatchard* specific ion-interaction theory [28] and the *Helgeson-Kirkham-Flowers* model [56] are implemented as well. In the present work, all calculations and optimizations have been performed using ChemSage [15, 55].

Sodium Chloride

The NaCl–H_2O system is definitely the most thoroughly investigated binary electrolyte system and has therefore often served as a reference for testing thermodynamic models. In several reviews [39, 42, 46, 57], various modifications of the *Pitzer* equations have been employed to correlate the thermodynamic properties over the entire experimentally accessible temperature, pressure, and concentration ranges. *Archer*'s model [39] uses as many as 57 adjustable parameters, but it covers the most extended temperature and pressure ranges ($250 < T/\mathrm{K} < 600$, $P \leq 1000\,\mathrm{bar}$) of these studies.

As explained above, the present model uses the classical *Pitzer* equations with the 9 parameters given in Table 1. The interaction parameters valid at 25°C and the heat capacity parameters together with the partial molar heat capacities at infinite dilution of the ions were taken from *Pitzer* and *Mayorga* [47] and *Criss* and *Millero* [49], respectively, and used without modification. The enthalpy parameters of *Silvester* and *Pitzer* [31] do not reproduce experimental values very well, probably due to the use of an A_L value that is different from *Bradley* and *Pitzer*'s [43] definition. Thus, the enthalpy parameters were re-fitted to apparent molar relative enthalpies of NaCl at 25°C taken from *Parker* [58] and the NBS tables [59]. Since the temperature dependence of solubility depends on the partial molar enthalpy of NaCl at saturation [60], which in turn is related to the derivative of the apparent molar relative enthalpy with respect to molality, a ten times higher weight was put on the two data points close to saturation.

Standard thermodynamic quantities (Table 2) were taken from

(a) CODATA [61] and *Criss* and *Millero* [49] for the aqueous ions,
(b) the F*A*C*T thermodynamic database [53] for ice, water, and steam (the latter is modelled as non-ideal gas using a second virial coefficient that is calculated from critical data according to the *Tsonopoulos* correlation [64]), and
(c) *Archer*'s work for NaCl [62] and $NaCl \cdot 2H_2O$ [39]; however, the standard enthalpies have been slightly adjusted (within the reported uncertainties) to reproduce the solubility at 25°C (in the case of NaCl) and the peritectic temperature (in the case of $NaCl \cdot 2H_2O$).

Table 1. Binary *Pitzer* parameters for osmotic coefficients, apparent molar relative enthalpies, and apparent molar heat capacities, all valid at 25°C; parameters without superscript were determined in the present study

	NaCl	Na_2CO_3	$NaHCO_3$	NaOH	HCl
$\beta^{(0)}_{MX}$	0.0765[a]	0.0362[c]	0.028[c]	0.0864[a]	0.1775[a]
$\beta^{(1)}_{MX}$	0.2664[a]	1.51[c]	0.044[c]	0.253[a]	0.2945[a]
C^{ϕ}_{MX}	0.00127[a]	0.0052[c]		0.0044[a]	0.0008[a]
$10^3 \cdot \beta^{(0)L}_{MX}$	0.7293	2.127	1.00[c]	0.5307	−0.3081[d]
$10^3 \cdot \beta^{(1)L}_{MX}$	0.9031	5.059	1.10[c]	0.7495	0.1419[d]
$2 \cdot 10^3 \cdot \lvert z_M z_X \rvert^{1/2} C^L_{MX}$	−0.1066	−0.3765		−0.1355	−0.06213[d]
$10^5 \cdot \beta^{(0)J}_{MX}$	−1.53[b]	−3.019[c]	−1.929[c]	−1.41[b]	−0.304[b]
$10^5 \cdot \beta^{(1)J}_{MX}$	−0.0365[b]	−15.425[c]	−3.562[c]	−3.76[b]	0.678[b]
$2 \cdot 10^5 \cdot \lvert z_M z_X \rvert^{1/2} C^J_{MX}$	0.1824[b]	0.460		0.134[b]	0.00372[b]

[a] *Pitzer* and *Mayorga* [47]; [b] *Criss* and *Millero* [49]; [c] *Peiper* and *Pitzer* [51]; [d] *Silvester* and *Pitzer* [31]

Table 2. Standard thermodynamic quantities of phase constituents ($P^\circ = 1$ bar)

	$\Delta_f H^\circ_{298}$/kJ · mol^{-1}	S°_{298}/J · K^{-1} · mol^{-1}	C°_{p298}/J · K^{-1} · mol^{-1}
$H_2O(g)$	−241.834[a]	188.834[a]	33.59[a]
$H_2O(l)$	−285.83[b]	69.95[b]	75.38[a]
$H^+(aq)$	0	0	0
$Na^+(aq)$	−240.34[b]	58.45[b]	43.01[c]
$OH^-(aq)$	−230.015[b]	−10.90[b]	−140.80[c]
$Cl^-(aq)$	−167.08[b]	56.60[b]	−126.32[c]
$HCO_3^-(aq)$	−689.93[b]	98.4[b]	−54.23[d]
$CO_3^{2-}(aq)$	−675.23[b]	−50.0[b]	−314.5[d]
$CO_2(aq)$	−413.26[b]	119.36[b]	185.73[e]
$H_2O(s)$	−292.816[a]	44.529[a]	36.246[a,f]
$NaCl(s)$	−411.22[g]	71.97[h]	50.23[h]
$NaCl \cdot 2H_2O(s)$	−997.69[g]	162.51[i]	137.0[i]
$Na_2CO_3 \cdot 10H_2O(s)$	−4080.068[g]	560.98[g]	550.31[j]
$Na_2CO_3 \cdot 7H_2O(s)$	−3199.133[g]	423.20[g]	420[j]
$Na_2CO_3 \cdot H_2O(s)$	−1430.980[g]	164.66[g]	145.6[j]

[a] F*A*C*T database [53]; [b] CODATA key value [61]; [c] *Criss* and *Millero* [49]; [d] *Peiper* and *Pitzer* [51]; [e] *Königsberger et al.* [19]; [f] constant value at $T < 0$°C; [g] this work; [h] *Archer* [62]; [i] *Archer* [39]; [j] *Vanderzee* [63]

Figure 1 shows apparent molar heat capacities of NaCl(*aq*) at various temperatures. In the present model (Eq. (22)) the heat capacities are treated as temperature independent; the change of the calculated values with temperature is solely due the temperature dependence of the *Debye-Hückel* coefficient A_J (dash-dot line for 75°C). In the two other extrapolation methods, Eq. (21) and particularly Eq. (20), the heat capacity exhibits a much more pronounced temperature dependence (dashed and dotted lines, respectively, for 75°C).

In Fig. 2, relative apparent molar enthalpies are compared with values calculated according to the *Pitzer* and SIT models. The SIT model gives a reasonable fit at lower molalities, but if all data points are included, the 1-parameter SIT model

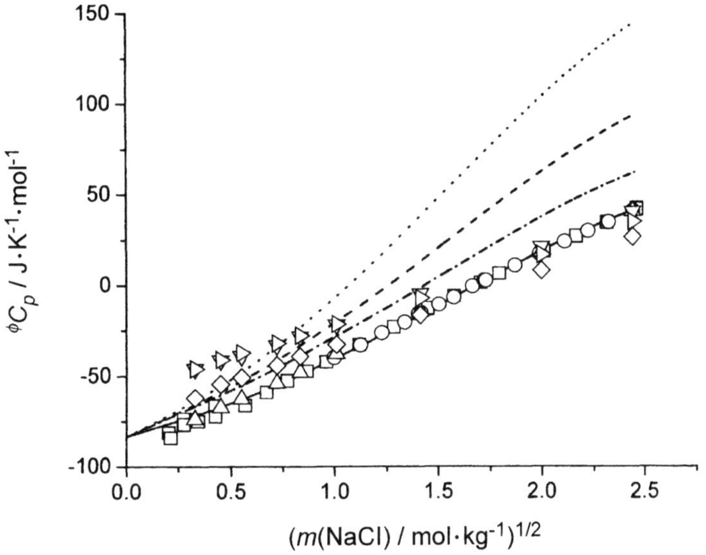

Fig. 1. Apparent molar heat capacities of NaCl(aq); experimental data at 25°C: squares [65], circles [66], up-triangles [67]; at 50°C: down triangles [67]; at 75°C: right-triangles [67]; at 100°C: diamonds [67]; calculated values at 25°C: solid line, present model; at 75°C: dash-dot line, present model; dashed line, according to Eq. (21); dotted line, according to Eq. (20)

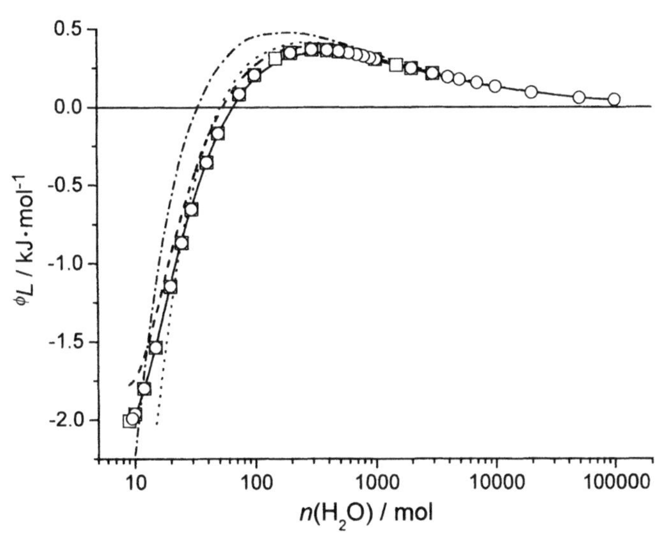

Fig. 2. Apparent molar relative enthalpy of NaCl(aq) at 25°C; experimental data: circles [58], squares [59]; calculated values: solid line, present model; dashed line [31]; dotted line, SIT model with lower weight on data at high molalities; dash-dot line, SIT model with equal weights on all data

is inferior to the more flexible, 3-parameter *Pitzer* model. In any case, the SIT model seems less suitable for representing these data to saturation molalities.

The present constant heat capacity *Pitzer* model satisfactorily reproduces enthalpies of solution of NaCl(s) measured recently between 25°C and 60°C [68] (Fig. 3). However, when the present model is used to extrapolate relative apparent

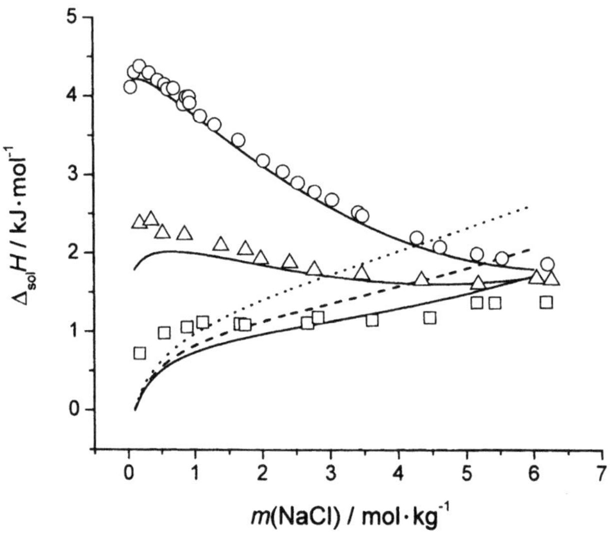

Fig. 3. Enthalpies of dissolution of NaCl(s) at 24.4°C (circles), 44.3°C (triangles), and 59.2°C (squares) [68]; calculated values: solid lines, present model at these temperatures; dashed line, according to Eq. (21) at 59.2°C; dotted line, according to Eq. (20) at 59.2°C

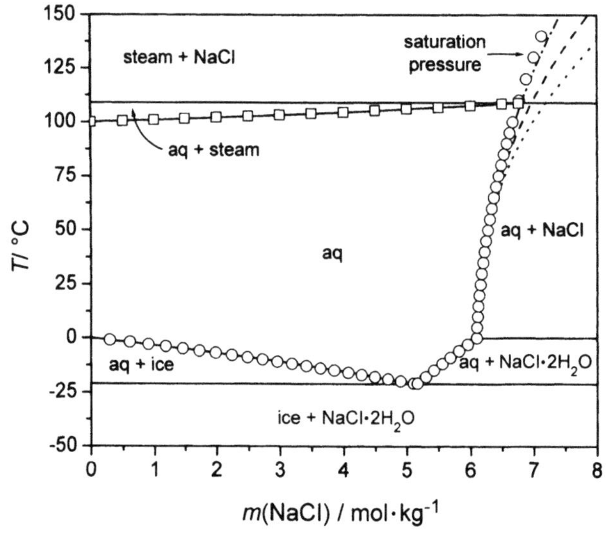

Fig. 4. Temperature-composition phase diagram of the NaCl–H_2O system at $P = 1.01325$ bar and at saturation pressure; experimental data: circles [70]; boiling points: squares, as tabulated by *Clarke* and *Glew* [57]; calculated values: solid and dash-dot lines, present model; dashed line, according to Eq. (21); dotted line, according to Eq. (20)

molar enthalpies to 100°C, deviations of about 20% from the values of Ref. [32] are noted. Similar deviations have been obtained by *Gates et al.* [69], although these authors used a more refined, temperature dependent heat capacity model based on their highly precise measurements.

Figure 4 shows the temperature-composition phase diagram of the NaCl–H_2O system. The present model reproduces experimental data for all equilibria between

Table 3. Ternary and neutral-species interaction parameters; the temperature independent values have been reported for 25°C

i	j	k	$^S\theta_{ij}$	ψ_{ijk}	λ_{ij}
H^+	Na^+	Cl^-	0.036[a]	−0.004[a]	
Cl^-	OH^-	Na^+	−0.05[a]	−0.006[a]	
				$-9.93(T/K)^{-1}+0.0273$[b]	
Cl^-	HCO_3^-	Na^+	0.0359[c]	−0.0143[c]	
Cl^-	CO_3^{2-}	Na^+	−0.053[c]	0.016[d]	
				$11.18(T/K)^{-1}-0.0235$[e]	
$CO_2(aq)$	Na^+				0.10[f]
$CO_2(aq)$	Cl^-				−0.005[f]

[a] *Pitzer* and *Kim* [74]; [b] *Pabalan* and *Pitzer* [6], see text; [c] *Peiper* and *Pitzer* [51]; [d] *Thurmond* and *Millero* [75]; [e] this study, see text; [f] *Harvie et al.* [4]

−21 and 110°C very well; even up to 150°C the agreement with solubility data [70] is quite reasonable. The two other extrapolation methods, Eq. (21) and particularly Eq. (20), which do not correspond to constant heat capacities, result in significant deviations above about 85 and 65°C respectively.

An independent test of the present model is provided by the prediction of equilibrium constants in a mixed electrolyte solution. The ionic product of water, defined by $K_w = m(H^+)m(OH^-)$, has been precisely measured in $NaCl(aq)$ by hydrogen electrode potentiometry in the temperature range from 25 to 300°C [71–73]. In these studies, the NaCl solution must actually be regarded as a four-component system (HCl–$NaOH$–$NaCl$–H_2O). In addition to the NaCl parameters, the binary *Pitzer* parameters for NaOH (see below) and HCl [2, 31, 47, 49] were extrapolated to high temperatures according to the present method (Eq. (22)). Also, four ternary mixing parameters valid at 25°C have been reported [74] (see Table 3) which were extrapolated according to Eq. (23). Most surprisingly, $\log K_w$ values calculated from the present model agree with measured values within the experimental uncertainties up to at least 250°C (Figs. 5 and 6). Presumably, there is some sort of error compensation involved at high temperatures, but the better predictive power of the present method (Eqs. (22) and (23)) compared to Eq. (20) is nonetheless obvious.

Sodium Carbonate, Hydrogen Carbonate, and Hydroxide

Vanderzee [80] and *Peiper* and *Pitzer* [51] have thoroughly analyzed the thermodynamic properties of aqueous solutions of sodium carbonate and its hydrolysis products hydrogen carbonate and hydroxide. In the early 1980's, when their work was done, experimental data were only available at 25°C. Very recently, *Polya et al.* (Oak Ridge National Laboratory) [45] have measured the enthalpies of dilution of $Na_2CO_3(aq)$ and $NaHCO_3(aq)$ at moderate molalities (up to $m = 1.5\,mol \cdot kg^{-1}$) from $T = 25$ to 250°C and at pressures of $P = 70$ and 400 bar. To correlate their data, *Polya et al.* [45] employed a comprehensive *Pitzer* model with *Peiper* and *Pitzer*'s [51] parameters for 25°C and 13 enthalpy parameters for each pressure.

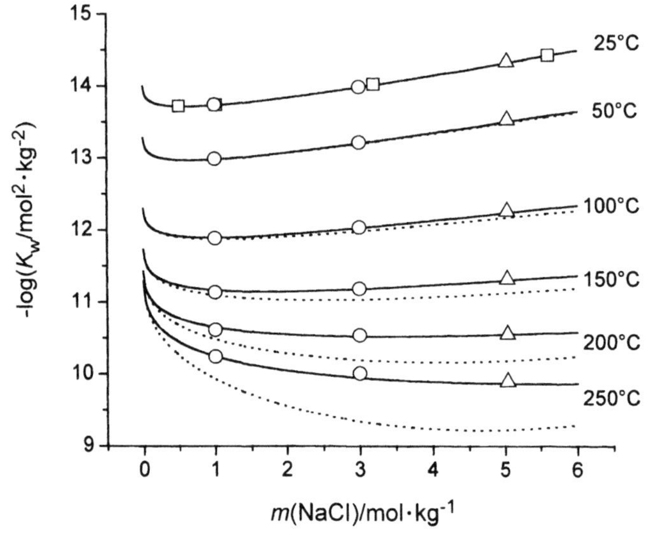

Fig. 5. Ionic product of water in NaCl(aq); experimental data: squares [71]; circles [72], triangles [73]; calculated values: solid lines, present model; dotted lines, according to Eq. (20)

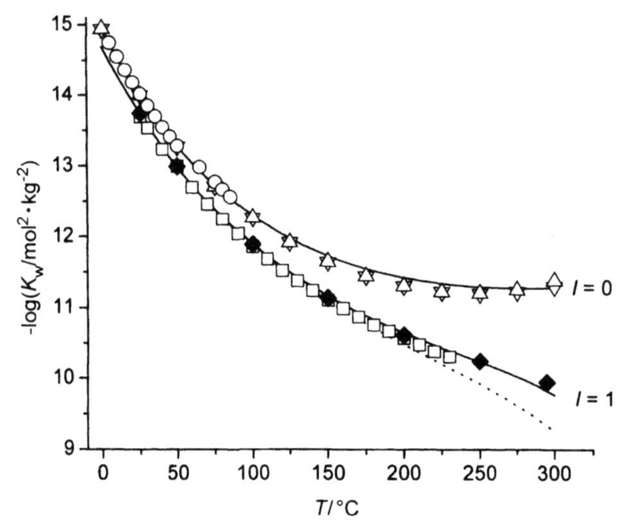

Fig. 6. Ionic product of water at $I = 0$ and $I = 1\,\mathrm{mol \cdot kg^{-1}}$ NaCl as a function of temperature; experimental values: diamonds [72], squares [76], circles [77], up-triangles [78], down-triangles [79]; calculated values: solid lines, present model; dotted line, according to Eq. (20)

Peiper and *Pitzer* [51] used the *Pitzer* model to derive binary interaction parameters together with their temperature dependencies and parameters for ternary interactions among the mentioned ions and with chloride. In view of the geochemical importance, *Monnin* and *Schott* [5] developed a model to calculate the solubilities of sodium carbonate minerals in brines. These authors employed *Peiper* and *Pitzer*'s aqueous model together with additional parameters taken from *Harvie et al.* [4] to derive standard enthalpies of formation and standard entropies of the solid phases from solubility data. Using these quantities together with *Peiper* and

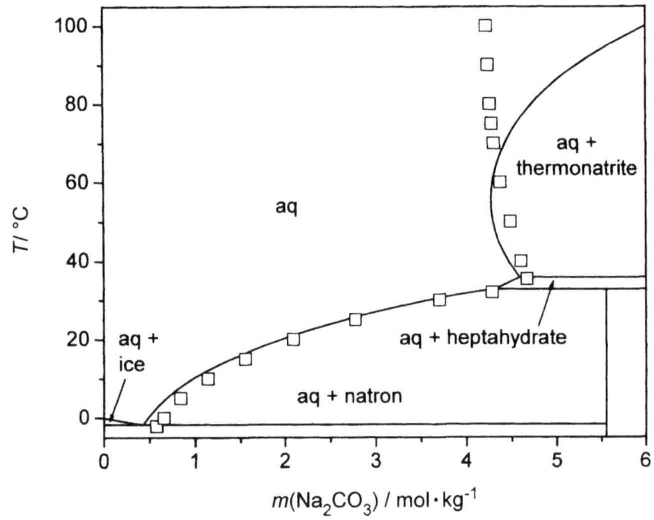

Fig. 7. Phase diagram of the Na_2CO_3–H_2O system; experimental data are from Ref. [81]; calculated values were obtained using data from Ref. [5]

Pitzer's temperature dependent interaction parameters according to Eq. (20) results in the Na_2CO_3–H_2O phase diagram shown in Fig. 7. Due to the poor agreement with experimental solubility data and the fact noticed earlier [19, 20] that *Monnin* and *Schott*'s quantities for the solid phases disagree with values carefully derived from calorimetric data [63, 82], a new assessment of this heterogeneous system seemed worthwhile.

As for NaCl, the present model uses the classical *Pitzer* equations with the parameters given in Table 1. The parameters valid at 25°C for Na_2CO_3 and $NaHCO_3$ were taken from *Peiper* and *Pitzer* [51], those for NaOH from *Pitzer* and *Mayorga* [47]. Enthalpy and heat capacity parameters for $NaHCO_3$ [51] were found satisfactory. For NaOH, the heat capacity parameters were taken from *Criss* and *Millero* [49], and the enthalpy parameters were re-determined to give a better fit to the data from Refs. [35, 58] at higher molalities (Fig. 8). Although the SIT model was appropriate to represent solubilities of sparingly soluble electrolytes in $NaClO_4$ media [18], it is obviously not suitable for representing apparent molar relative enthalpies of NaOH(*aq*) at higher molalities. The Na_2CO_3 enthalpy parameters reported by *Peiper* and *Pitzer* [51] reproduce experimental values only at low molalities, which is not surprising as these authors used only two parameters. Therefore, three parameters were fitted to the apparent molar relative enthalpies of Na_2CO_3 at 25°C given in Table 3 of Ref. [83], which are essentially the same as those reported in the NBS tables [59]. For the same reason as in the case of NaCl, a higher weight was given to the data points closest to saturation. To model the apparent molar heat capacities of Na_2CO_3, the $\beta^{(0)J}$ and $\beta^{(1)J}$ parameters given by *Peiper* and *Pitzer* [51] were retained. These parameters were derived from measurements performed up to 1 mol · kg^{-1} [84]. In addition, a C^J parameter was estimated in order to reproduce data at higher molalities measured by *Chernen'kaya* [85] (Fig. 9).

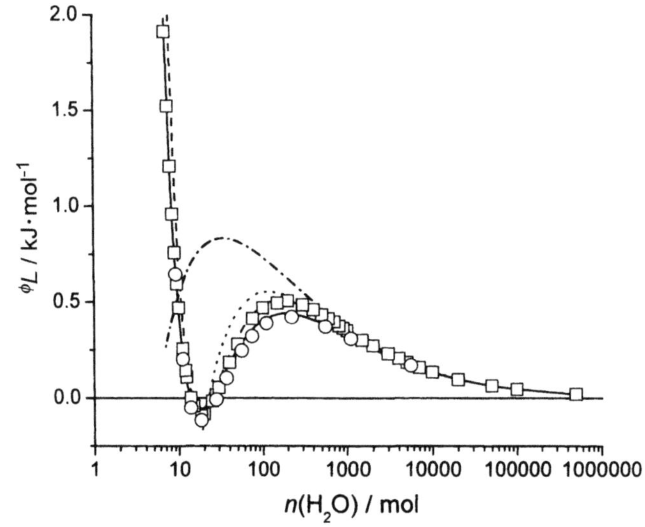

Fig. 8. Apparent molar relative enthalpy of NaOH(*aq*) at 25°C; experimental data: circles [35], squares [58]; calculated values: solid line, present model; dashed line [31]; dotted line, SIT model with lower weight on data at high molalities; dash-dot line, SIT model with equal weights on all data

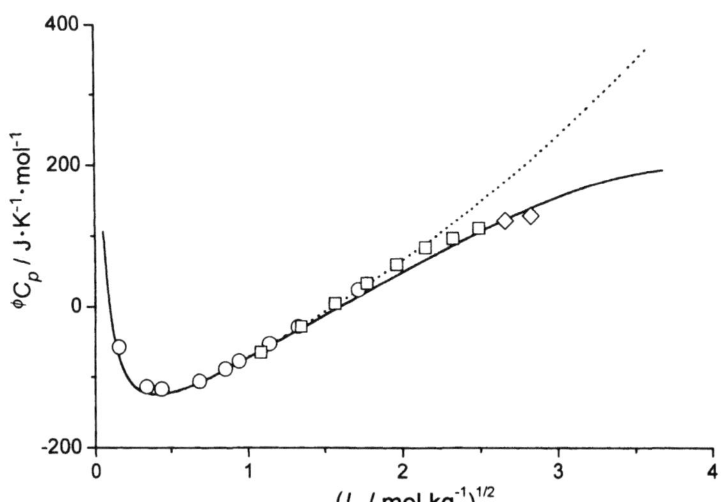

Fig. 9. Apparent molar heat capacity of Na_2CO_3(*aq*) at 25°C; experimental data: circles [84]; squares [85]; diamonds [86]; calculated values: solid line, present model; dotted line [51]

It should be noted that equilibrium enthalpy and heat capacity data were used in these optimizations. Owing to hydrolysis, hydrogen carbonate and hydroxide ions are formed at equilibrium. Since the degree of hydrolysis increases with dilution, the contribution of the enthalpy of hydrolysis becomes larger at low molalities as can be seen from Fig. 10. The effect of temperature on the chemical equilibria leads to a so-called 'relaxation' contribution to the measured heat capacity of the solution which results in an increase of $^{\phi}C_p$ at low ionic strengths (Fig. 9) as has been discussed *e.g.* by *Peiper* and *Pitzer* [51] and *Vanderzee* [80]. *Hepler* and

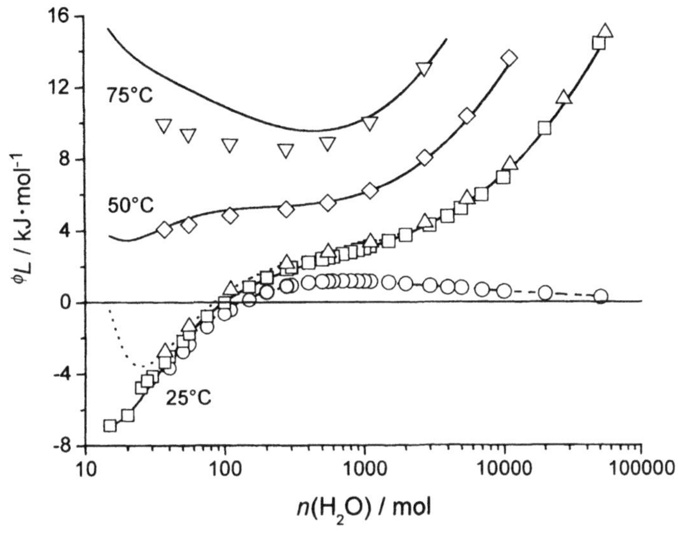

Fig. 10. Apparent molar relative enthalpies of $Na_2CO_3(aq)$; experimental data at 25°C: squares [59, 83], up-triangles [45]; circles, data corrected with respect to hydrolysis [83]; at 50°C: diamonds [45]; at 75°C: down-triangles [45]; calculated values: solid lines, present model; dashed line, present model, hydrolysis suppressed; dotted line, enthalpy parameters from Ref. [45]

Hovey [87] show that the relaxation contribution is always positive and give many references to systems for which this effect has been observed. However, an analytical expression of the relaxation contribution is different for nearly every system that has been investigated and sometimes quite cumbersome to derive.

In ChemSage, heat capacities and enthalpies are calculated by appropriate numerical differentiation of equilibrium *Gibbs* energies with respect to temperature. In this way, chemical reactions and shifts of equilibria with temperature, and thus the effects mentioned above, are automatically accounted for. If one wants to calculate the properties of hypothetical, pure electrolytes, the effects of equilibria and relaxation can easily be suppressed by disregarding the reaction products in the calculation. Therefore, ChemSage or F*A*C*T are truly 'facilities for the analysis of chemical thermodynamics' that allow the user to deal with these effects correctly without the need of performing complicated thermodynamic derivations.

The standard enthalpies of formation and standard entropies of sodium carbonate decahydrate (natron), heptahydrate, and monohydrate (thermonatrite) were optimized with respect to solubility data. The *Bayes* formalism [88, 89] used in the ChemSage optimizer [15] permits to weight parameters (in this case, the standard thermodynamic quantities of the solids) and experimental data (solubilities) independently from each other. Thereby, the carefully evaluated calorimetric quantities reported by *Vanderzee* [63, 82] were used as *a priori* parameters and their weights were chosen such that they changed essentially within their reported uncertainties. With an ordinary least-squares optimization, the changes in the parameter values would be much larger.

The phase diagram thus calculated agrees very reasonably with experimental data given by *Linke* and *Seidell* [81] (Fig. 11). Also, heptahydrate and thermonatrite solubilities are well reproduced in the metastable range, although these data

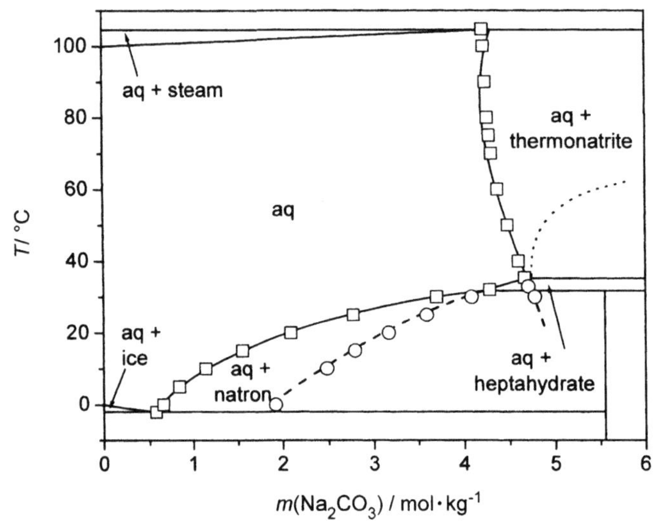

Fig. 11. Phase diagram of the Na_2CO_3–H_2O system; experimental data: squares, stable equilibria; circles, metastable equilibria [81]; calculated curves according to the present model: solid lines, stable equilibria; dashed lines, metastable equilibria; dotted line: thermonatrite solubility calculated from the present model but with $C^J = 0$, *i.e.* with *Peiper* and *Pitzer*'s [51] heat capacity parameters only

were not used in the optimizations. Figure 11 also illustrates the importance of correct apparent molar heat capacity values close to saturation. Obviously, *Peiper* and *Pitzer*'s [51] model ($C^J = 0$), which wrongly extrapolates to high molalities (dotted line in Fig. 9), results in a completely wrong solubility prediction (dotted line in Fig. 11). It should also be noted that C^J and the standard thermodynamic quantities of thermonatrite are strongly correlated and it is quite difficult to determine optimal values. This explains that the change in the *a priori* parameters for thermonatrite had to be larger than the experimental uncertainties in order to reproduce experimental solubilities. However, there is still some discrepancy at temperatures above 90°C where calculated solubilities start to increase again.

The present model was also used to predict relative apparent molar enthalpies of $Na_2CO_3(aq)$ at higher temperatures (Fig. 10). Whereas the agreement with the data measured by *Polya et al.* [45] is nearly perfect at 50°C, it becomes worse at 75°C, similar to the findings for NaCl. Unfortunately, *Polya et al.* limited their measurements to $m = 1.5\,\mathrm{mol \cdot kg^{-1}}$, whereas solubilities of thermonatrite range beyond $4\,\mathrm{mol \cdot kg^{-1}}$. Almost inevitable with the *Pitzer* model, extrapolations of enthalpies and heat capacities to higher molalities are rather poor. Particularly, using the enthalpy parameters of Ref. [45] for 25°C, the partial molar enthalpy of $Na_2CO_3(aq)$ changes its sign close to saturation, thus leading to a strong increase rather than to the slight decrease observed experimentally for thermonatrite solubilities just above 40°C. It must be concluded that despite the high quality of the ORNL data, the *Pitzer* model based on them fails completely to predict sodium carbonate solubilities. As has been noted by *Peiper* and *Pitzer* [51] two decades ago, measurements of heat capacities and enthalpies of dilution up to saturation are very desirable; however, they are still lacking for this important system.

Solubility Predictions for Multicomponent Electrolyte Systems

The data obtained in this work were used to predict solubilities in the ternary systems Na_2CO_3–NaCl–H_2O and Na_2CO_3–NaOH–H_2O and in the quaternary system Na_2CO_3–NaCl–NaOH–H_2O for $0°C \leq T \leq 100°C$.

The NaCl–NaOH–H_2O system

Pabalan and *Pitzer* [6] considered it necessary to introduce a temperature dependent ternary interaction parameter $\psi_{\mathrm{Na,OH,Cl}} = -9.93(T/\mathrm{K})^{-1} + 0.0273$ (which results in $\psi_{\mathrm{Na,OH,Cl}} = -0.006$ for 25°C [74]) in order to reproduce NaCl solubilities in NaOH solutions at temperatures different from 25°C. If their $\psi_{\mathrm{Na,OH,Cl}}$ is used with the present model, the results between 0 and 100°C are of similar quality as in *Pabalan* and *Pitzer*'s work [6]. This temperature dependent $\psi_{\mathrm{Na,OH,Cl}}$ parameter also influences pK_w values calculated at higher temperatures (see Fig. 5). The results obtained are slightly worse than with the temperature independent $\psi_{\mathrm{Na,OH,Cl}}$ parameter [74] but still in very good agreement with experimental values.

The Na_2CO_3–NaOH–H_2O system

Solubilities are in general well reproduced, particularly at higher temperatures (Fig. 12). At low temperatures (Fig. 13), the natron solubilities agree very well with experimental data, whereas measured solubilities of heptahydrate and thermonatrite are generally higher than the calculated values. This may be either due to experimental difficulties (*e.g.* supersaturation) or model imperfections (*e.g.* wrong extrapolation of the *Gibbs* energies of these phases to low temperatures). However, metastable solubilities in water are very well modelled (Fig. 11), and changing ternary parameters would also affect natron solubilities. Thus, it is suggested that more solubility data are measured for this system at low temperatures.

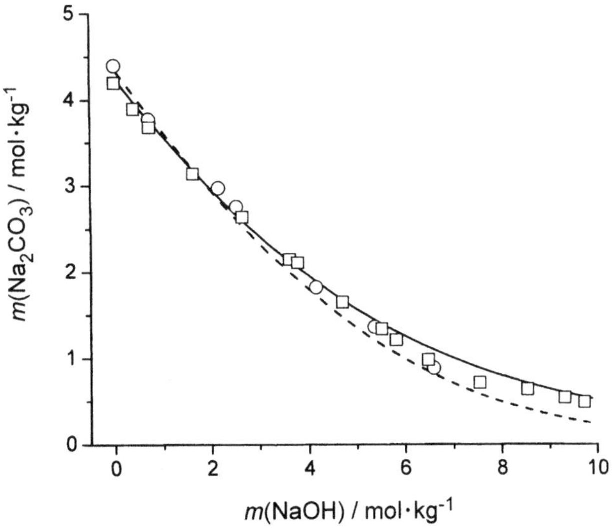

Fig. 12. Solubility of thermonatrite in NaOH solutions as calculated from the present model; experimental data were taken from Ref. [81]; circles, dashed line, 60°C; squares, solid line, 100°C

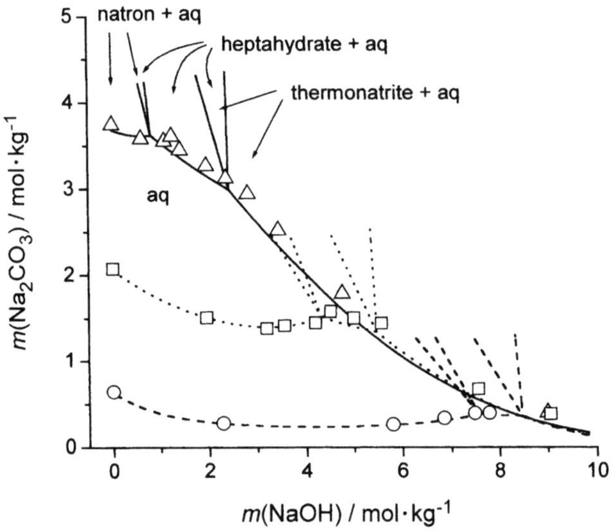

Fig. 13. Isothermal phase diagrams of the Na_2CO_3–NaOH–H_2O system; experimental data from Ref. [81]: circles, 0°C; squares, 20°C; triangles, 30°C; calculated values from the present model: the curves for the corresponding temperatures are dashed, dotted, and solid, respectively; one-, two-, and three-phase fields are indicated at the 30°C phase diagram

It should be noted that these phase diagrams, which have been conveniently generated using the so-called two-dimensional phase mapping feature of ChemSage, are of Type III in *Pelton* and *Schmalzried*'s [90, 91] classification. This means that two molar quantities are plotted against each other, whereas thermodynamic potentials like temperature and pressure are kept constant. In this type of phase diagrams, one-, two-, and three-phase fields are all two-dimensional. In the literature, however, usually only the solubility curves are depicted, whereas the other phase fields are omitted.

The Na_2CO_3–NaCl–H_2O system

Solubilities calculated with the temperature independent ternary parameters given in Table 3 are already reasonable, although they are a bit lower than experimental values, especially at higher temperatures. If, similar to NaCl–NaOH–H_2O, a temperature dependent parameter $\psi_{\mathrm{Na,CO_3,Cl}} = 11.18(T/\mathrm{K})^{-1} - 0.0235$, which essentially retains the 25°C value, is introduced, the agreement with experimental data can be improved (Figs. 14 and 15). Surprisingly, there is a lack of solubility data for this important system, so that the measurement of new solubility data is strongly recommended, particularly for temperatures above 25°C.

The Na_2CO_3–NaCl–NaOH–H_2O system

Phase diagrams for this system at NaCl saturation agree reasonably well with experimental data [81], despite the high ionic strengths of these mixed electrolyte solutions (Fig. 16). No further parameters have been adjusted; the results can therefore be regarded as a true prediction of solubilities in a multicomponent

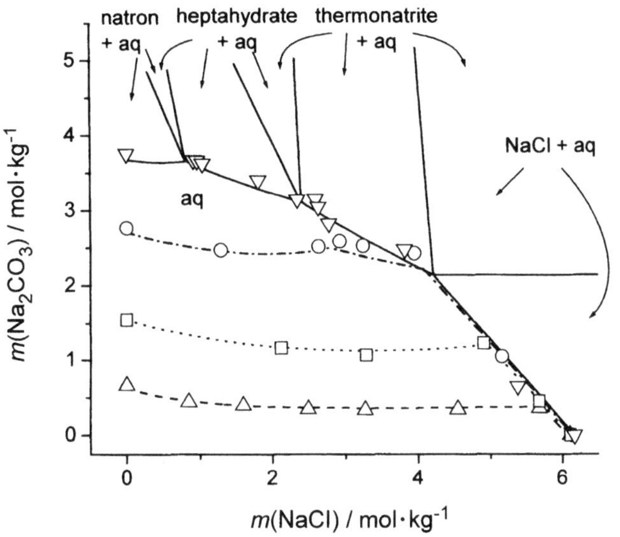

Fig. 14. Isothermal phase diagrams of the Na_2CO_3–NaCl–H_2O system; experimental data from Ref. [81]: up-triangles, 0°C; squares, 15°C; circles, 25°C; down-triangles, 30°C; calculated values from the present model: the curves for the corresponding temperatures are dashed, dotted, dash-dot, and solid, respectively; one-, two-, and three-phase fields are indicated at the 30°C phase diagram

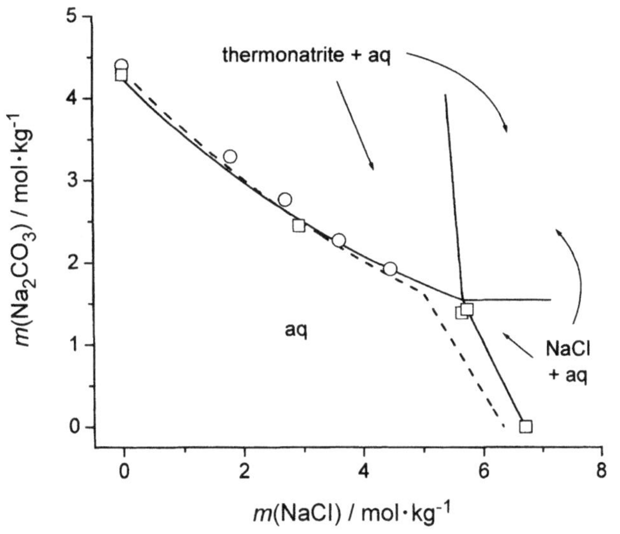

Fig. 15. Isothermal phase diagrams of the Na_2CO_3–NaCl–H_2O system as calculated from the present model; experimental data were taken from Ref. [81]; circles, dashed line, 60°C; squares, solid line, 100°C

electrolyte system. However, there are some discrepancies that can be analyzed by investigating the effects of changing thermodynamic data on calculated solubilities. For instance, whereas thermonatrite solubilities agree perfectly with experimental data at 60°C, making this phase less stable by about $600\,J \cdot mol^{-1}$ would result in very good agreement at lower temperatures. However, this change in the standard *Gibbs* energy is not compatible with solubilities in pure water. At higher

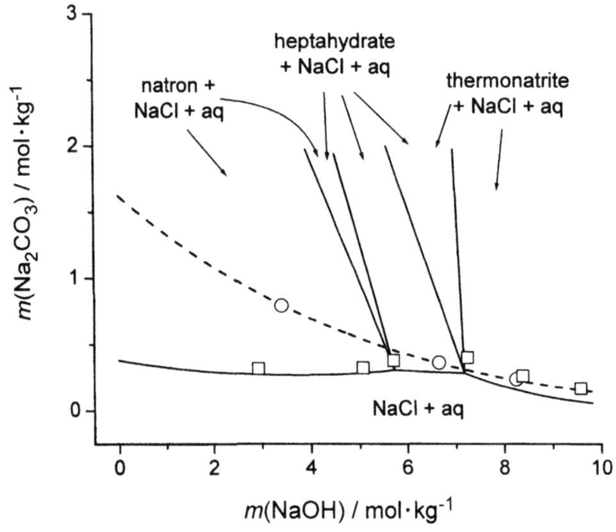

Fig. 16. Isothermal phase diagrams of the $Na_2CO_3-NaCl-NaOH-H_2O$ system at NaCl saturation as calculated from the present model; experimental data are from Ref. [81]; circles, dashed line, 60°C; squares, solid lines, 0°C

temperatures, assigning a value to ψ_{Na,OH,CO_3} (which was not needed for the Na_2CO_3–NaOH–H_2O subsystem) may be more appropriate. This obviously points at the limitations of the present method with its temperature independent heat capacities for aqueous and solid phases. Using the flexibility of the *Pitzer* model, a full optimization of the system could be performed (*i.e.* by taking all available experimental information into account). This would lead to complicated temperature functions of standard *Gibbs* energies and *Pitzer* parameters which is, however, beyond the scope of this study.

Discussion and Conclusions

Although *Pitzer* models are constantly improved for a limited number of electrolytes, there is a need for simple models which can use the little information that is available for the vast majority of electrolytes. It has been shown that *Pitzer* parameters for the osmotic coefficient, apparent molar relative enthalpy, and heat capacity, all valid for 25°C, can be employed to predict solubilities in the temperature range from 0 to 100°C. It is most important, however, that the parameters are (*i*) adjusted to data measured up to saturation and then (*ii*) properly extrapolated to higher temperatures. Of the three methods examined here, Eqs. (22) and (23) proposed in this work were found most satisfactory. This new method corresponds to constant heat capacities of the solution, apart from the temperature dependence of the *Debye-Hückel* coefficient A_J. Previously, constant standard heat capacities of reaction have been found appropriate to reproduce standard equilibrium constants very accurately in the temperature range from 0 to 100°C or even higher [19, 92]. Also, the so-called density model, in its modification by *Anderson et al.* [93], is essentially a constant heat capacity model at these temperatures.

Solubility predictions for complex systems are quite reasonable when temperature independent ternary *Pitzer* parameters are used, but they can certainly be improved when some temperature dependence of the ψ parameters is introduced. As the latter is usually fitted to solubility data [6], one must be careful not to use the same data for parametrization and then for proving the predictive capabilities of the model. In general, solubilities in complex media can be used to validate models, as calculated solubility values are very sensitive to changes in the ternary interaction parameters.

Lu et al. [94, 95] have recently proposed a *semi*-empirical model that uses ion-specific parameters and a universal relation for the influence of temperature on some model parameters which is obtained by a least squares fit to experimental activity and osmotic coefficients of 13 electrolytes between 50 and 300°C. Also, four temperature coefficients for each solubility product have been fitted to salt solubilities in water. These authors claim that their model can predict osmotic and activity coefficients, apparent molar enthalpies, and solubilities (the latter in mixed electrolyte solutions) over wide ranges of temperatures. In the present communication, the word 'prediction' has a different meaning, as no fitting to properties as a function of temperature is involved.

The ChemSage software used in this work is a true facility for the analysis of chemical thermodynamics. It is capable of calculating solubilities in complex media, integral and partial thermodynamic quantities of phases, and complete phase diagrams quickly and conveniently. This software is also invaluable for studying the interrelations and correlations among the various interaction parameters, thermodynamic quantities of solids, and experimental data. Thereby, different scenarios can be designed that help to identify the conditions for which new solubility measurements should be performed. Such suggestions have particularly been made regarding the measurement of sodium carbonate solubilities in various salt solutions. For this task, a recently developed solubility apparatus by *Capewell et al.* [96] could be favourably used.

The results of the present work make it seem feasible that the currently available information on *Pitzer* parameters at 25°C can be employed to establish databases for solubility predictions in complex systems from 0°C to 100°C.

Acknowledgements

I am grateful to Profs. *Peter May* and *Glenn Hefter* (Murdoch University, Western Australia) for stimulating discussions on solution chemistry and thermodynamic modelling. Prof. *May* also pointed out the significance of the density model to the present work.

References

[1] Primas H (1983) Chemistry, Quantum Mechanics and Reductionism, 2nd edn. Springer, Berlin
[2] Pitzer KS (1991) In: Pitzer KS (ed) Activity Coefficients in Electrolyte Solutions, 2nd edn. CRC Press, Boca Raton, p 75
[3] Harvie CE, Weare JH (1980) Geochim Cosmochim Acta **44**: 981
[4] Harvie CE, Møller N, Weare JH (1984) Geochim Cosmochim Acta **48**: 723
[5] Monnin C, Schott J (1984) Geochim Cosmochim Acta **48**: 571
[6] Pabalan RT, Pitzer KS (1987) Geochim Cosmochim Acta **51**: 2429

[7] Pabalan RT, Pitzer KS (1991) In: Pitzer KS (ed) Activity Coefficients in Electrolyte Solutions, 2nd edn. CRC Press, Boca Raton, p 435
[8] Møller N (1988) Geochim Cosmochim Acta **52**: 821
[9] Greenberg JP, Møller N (1989) Geochim Cosmochim Acta **53**: 2503
[10] Spencer RJ, Møller N, Weare JH (1990) Geochim Cosmochim Acta **54**: 575
[11] Monnin C (1995) J Chem Eng Data **40**: 828
[12] Monnin C (1999) Chem Geol **153**: 187
[13] Marion GM, Farren RE (1999) Geochim Cosmochim Acta **63**: 1305
[14] Kaufman L, Bernstein H (1970) Computer Calculation of Phase Diagrams. Academic Press, New York
[15] Königsberger E, Eriksson G (1995) CALPHAD **19**: 207
[16] Gamsjäger H, Königsberger E, Preis W (1998) Pure Appl Chem **70**: 1913
[17] Gamsjäger H, Preis W, Königsberger E, Magalhaes MCF, Brandao P (1999) J Solution Chem **28**: 711
[18] Preis W, Königsberger E, Gamsjäger H (2000) J Solution Chem **29**: 605
[19] Königsberger E, Königsberger L-C, Gamsjäger H (1999) Geochim Cosmochim Acta **63**: 3105
[20] Königsberger E, Königsberger L-C, Gamsjäger H (2001) Geochim Cosmochim Acta **65**: 183
[21] Königsberger E, Gamsjäger H (1987) Ber Bunsenges Phys Chem **91**: 785
[22] Königsberger E, Schmidt P, Gamsjäger H (1992) J Solution Chem **21**: 1195
[23] Nguyen AM, Königsberger E, Marhold H, Gamsjäger H (1993) Monatsh Chem **124**: 1011
[24] Gamsjäger H, Marhold H, Königsberger E, Tsai YJ, Kolmer H (1995) Z Naturforsch **50a**: 59
[25] Königsberger E, Tran-Ho L-C, Gamsjäger H (1998) Monatsh Chem **129**: 1061
[26] Gamsjäger H, Königsberger E, Preis W (2000) Aquatic Geochem **6**: 119
[27] Königsberger E, Eriksson G (1999) J Solution Chem **28**: 721
[28] Grenthe I, Plyasunov AV, Spahiu K (1997) In: Grenthe I, Puigdomenech I (eds) Modelling in Aquatic Chemistry, Nuclear Energy Agency. OECD, Paris, p 325
[29] Pitzer KS (1973) J Phys Chem **77**: 268
[30] Mesmer RE, Palmer DA, Simonson JM, Holmes HF, Ho PC, Wesolowski DJ, Gruszkiewicz MS (1997) Pure Appl Chem **69**: 905
[31] Silvester LF, Pitzer KS (1978) J Solution Chem **7**: 327
[32] Busey RH, Holmes HF, Mesmer RE (1984) J Chem Thermodyn **16**: 343
[33] Holmes HF, Mesmer RE (1986) J Chem Thermodyn **18**: 263
[34] Hovey JK, Pitzer KS, Rard JA (1993) J Chem Thermodyn **25**: 173
[35] Simonson JM, Mesmer RE, Rogers PSZ (1989) J Chem Thermodyn **21**: 561
[36] Holmes HF, Busey RH, Simonson JM, Mesmer RE, Archer DG, Wood RH (1987) J Chem Thermodyn **19**: 863
[37] Ananthaswamy J, Atkinson G (1985) J Chem Eng Data **30**: 120
[38] Archer DG (1991) J Phys Chem Ref Data **20**: 509
[39] Archer DG (1992) J Phys Chem Ref Data **21**: 793
[40] Clegg SL, Rard JA, Pitzer KS (1994) J Chem Soc Faraday Trans **90**: 1875
[41] Pitzer KS, Wang P, Rard JA, Clegg SL (1999) J Solution Chem **28**: 265
[42] Silvester LF, Pitzer KS (1977) J Phys Chem **81**: 1822
[43] Bradley DJ, Pitzer KS (1979) J Phys Chem **83**: 1599
[44] Archer DG, Wang P (1990) J Phys Chem Ref Data **19**: 371
[45] Polya DA, Woolley EM, Simonson JM, Mesmer RE (2001) J Chem Thermodyn **33**: 205
[46] Pitzer KS, Peiper JC, Busey RH (1984) J Phys Chem Ref Data **13**: 1
[47] Pitzer KS, Mayorga G (1973) J Phys Chem **77**: 2300
[48] Pitzer KS, Mayorga G (1974) J Solution Chem **3**: 539

[49] Criss CM, Millero FJ (1996) J Phys Chem **100**: 1288
[50] Criss CM, Millero FJ (1999) J Solution Chem **28**: 849
[51] Peiper JC, Pitzer KS (1982) J Chem Thermodyn **14**: 613
[52] Roy RN, Gibbons JJ, Williams R, Godwin L, Baker G, Simonson JM, Pitzer KS (1984) J Chem Thermodyn **16**: 303
[53] Bale CW, Pelton AD (1999) FACT-Win – User Manual. CRCT, Ecole Polytechnique de Montréal, Québec, Canada (http://www.crct.polymtl.ca)
[54] Pelton AD (2000) personal communication
[55] Eriksson G, Hack K (1990) Metall Trans **21B**: 1013
[56] Helgeson HC, Kirkham DH, Flowers GC (1981) Am J Sci **281**: 1249
[57] Clarke ECW, Glew DN (1985) J Phys Chem Ref Data **14**: 489
[58] Parker VB (1965) Thermal Properties of Aqueous Uni-univalent Electrolytes, US Nat Bur Standards, NSRDS-NBS 2
[59] Wagman DD, Evans WH, Parker VB, Schumm RH, Halow I, Bailey SM, Churney KL, Nuttall RL (1982) J Phys Chem Ref Data **11** (Supplement 2)
[60] Williamson AT (1944) Trans Faraday Soc **40**: 421
[61] Cox JD, Wagman DD, Medvedev VA (1989) CODATA Key Values for Thermodynamics. Hemisphere Publishing, New York
[62] Archer DG (1997) J Chem Eng Data **42**: 281
[63] Vanderzee CE (1982) J Chem Thermodyn **14**: 219
[64] Tsonopoulos C (1974) AIChE J **20**: 273
[65] Tanner JE, Lamb FW (1978) J Solution Chem **7**: 303
[66] Perron G, Roux AH, Desnoyers JE (1981) Can J Chem **59**: 3049
[67] Saluja PPS, Jobe DJ, LeBlanc JC, Lemire RJ (1995) J Chem Eng Data **40**: 398
[68] Hubert N, Solimando R, Pere A, Schuffenecker L (1997) Thermochim Acta **294**: 157
[69] Gates JA, Tillett DM, White DE, Wood RH (1987) J Chem Thermodyn **19**: 131
[70] Cohen-Adad R, Lorimer JW (1991) Alkali Metal and Ammonium Halides in Water and Heavy Water (Binary Systems), IUPAC Solubility Data Series, vol 47. Pergamon Press
[71] Kron I, Marshall SL, May PM, Hefter G, Königsberger E (1995) Monatsh Chem **126**: 819
[72] Busey RH, Mesmer RE (1976) J Solution Chem **5**: 147
[73] Palmer DA, Drummond SE (1988) J Solution Chem **17**: 153
[74] Pitzer KS, Kim JJ (1974) J Am Chem Soc **96**: 5701
[75] Thurmond V, Millero FJ (1982) J Solution Chem **11**: 447
[76] Becker P, Bilal BA (1985) J Solution Chem **14**: 367
[77] Bezboruah CP, Filomen J, Camoes GFC, Covington AK, Dobson JV (1973) J Chem Soc Faraday Trans 1 **69**: 949
[78] Chen X, Oscarson JL, Gillespie SE, Cao H, Izatt RM (1994) J Solution Chem **23**: 747
[79] Sweeton FH, Mesmer RE, Baes CF Jr (1974) J Solution Chem **3**: 191
[80] Vanderzee CE (1982) J Chem Thermodyn **14**: 1051
[81] Linke WF and Seidell A (1965) Solubilities. Inorganic and Metal-Organic Compounds, vol 2. American Chemical Society
[82] Berg RL, Vanderzee CE (1978) J Chem Thermodyn **10**: 1113
[83] Berg RL, Vanderzee CE (1978) J Chem Thermodyn **10**: 1049
[84] Perron G, Desnoyers JE, Millero FJ (1975) Can J Chem **53**: 1134
[85] Chernen'kaya EI (1971) Zh Prikl Khim **44**: 1543
[86] Aseyev GG (1996) Thermal Properties of Electrolyte Solutions. Begell House, New York
[87] Hepler LG, Hovey JK (1996) Can J Chem **74**: 639
[88] Königsberger E, Gamsjäger H (1990) Monatsh Chem **121**: 119
[89] Königsberger E (1991) CALPHAD **15**: 73
[90] Pelton AD, Schmalzried H (1973) Met Trans **4**: 1395
[91] Königsberger E, Gamsjäger H (1991) Ber Bunsenges Phys Chem **95**: 734

[92] Puigdomenech I, Rard JA, Plyasunov AV, Grenthe I (1997) In: Grenthe I, Puigdomenech I (eds) Modelling in Aquatic Chemistry. Nuclear Energy Agency, OECD, Paris, p 427
[93] Anderson GM, Castet S, Schott J, Mesmer RE (1991) Geochim Cosmochim Acta **55**: 1769
[94] Lu X, Zhang L, Wang Y, Shi J, Maurer G (1996) Ind Eng Chem Res **35**: 1777
[95] Lu X, Zhang L, Wang Y, Shi J (1996) Fluid Phase Equil **116**: 201
[96] Capewell SG, Hefter GT, May PM (1999) Rev Sci Instrum **70**: 1481

Received April 20, 2001. Accepted May 2, 2001

Invited Review

Preferential Solvation in Mixed Solvents X. Completely Miscible Aqueous Co-Solvent Binary Mixtures at 298.15 K

Yizhak Marcus*

Department of Inorganic and Analytical Chemistry, The Hebrew University of Jerusalem, Jerusalem 91904, Israel

Summary. The *Kirkwood-Buff* integrals for 18 completely miscible aqueous co-solvent binary mixtures have been recalculated from thermodynamic data, and the volume-corrected preferential solvation parameters derived from them are presented. Also presented are these latter quantities for 15 additional such mixtures, for which the volume correction has not been applied previously. The self-interaction of the water, the mutual interaction of the water and the co-solvent, and the self-interaction of the co-solvent at infinite dilution derived from these integrals and parameters are then discussed. The systems studied include aqueous hydrogen peroxide, methanol, ethanol, 1- and 2-propanol, 2-methyl-2-propanol, 2,2,2-trifluoroethanol, 1,1,1,3,3,3-hexafluoro-2-propanol, ethane-1,2-diol, glycerol, 2-methoxyethanol (at 313 and 343 K), 2-ethoxyethanol, 2-butoxyethanol, 2-aminoethanol, N-methyl- and N,N-dimethyl-2-aminoethanol, tetrahydrofuran, 1,4-dioxane, acetone, formic, acetic, and propanoic acids, piperidine, pyridine, acetonitrile, formamide, N-methyl- and N,N-dimethylformamide, N-methylacetamide, N-methylpyrrolidin-2-one (at 303 K), hexamethyl phosphoric triamide, dimethylsulfoxide, and tetramethylenesulfone (at 303 K).

Keywords. Binary aqueous solvents; Preferential solvation; *Kirkwood-Buff* integrals.

Introduction

A fairly large number of preferential solvation studies employed the *Kirkwood-Buff* integrals, G_{ij},

$$G_{ij} = \int_0^\infty (g_{ij} - 1)4\pi r^2 dr \tag{1}$$

where g_{ij} is the pair correlation function, denoting the probability of finding a molecule of species i in a volume element at the distance r from the center of a molecule of species j. The product $\rho_i \cdot G_{ij}$, where ρ_i is the number density of molecules of species i in the bulk of the mixture, represents the excess, if positive, or deficiency, if negative, of molecules of species i in the space around a given

* E-mail: ymarcus@vms.huji.ac.il

molecule of species j. An excess means attractive interactions between molecules of i and j, *i.e.* preferential mutual solvation, whereas a deficiency means that self-interactions of i-i and j-j are preferred to mutual interactions. These integrals have been obtained for binary aqueous solvent mixtures (where in the following subscript W signifies water and subscript S the co-solvent; i and j represent any component) mainly from thermodynamic data [1–20] (where Ref. [20] constitutes Part 9 of this series), but also from structural fluctuation measurements [21, 22], and have been employed to describe preferential solvation.

From thermodynamic data the integrals are calculated using the following expressions:

$$G_{WS} = RT(\kappa_T - V_W V_S / VD) \tag{2}$$

$$G_{WW} = RT(\kappa_T + (x_S/(1-x_S))(V_S^2/VD)) - V/(1-x_S) \tag{3}$$

$$G_{SS} = RT(\kappa_T + ((1-x_S)/x_S)(V_W^2/VD)) - V/x_S \tag{4}$$

where κ_T is the isothermal compressibility of the mixture, x_s is the mole fraction of the co-solvent, V_W and V_S are the partial molar volumes of the components in the mixture, and V is the molar volume of the mixture. The function D is (assuming water to be the by far more volatile component):

$$D = RT(1 + (1-x_S)(\partial \ln f_W / \partial(1-x_S))_{T,P}) \tag{5}$$

where $f_W = p_W/(1-x_S)p_W^\circ$ is the activity coefficient of water and p_W and p_W° are its fugacities above the mixture and pure water, respectively. Partial pressures can be used instead of fugacities if they are sufficiently low; otherwise, they must be corrected for the vapor virial coefficients. A corresponding expression involving $x_S(\partial \ln f_S/\partial x_S)_{T,P}$ can, of course, be employed too. A less desirable alternative is the use of the second derivative of the excess *Gibbs* energy of the mixture, G^E,

$$D = RT + x_S(1-x_S)(\partial^2 G^E / \partial x_s^2)_{T,P} \tag{6}$$

because of the double derivation, so that very accurate G^E values have to be employed. On the other hand, the partial molar volumes can be readily calculated sufficiently accurate from those of the pure components and the excess molar volume of the mixtures, V^E (that is invariably negative for aqueous mixtures), and its first derivative with respect to the composition. Values of the isothermal compressibility of the mixtures, κ_T, as a function of the composition are not generally known. Instead, since their contribution to G_{ij} is quite small, linearly interpolated values between the κ_T of the pure components or values of the adiabatic compressibility of the mixtures, κ_S, available from measurements of ultrasound speed, can be employed. The latter values might be increased by 5 to 10% to approximate the κ_T values: $\kappa_T = \kappa_S + VT/\alpha_P^2 C_P$, where α_P and C_P are the isobaric expansivity and molar heat capacity.

In only a fraction of these studies preferential solvation parameters, δx_{ij}, or local compositions around molecules of the components, x_{ij}^L, have been calculated. In some of the latter investigations, only the limiting preferential solvation parameters, δx_{ij}°, were reported. In these quantities the first subscript index pertains to a component in the environment of a molecule of the component designated by the second index for both $i = j$ and $i \neq j$. The preferential solvation takes place in a concentric region around each molecule in the mixture, designated as the correlation volume, V_{cor} (the region where $g_{ij} \neq 1$, see Eq. (1)).

The preferential solvation parameters and local compositions, δx_{ij} and x^{L}_{ij}, are related to the *Kirkwood-Buff* integrals and to V_{cor} as follows:

$$\delta x_{WW} = x^{L}_{WW} - x_{W} = x_{W}x_{S}(G_{WW} - G_{SW})/(x_{W}\ G_{WW} + x_{S}\ G_{WS} + V_{cor,W}) \quad (7a)$$

$$\delta x_{WS} = x^{L}_{WS} - x_{W} = x_{W}x_{S}(G_{WS} - G_{SS})/(x_{W}\ G_{WS} + x_{S}\ G_{SS} + V_{cor,S}) \quad (7b)$$

It should also be remembered that $\delta x_{SW} = -\delta x_{WW}$ and $\delta x_{SS} = -\delta x_{WS}$. Positive values of δx_{WS} denote favoured mutual preferential solvation of water and the co-solvent, concomitant with disfavoured self-interaction of the co-solvent molecules; positive values of δx_{WW} denote preferential self-interaction of the water molecules and are concomitant with disfavoured mutual interaction of water with the co-solvent molecules.

The correlation volume, V_{cor}, is calculated for consecutive solvation shells, taking into account partial penetration of molecules from farther shells into nearer ones as well as the preferential solvation in these shells. Hence, the calculation must be carried out iteratively [10, 11, 15, 17, 20].

$$V_{cor,W}/\mathrm{cm}^3 \cdot \mathrm{mol}^{-1} = 2522.7 \cdot (-0.085m + (0.1363/2)V_W^{1/3} + 0.1363 \cdot (m - 0.5) \cdot (x^{L}_{WW}V_W + (1 - x^{L}_{WW})V_S)^{1/3})^3 \quad (8a)$$

$$V_{cor,S}/\mathrm{cm}^3 \cdot \mathrm{mol}^{-1} = 2522.7 \cdot (-0.085m + (0.1363/2)V_S^{1/3} + 0.1363 \cdot (m - 0.5) \cdot (x^{L}_{WS}V_W + (1 - x^{L}_{WS})V_S)^{1/3})^3 \quad (8b)$$

The numerical coefficients arise from the relationship between the molar volume and the mean molecular diameter, *d*, of the solvent according to *Kim* [71]:

$$d/\mathrm{nm} = -0.085 + 0.1363 \cdot (V/\mathrm{cm}^3 \cdot \mathrm{mol}^{-1})^{1/3} \quad (8c)$$

m designates the sequential number of the shell to the limits of which V_{cor} is calculated. The larger the values of V_{cor} in the denominator of Eq. (7), the smaller are the δx_{ij} values, becoming indistinguishable from zero when sufficiently many solvation shells are included (large correlation volumes). The partial molar volumes in Eq. (8) can be replaced by those of the pure components without loss of accuracy.

A feature to be noted in the application of Eq. (7) is that even for ideal mixtures (where $G^E = 0, V^E = 0$, and $G_{ii} + G_{jj} = 2G_{ij}$) the δx_{ij} are non-zero, as if preferential solvation existed in them. This arises from the differences of the molar volumes of the pure components, resulting in *Kirkwood-Buff* integral curves $G_{ii}(x_S)$ and $G_{jj}(x_S)$ that are parallel but shifted by $|V^{\circ}_{j} - V^{\circ}_{i}|$ from the $G_{ij}(x_S)$ curve and satisfying $G_{ii} + G_{jj} - 2G_{ij} = 0$. To overcome this, *Matteoli* [72] suggested the use of volume-corrected values, $\Delta G_{ij} = G_{ij} - G^{id}_{ij}$, where the ideal G^{id}_{ij} are calculated from Eqs. (2)–(4) by setting $D = RT$ and $V^E = 0$. Then, volume corrected preferential solvation parameters $\delta x'_{ij}$ are calculated from Eq. (7) using ΔG instead of G. Due to the differences in the volumes of the components, the actual values of the local mole fractions and the preferential solvation parameters still depend on G rather than on ΔG, but the interpretation in terms of intermolecular interactions should be done with regard to $\delta x'_{ij}$. Values of the non-corrected δx_{ij} for ideal systems with various molar volumes of the co-solvent are shown in Fig. 1.

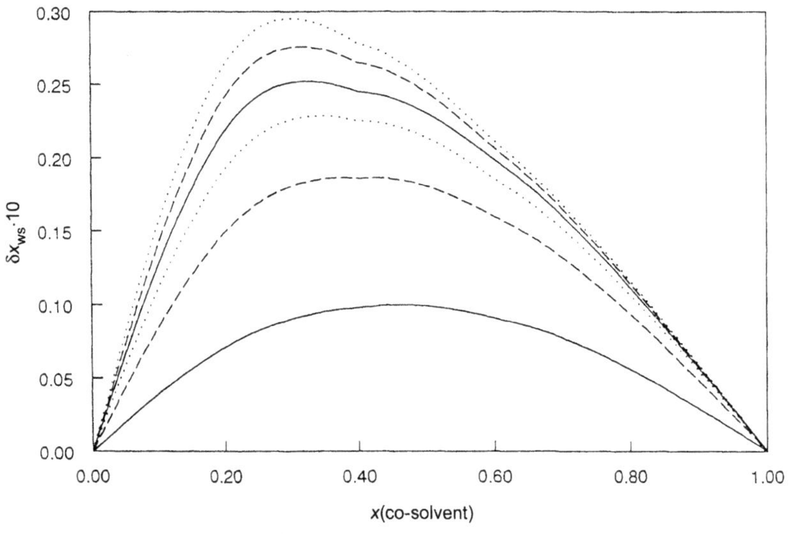

Fig. 1. Preferential solvation parameters, δx_{WS}, in ideal systems with $V_W^\circ = 18.07\,cm^3\,mol^{-1}$ (*i.e.* that for water) and increasing values of V_S°: 30, 50, 70, 90, 120, and $180\,cm^3 \cdot mol^{-1}$ (from the bottom)

The list of completely water-miscible co-solvents of water for which the *Kirkwood-Buff* integrals of their aqueous mixtures have been reported is shown in Table 1. In the following, co-solvents are designated by the abbreviations listed in this table. Most of them have been studied at 298.15 K, but some at other temperatures (*e.g. MEA*, *DEA*, and *TEA* at 273.15 K and *TMS* at 303.15 K) or at several temperatures (*e.g.* MeOH and MeCN). A few co-solvents only partially miscible with water at 298.15 K have also been studied in this respect: 1-butanol [4], 2-butanol [4], 2-methyl-1-propanol [4], propylene carbonate [10], and triethylamine [10].

It is the purpose of this paper to present and discuss the $\delta x'_{ij}$ values of those binary aqueous solvent systems that have already been subjected to *Kirkwood-Buff* integral calculations for 298.15 K (Table 1), but for which even the non-corrected δx_{WS} preferential solvation parameters have not been reported over the entire composition range by taking into account the correlation volumes V_{cor}. These systems include aqueous mixtures of EtOH, *n*-PrOH, *t*-BuOH, *TFE*, *HF-i*-PrOH, MeOEtOH, EtOEtOH, BuOEtOH, *MEA*, Me_2NEtOH, *THF*, Dioxane, Me_2CO, *Pip*, *Py*, and *DMSO*. For the sake of completeness, N-methylethanolamine (MeNHEtOH) and N-methylpyrrolidin-2-one (N*MPy*), for which no *Kirkwood-Buff* integrals have been reported so far, are included. For those systems where δx_{ij} has been presented previously but not corrected for the volume effect, new $\delta x'_{ij}$ values have been calculated; these are also presented. As mentioned above, these parameters are a direct quantitative measure of how much (in terms of mole fractions) of the designated component i is accumulated near (if positive) or banished away from (if negative) the component j. Altogether, figures for 33 systems are shown. In a subsequent paper, other aqueous co-solvent mixtures of interest will be dealt with.

Table 1. Aqueous solvent mixtures for which *Kirkwood-Buff* integrals have been obtained at 298.15 K (except where noted otherwise in the text)

Co-solvent	Abbreviation	References	$G_{SS}^{\infty}/cm^3 \cdot mol^{-1}$
Hydrogen peroxide	H_2O_2	8, 10	−53
Methanol	MeOH	2, 4, 13, 17, 18	−45
Ethanol	EtOH	1, 2, 4, 13, 18, 22	−50
1-Propanol	*n*-PrOH	4, 12, 13, 18	−46
2-Propanol	*i*-PrOH	10, 18	40 (?)
2-Methyl-2-propanol	*t*-BuOH	3, 4, 13, 18	−132
2,2,2-Trifluoroethanol	*TFE*	9	−61
1,1,1,3,3,3-Hexafluoro-2-propanol	*HF-i*-PrOH	9	−89
Ethane-1,2-diol	*EG*	12, 19	−98
Glycerol	*Gly*	20	−95
2-Methoxyethanol	MeOEtOH	12	−230
2-Ethoxyethanol	EtOEtOH	21	−107
2-Butoxyethanol	BuOEtOH	21	< −500 (?)
2-Aminoethanol	*MEA*	12, 14, 15	−116
Diethanolamine	*DEA*	14, 15	
Triethanolamine	*TEA*	14, 15	
N,N-Dimethyl-2-aminoethanol	Me_2NEtOH	4	−157
Tetrahydrofuran	*THF*	4, 6	−35
1,4-Dioxane	Dioxane	4	−127
Acetone	Me_2CO	4, 8	−76
Formic acid	HCOOH	10	−140
Acetic acid	MeCOOH	10	−8
Propanoic acid	EtCOOH	10	50
Ethane-1,2-diamine	*En*	12	
Piperidine	*Pip*	4	−26
Pyridine	*Py*	4	−146
Acetonitrile	MeCN	4, 7, 1	155
Formamide	*FA*	10	>600 (?)
N-Methylformamide	*NMF*	10	−129
N,N-Dimethylformamide	*DMF*	10, 14	−162
N-Methylacetamide	*NMA*	10, 19	−132
Hexamethylphosphoric triamide	*HMPT*	10	−370
Dimethylsulfoxide	*DMSO*	4, 7	−76
Tetramethylenesulfone	*TMS*	6, 10	5

Sources of Data

Aqueous EtOH

Activity coefficient expressions of both components are available from *Hansen* and *Miller* [23] for 298.15 K over the entire composition range in good agreement with earlier [24, 25] and later [26] data used by others who have calculated the *Kirkwood-Buff* integrals [1–4]. The partial molar volumes are available from the data of *Benson* and *Kiyohara* [27] and adiabatic compressibilities from *D'Arrigo* and *Paparelli* [28] and from *Omori* [29].

Aqueous n-PrOH

Activity coefficients of both components at 298.15 K are available from *Matteoli* and *Lepori* [30] in good agreement with earlier data of *Butler* [31], but leading to G^E values some 8% higher than those interpolated from *Westmeier*'s data [32] at 293.15 and 303.15 K that have not been used. However, the corresponding V^E data of *Westmeier* [32] have been employed to yield the partial molar volumes. The isothermal compressibility was calculated approximately from the compression data of *Moriyoshi* and *Uosaki* [33] as $\kappa_T \approx \Delta V/V_{(P=0.1\,\mathrm{MPa})}\Delta P$.

Aqueous t-BuOH

The G^E values, interpolated for 298.15 K from *Westmeier*'s data [34] at 293.15 and 303.15 K, have not been used, since activity coefficient data for 298.15 K from *Brown* and *Ives* [35] were available that needed to be differentiated only once. However, *Westmeier*'s data [34] for V^E, likewise interpolated, were used to obtain the partial molar volumes. Adiabatic compressibilities were taken from *Sehgal et al.* [36].

Aqueous TFE

The G^E values of *Cooney* and *Morcom* [37] and the V^E values of *Rochester* and *Symonds* [38] at 298.15 K were used. The estimated [39] value of the isothermal compressibility of *TFE* was used to interpolate linearly the values for the aqueous mixtures.

Aqueous HF-i-PrOH

The G^E values of *Blandamer et al.* [9] and the the V^E values of *Rochester* and *Symonds* [38] for 298.15 K were used. No compressibility data were found, but in view of the similarity of the κ_T of *TFE* and EtOH, the value for 2-propanol was used for that of *HF-i*-PrOH with linear interpolation for the values of the aqueous mixtures.

Aqueous MeOEtOH

The activity coefficients of water at 343.15 and 363.15 K were available from *Chiavone-Filho et al.* [40] and were extrapolated to 313.15 K. Excess molar volumes, V^E, and adiabatic compressibility data at 283.15, 298.15, and 313.15 K were available from *Pagé et al.* [41] and were extrapolated to 343.15 K. Therefore, two sets of *Kirkwood-Buff* integrals had to be calculated (for 313.15 and 343.15 K) with the uncertainty of the temperature extrapolations kept in mind.

Aqueous EtOEtOH

Here again the activity coefficients of water at 343.15 and 363.15 *K* were available from *Chiavone-Filho et al.* [40], but in this case there were also H^E data at 298.15 and 303.15 K from *Tamura et al.* [42] that facilitated the extrapolation of G^E to

298.15 K. Data for V^E and κ_T at 298.15 K were available from the latter authors and of V^E also from *Douheret* and *Pal* [43] in reasonable agreement.

Aqueous BuOEtOH

There are older G^E data for 298.15 K and V^E data at 300.15 K from *Scatchard* and *Wilson* [44], but the G^E do not agree with the more recent data of *Koga* [45] whose activity coefficients of water have also been used here. The V^E data of *Douheret* and *Pal* [43] for 298.15 K are in agreement with those of *Scatchard* and *Wilson* [44] at 2 K higher, and both sets could be used. Adiabatic compressibilities data, κ_S, that agreed sufficiently well with the estimated [39] κ_T value of pure BuOEtOH interpolated linearly the values for the aqueous mixtures were taken from *Lara* and *Desnoyers* [46].

Aqueous MEA

Previously, the 273.15 K data were used [14], but now the 298.15 K G^E and V^E data of *Touhara et al.* [47] could be employed. *Hawrylak et al.* [48] provided both V^E and κ_S data at 298.15 K, the former being in agreement with those of *Touhara et al.*

Aqueous MeNHEtOH

The 298.15 K G^E and V^E data of *Touhara et al.* [47] could be employed. No compressibility data were found; therefore, the means of those for *MEA* and Me_2NEtOH from *Hawrylak et al.* [48] were used.

Aqueous Me_2NEtOH

The same sources of data as for *MEA* were used.

Aqueous THF

The water activity coefficient data of *Treiner et al.* [49] and V^E data of *Kiyohara* and *Benson* [50] for 298.15 K were used. Values of κ_S from *Aminabhavi et al.* [51], who provided also V^E data in agreement with those of *Kiyohara* and *Benson* [50], were employed.

Aqueous Diox

Data for G^E at 298.15 K were available from *Goats* and *Sullivan* [52] and were combined with the fewer values calculated from the activities reported by *Vierk* [53] and *Bacarella et al.* [54] Also used were the V^E from *Rebollo et al.* [55] and κ_S data were taken from *Aminabhavi et al.* [51].

Aqueous Me_2CO

Data for G^E at 298.15 K were available from *Orye* and *Prausnitz* [56], but the data of *Villamañan* and *van Ness* [57] obtained at higher temperatures could be extrapolated to 298.15 K by means of the expressions provided by these authors. The two sets are in reasonable agreement, but the latter was used. V^E data were calculated from the densities provided for 298.15 K by *Noda et al.* [58], and compressibility data, κ_T, were taken from *Egorov et al.* [59].

Aqueous Pip

The G^E and V^E data of *Nakanishi et al.* [60] for 298.15 K were employed as well as κ_S values for 293.15 K of *Patil* [61] for the pure solvent that were used to interpolate linearly the values for the aqueous mixtures.

Aqueous Py

The water activity coefficient data of *Chan* and *van Hook* [62] and the G^E data of *Abe et al.* [63] as well as the V^E data of *Abe et al.* [63] for 298.15 K were employed, as were the κ_S data of *Sharma* and *Singh.* [64].

Aqueous NMPy

The G^E data of *Zielenkiewicz* [65] for 303.15 K had to be used, and it was assumed that the V^E data of *Davis* [66] for 298.15 K would still give valid values of the *Kirkwood-Buff* integrals. The κ_S data of *Pal et al.* [67] were employed.

Aqueous DMSO

The activity coefficient data of *Lam* and *Benoit* [68] and those of *Chan* and *van Hook* [69] in agreement with them as well as the V^E data of *de Visser et al.* [70] for 298.15 K and the κ_S data of *Aminabhavi et al.* [51] were employed.

Results

Aqueous EtOH

The *Kirkwood-Buff* integral curves $G_{WW}(x_S)$, $G_{WS}(x_S)$, and $G_{SS}(x_S)$ resemble closely those reported by *Matteoli* and *Lepori* [4] and need not be repeated here. They do not show the rather irregular behavior reported by *Blandamer et al.* [9], however. The resulting volume-corrected preferential solvation parameters for the first solvation shell, $\delta x'_{WW}$ and $\delta x'_{WS}$, are shown in Fig. 2a. The curves show that water tends to aggregate near a water molecule, with a maximum of $\delta x'_{WW} = 0.16$ near $x_S = 0.45$, and that it tends to avoid slightly the vicinity of ethanol, with a minimum of $\delta x'_{WS} = -0.04$ near $x_S = 0.35$.

Aqueous n-PrOH

The *Kirkwood-Buff* integral curves $G_{WW}(x_S)$, $G_{WS}(x_S)$, and $G_{SS}(x_S)$ resemble closely those reported by *Matteoli* and *Lepori* [4] and need not be repeated here. The resulting volume-corrected preferential solvation parameters for the first and second solvation shells, $\delta x'_{WW}$ and $\delta x'_{WS}$, are shown in Fig. 2b. The curves show that

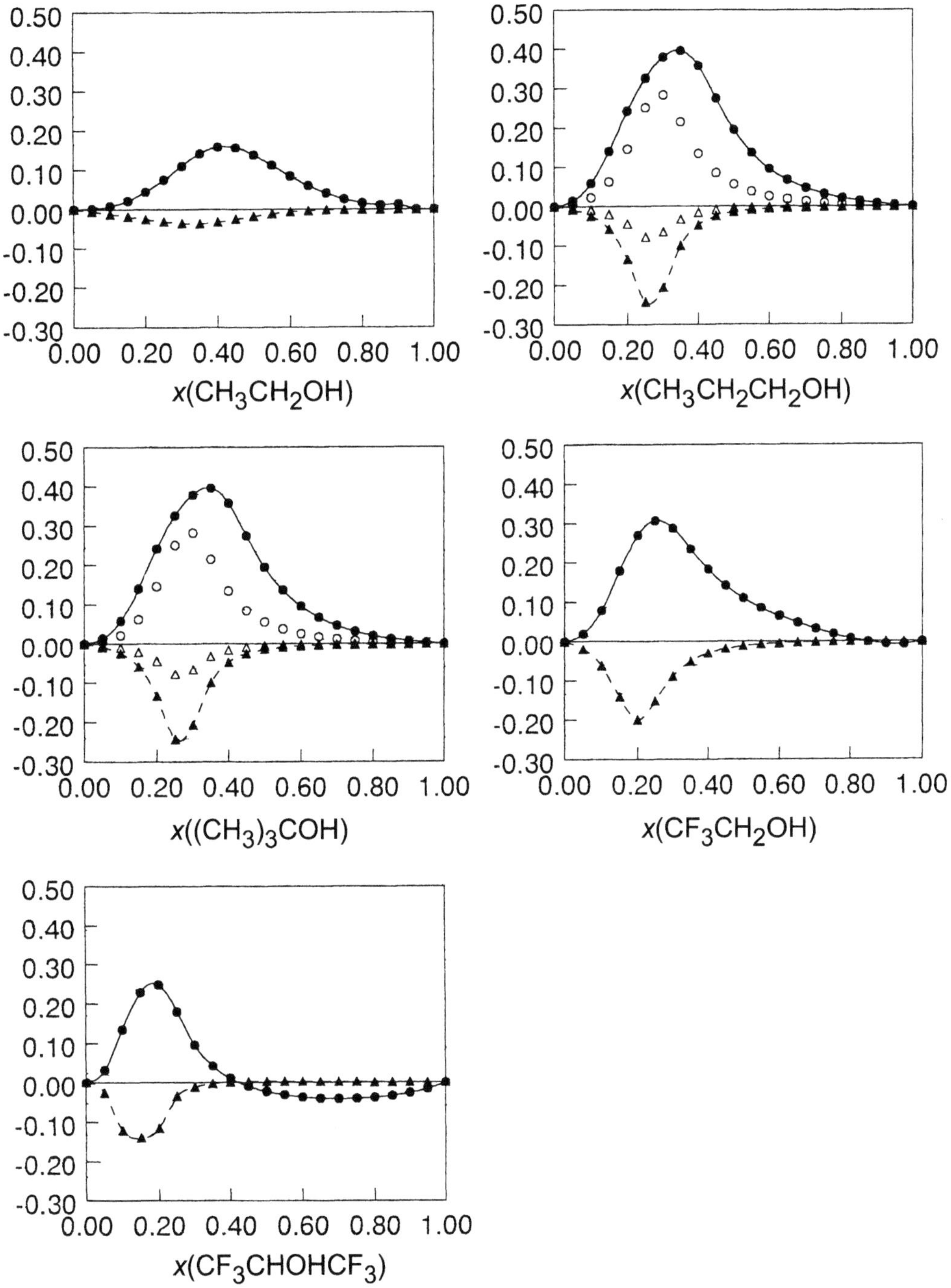

Fig. 2. Volume-corrected preferential solvation parameters $\delta x'_{WW}$ (●) for water-water interactions and $\delta x'_{WS}$ (▲) for water – co-solvent interactions in the first solvation shell at 298.15 K of aqueous a) ethanol, b) 1-propanol (empty symbols for second solvation shell), c) 2-methyl-2-propanol (empty symbols for second solvation shell), d) 2,2,2-trifluoroethanol, and e) 1,1,1,3,3,3-hexafluoro-2-propanol

water tends much more than in the case of aqueous ethanol to aggregate near a water molecule, with a maximum of $\delta x'_{WW} = 0.42$ in the first solvation shell and $\delta x'_{WW} = 0.22$ in the second, near $x_S = 0.30$. Even in the third solvation shell the maximal $\delta x'_{WW}$ value is above 0.1 (not shown), indicating that this system is not far from microheterogeneity. The avoidance of water in the vicinity of *n*-propanol molecule is correspondingly pronounced, the minima of $\delta x'_{WS}$ are -0.28 and -0.09 in the first and second solvation shells near $x_S = 0.25$.

Aqueous t-BuOH

In this case, the *Kirkwood-Buff* integrals differ in magnitude from those reported by *Matteoli* and *Lepori* [4] and do not show the irregularities noted by those authors in very dilute (*i.e.* water-rich) mixtures. The three G_{WS}, G_{WW}, and G_{SS} curves are shown in Fig. 3a. The resulting volume-corrected preferential solvation parameters for the first and second solvation shells, $\delta x'_{WW}$ and $\delta x'_{WS}$, are shown in Fig. 2c. Water tends to aggregate near a water molecule as much as in the case of aqueous *n*-propanol, with a maximum of $\delta x'_{WW} = 0.40$ in the first solvation shell and 0.28 in the second, near $x_S = 0.35$ and 0.30, respectively, and in the third solvation shell (not shown) it still is 0.15. The avoidance of water of the vicinity of a *t*-butanol molecule is also pronounced, the minima of $\delta x'_{WS}$ are -0.24 and -0.08 in the first and second solvation shells near $x_S = 0.25$.

Aqueous TFE

The *Kirkwood-Buff* integral curves $G_{WS}(x_S)$ and $G_{SS}(x_S)$ resemble closely those reported by *Balandamer et al.* [9], but the $G_{WW}(x_S)$ curve is more regular and doesn't show the second hump seen in the latter authors' results. The resulting $\delta x'_{WW}$ and $\delta x'_{WS}$, are shown in Fig. 2d. They lie in between those of ethanol and *n*-propanol, with a maximal $\delta x'_{WW} = 0.31$ near $x_S = 0.25$ and a minimal $\delta x'_{WS} = -0.20$ near $x_S = 0.20$.

Aqueous HF-i-PrOH

In this case, the *Kirkwood-Buff* integral curves $G_{WW}(x_S)$ and $G_{WS}(x_S)$ resemble closely those reported by *Balandamer et al.* [9], but the $G_{SS}(x_S)$ curve is considerably lower ($\sim$250 against $\sim$950 cm^3 mol^{-1}), although of the same general shape. The resulting $\delta x'_{WW}$ and $\delta x'_{WS}$, are shown in Fig. 2e and are somewhat lower (in absolute value) than those for aqueous 2,2,2-trifluoroethanol.

Aqueous MeOEtOH

The *Kirkwood-Buff* integrals $G_{WW}(x_S)$, $G_{WS}(x_S)$, and $G_{SS}(x_S)$ for 313.15 and 343.15 K are shown in Fig. 3b. It has to be noted that the temperature has a negligible effect on $G_{WS}(x_S)$ and $G_{SS}(x_S)$, but does affect $G_{WW}(x_S)$, the curve at 343.15 K being broader and somewhat lower than at 313.15 K. The resulting $\delta x'_{WS}$ are, correspondingly, practically identical at the two temperatures (Fig. 4a), but the

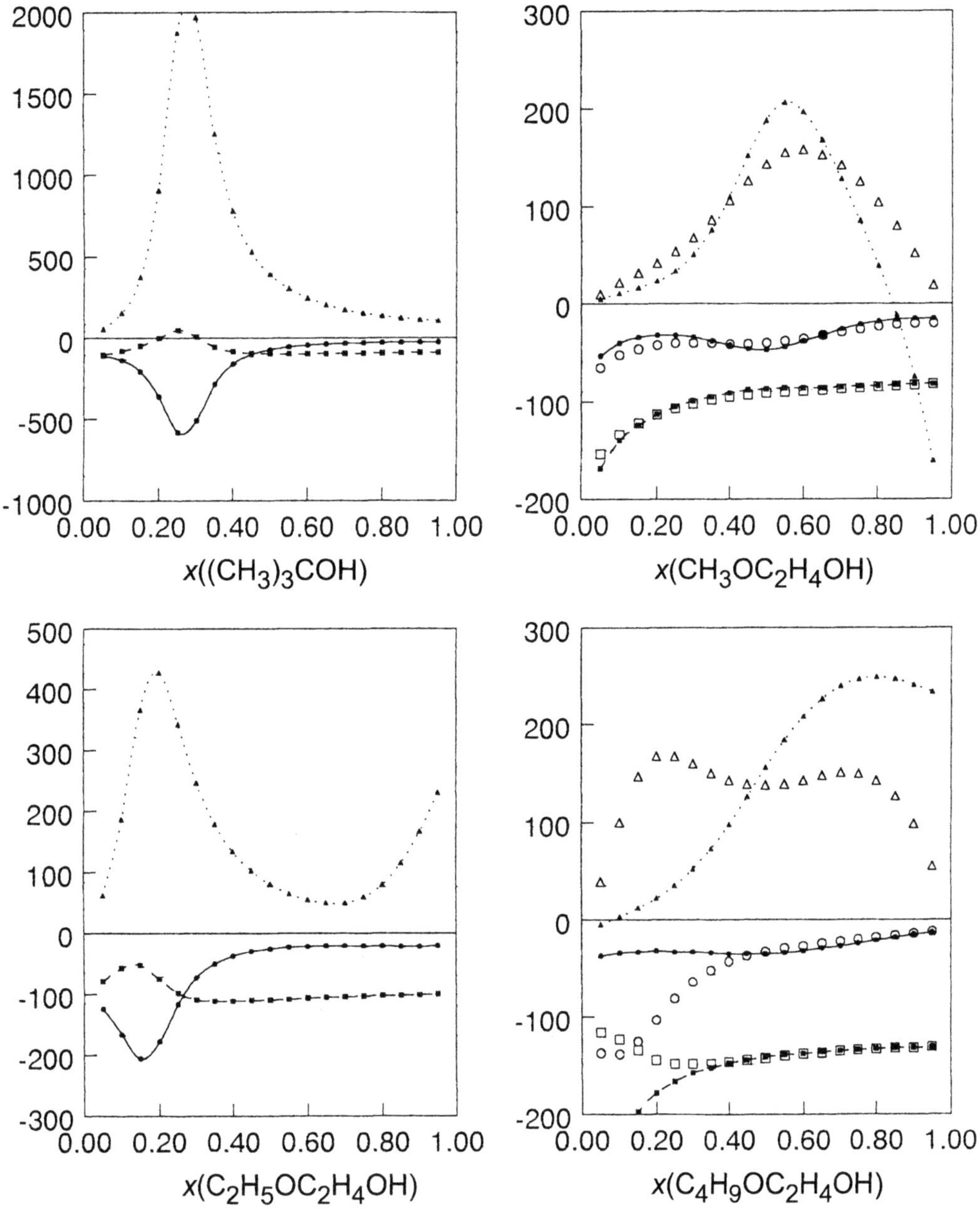

Fig. 3. *Kirkwood-Buff* integrals in $cm^3 \cdot mol^{-1}$ at 298.15 K (except frame b); G_{WW} (▲ and dotted line), G_{SS} (● and continuous line), and G_{SW} (■ and dashed line) of aqueous a) 2-methyl-2-propanol, b) 2-methoxyethanol (filled symbols at 313.15 K, empty symbols at 343.15 K), c) 2-ethoxyethanol, and d) 2-butoxyethanol (filled symbols from *Koga*'s data [45] and empty symbols from *Scatchard* and *Wilson*'s data [44]

$\delta x'_{WW}$ curves differ, that at 343.15 K being again broader and somewhat lower than that at 313.15 K. At the latter temperature, the maximal $\delta x'_{WW} = 0.10$ and the minimal $\delta x'_{WS} = -0.008$, both near $x_S = 0.50$. These values are smaller in absolute value than those for ethanol, showing the effect of the additional (etheral) oxygen atom on the interactions.

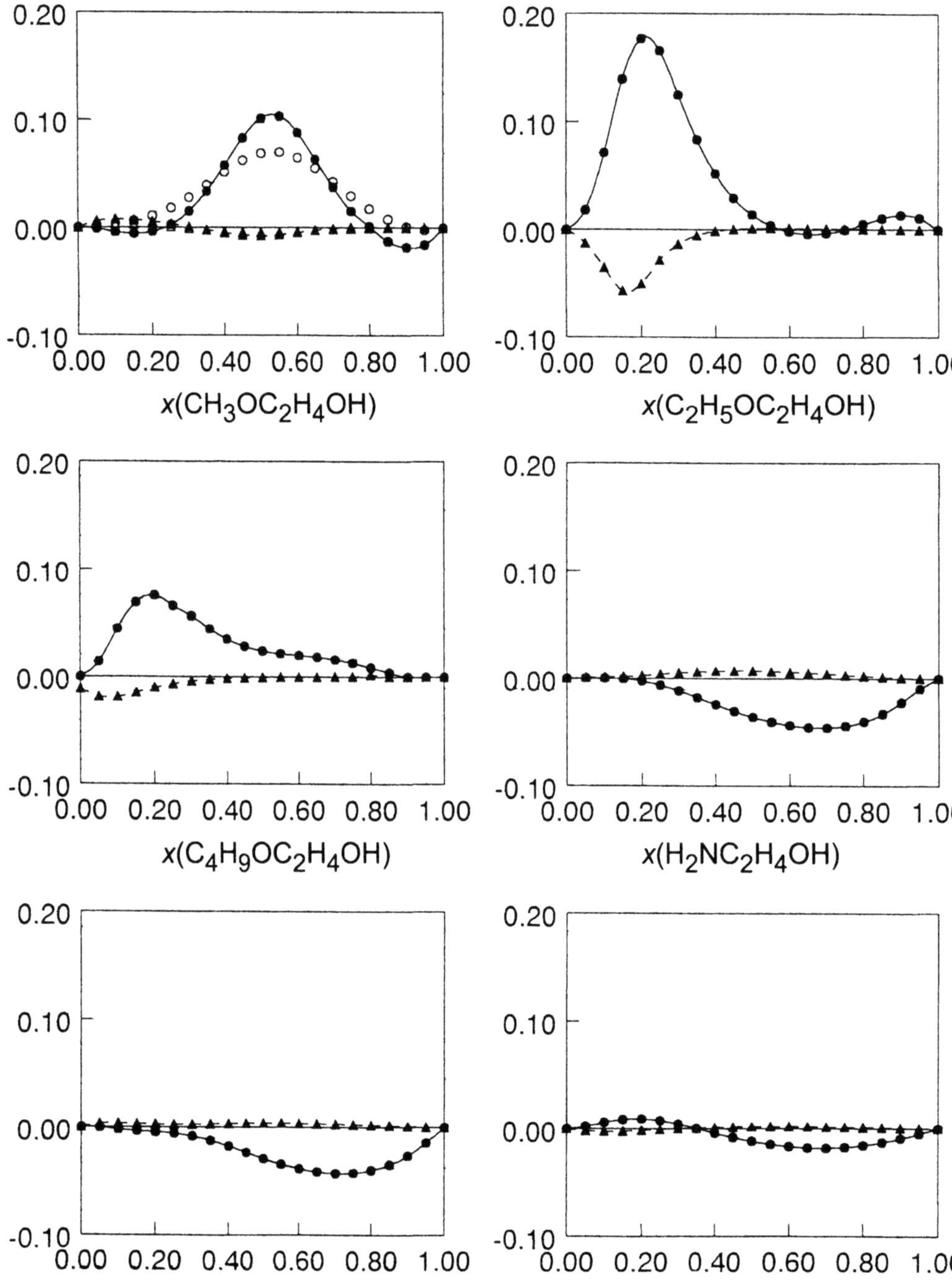

Fig. 4. Volume-corrected preferential solvation parameters $\delta x'_{WW}$ (●) for water-water interactions and $\delta x'_{WS}$ (▲) for water – co-solvent interactions in the first solvation shell at 298.15 K of aqueous a) 2-methoxyethanol (filled symbols at 313.15 K, empty symbols at 343.15 K), b) 2-ethoxyethanol, c) 2-butoxyethanol (from *Koga*'s data [45]), d) 2-aminoethanol, e) 2-N-methylaminoethanol, and f) 2-N,N-dimethylaminoethanol

Aqueous EtOEtOH

The *Kirkwood-Buff* integrals $G_{WW}(x_S)$, $G_{WS}(x_S)$, and $G_{SS}(x_S)$, are shown in Fig. 3c. The upturn of the $G_{WW}(x_S)$ curve at $x_S > 0.7$ is probably an artifact due to inaccuracy of the G^E extrapolation to 298.15 K. Otherwise, the curves show the expected behavior, exhibiting more extreme values than the corresponding curves for 2-methoxyethanol in the more water-rich mixtures, the ethyl group being more hydrophobic than the methyl group. The *Kirkwood-Buff* integral curves obtained by *Kato* from *Rayleigh-Brillouin* scattering [21] are quite different, however, and are deemed not to be correct. The calculated $\delta x'_{WW}$ and $\delta x'_{WS}$ are shown in Fig. 4b. Their extrema are larger than those of 2-methoxyethanol and in more water-rich mixtures: the maximal $\delta x'_{WW} = 0.18$ near $x_S = 0.20$ and the minimal $\delta x'_{WS} = -0.06$ near $x_S = 0.15$.

Aqueous BuOEtOH

The *Kirkwood-Buff* integrals $G_{WW}(x_S)$, $G_{WS}(x_S)$, and $G_{SS}(x_S)$ obtained from both the *Koga* [45] activity coefficient and the *Scatchard* and *Wilson* [44] G^E data are shown in Fig. 3d. Although the $G_{WS}(x_S)$ and $G_{SS}(x_S)$ curves for $x_S \geq 0.4$ from both sources agree with each other, they disagree for more water-rich mixtures, and the two $G_{WW}(x_S)$ curves are completely different. The expected gradual change from methoxy- through ethoxy- to butoxyethanol is not observed for either set of curves for butoxyethanol, although that based on the *Scatchard* and *Wilson* data [44] conform to the qualitative trend (but the extremal values are much lower than expected for the more hydrophobic butyl group), whereas those based on the *Koga* data [45] do not. The *Kirkwood-Buff* integrals reported by *Kato* from *Rayleigh-Brillouin* scattering [21] at 294 K reach only to $x_S = 0.2$ and show very large extrema (*e.g.* $G_{SS}(x_S = 0.05) \approx 13000\,\text{cm}^3 \cdot \text{mol}^{-1}$), completely at odds with those of other homogeneous aqueous mixtures. They would signify phase separation at 294 K that actually takes place only at the considerably higher temperature of about 323 K. The derived $\delta x'_{WW}$ and $\delta x'_{WS}$ values based on the *Scatchard* and *Wilson* data [44] are shown in Fig. 4c. Those based on the *Koga* data [45] (not shown) exhibit negative $\delta x'_{WW}$ and positive $\delta x'_{WS}$ values in water-rich mixtures that are completely counter-intuitive and contrary to the behaviour of other systems. The $\delta x'_{WW}$ and $\delta x'_{WS}$ values are lower than expected but of correct sign and shape and have the maximal $\delta x'_{WW} = 0.075$ near $x_S = 0.20$ and the minimal $\delta x'_{WS} = -0.019$ near $x_S = 0.15$, *i.e.* at more water-rich compositions that for the methoxy- and ethoxy-ethanols as expected.

Aqueous MEA

The *Kirkwood-Buff* integrals $G_{WW}(x_S)$, $G_{WS}(x_S)$, and $G_{SS}(x_S)$ at 298.15 K (not shown) resemble those obtained by the present author [14] at 273.15 K (shown there) but are shifted slightly towards more positive values, and also those reported by *Matteoli* and *Lepori* for 298.15 K [4] except $G_{WW}(x_S)$, these authors' curve turning upward whereas the present one turns downward at $x_S > 0.5$. The calculated $\delta x'_{WW}$ and $\delta x'_{WS}$, are shown in Fig. 4d, agreeing well with those reported for 273.15 K

[14] and are seen to be quite small (the minimal $\delta x'_{WW} = -0.046$ and the maximal $\delta x'_{WS} = 0.007$). The main feature for the system involving the very hydrophilic 2-aminoethanol is that $\delta x'_{WW}$ is negative and $\delta x'_{WS}$ is positive, signifying strong attraction between water and the co-solvent.

Aqueous MeNHEtOH

The *Kirkwood-Buff* integrals $G_{WW}(x_S)$, $G_{WS}(x_S)$, and $G_{SS}(x_S)$ (Fig. 6a) resemble closely those for 2-aminoethanol, but the distances between the curves are somewhat larger, reflecting the larger molar volume of the methyl derivative. The resulting $\delta x'_{WW}$ and $\delta x'_{WS}$ (Fig. 4e) are also very similar to those for 2-aminoethanol, but the extrema are even smaller (the minimal $\delta x'_{WW} = -0.043$ and the maximal $\delta x'_{WS} = 0.004$), signifying the expected somewhat lower hydrophilicity of the methyl derivative.

Aqueous Me_2NEtOH

The *Kirkwood-Buff* integrals $G_{WW}(x_S)$, $G_{WS}(x_S)$, and $G_{SS}(x_S)$ (not shown) are similar to those of the un- and singly-substituted aminoethanols with another widening of the spacing between the curves. The agreement with the curves reported by *Matteoli* and *Lepori* [4] is similar to that noted above for 2-aminoethanol. The resulting $\delta x'_{WW}$ and $\delta x'_{WS}$, shown in Fig. 4f, follow the trend set by the un- and singly-substituted aminoethanols, with $\delta x'_{WW}$ now showing slightly positive values at water-rich compositions, demonstrating that the self association of water in this region overcomes the mutual association due to the two methyl groups.

Aqueous THF

The *Kirkwood-Buff* integrals $G_{WS}(x_S)$, and $G_{SS}(x_S)$, resemble closely those reported by *Matteoli* and *Lepori* [4], but $G_{WW}(x_S)$ of these authors shows a single maximum of $\sim$3500 $cm^3 \cdot mol^{-1}$ at $x_S \approx 0.5$, whereas the present calculation gives a double-humped curve with maxima of 850 and 450 $cm^3 \cdot mol^{-1}$ at $x_S = 0.25$ and 0.75, respectively. Still, even with these lower G_{WW} values, considerable water-water association takes place, the maximal $\delta x'_{WW}$ being 0.33 near $x_S = 0.30$ as shown in Fig. 5a for the first solvation shell. This self association persists to farther shells, as shown in Fig. 5a also for the second one, and even in the fifth shell it is not completely negligible (maximal $\delta x'_{WW} = 0.028$). The mutual association of water with tetrahydrofuran is correspondingly depressed, with the minimal $\delta x'_{WS} = -0.20$ at $x_S = 0.25$, and this persists again to farther solvation shells.

Aqueous Diox

All *Kirkwood-Buff* integrals resemble closely those reported by *Matteoli* and *Lepori* [4] and need not be shown here. The $\delta x'_{WW}$ values show even larger self-association of water than in aqueous tetrahydrofuran, the maximal $\delta x'_{WW} = 0.57$ at $x_S = 0.65$,

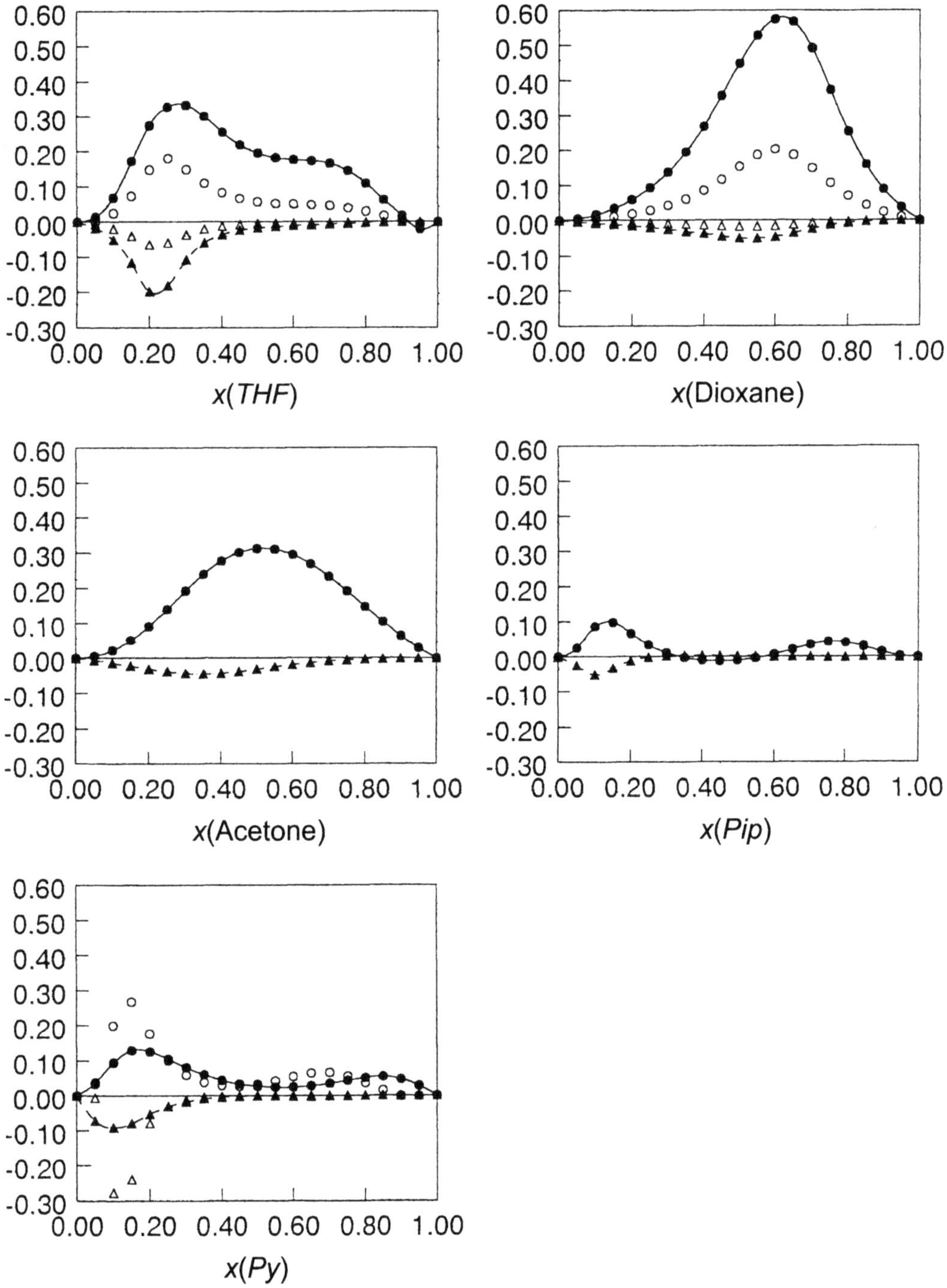

Fig. 5. Volume-corrected preferential solvation parameters $\delta x'_{WW}$ (●) for water-water interactions and $\delta x'_{WS}$ (▲) for water – co-solvent interactions in the first solvation shell at 298.15 K of aqueous a) tetrahydrofuran (empty symbols for second solvation shell), b) 1,4-dioxane (empty symbols for second solvation shell), c) acetone, d) piperidine, and e) pyridine (filled symbols from *Chan*'s data [62] and empty symbols from *Abe*'s data [63]

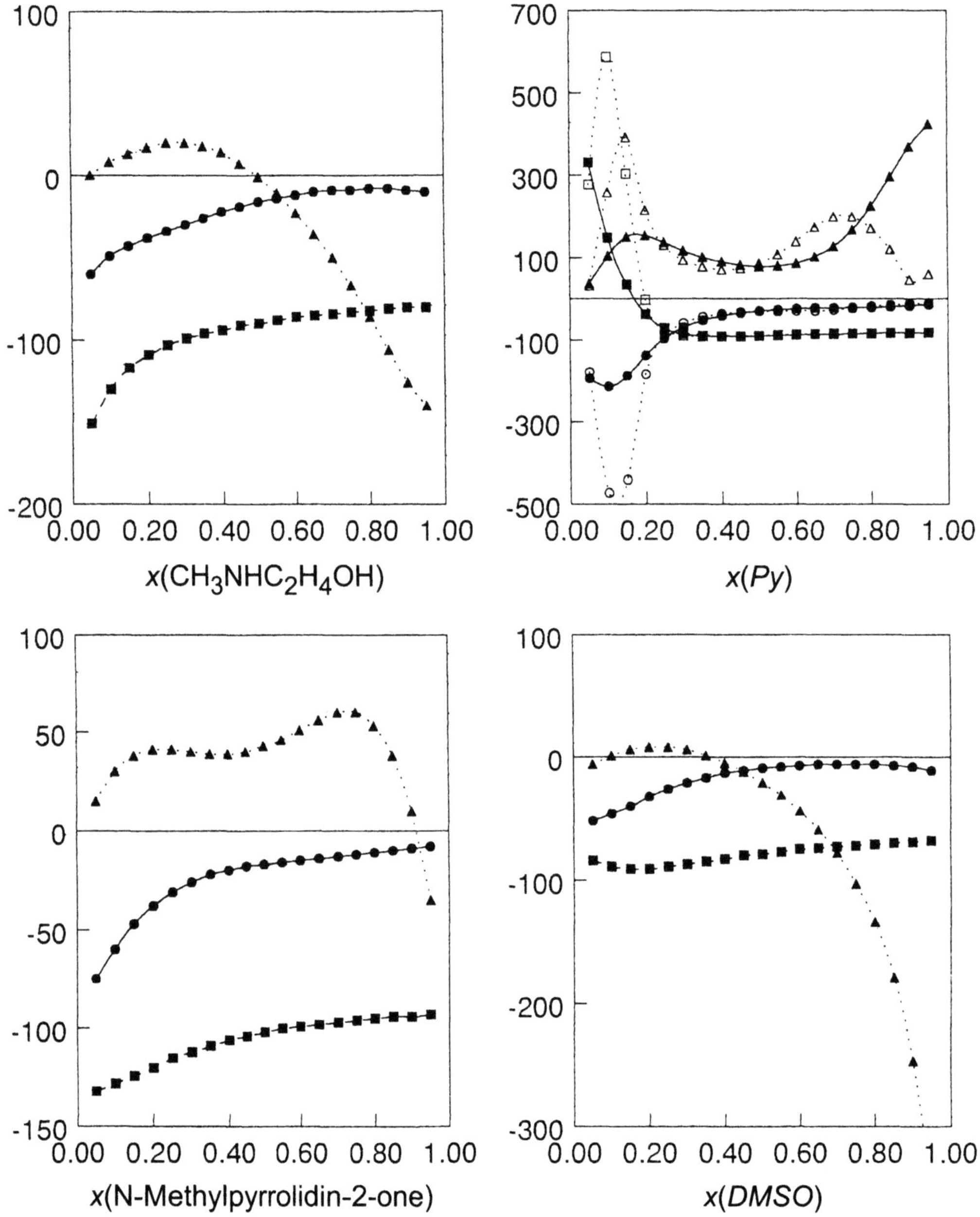

Fig. 6. *Kirkwood-Buff* integrals in $cm^3 \cdot mol^{-1}$ at 298.15 K (except frame b); G_{WW} (▲ and dotted line), G_{SS} (● and continuous line), and G_{SW} (■ and dashed line) of aqueous a) 2-N-methylaminoethanol, b) pyridine (filled symbols from *Chan*'s data [62] and empty symbols from *Abe*'s data [63], c) N-methylpyrrolidin-2-one, and d) dimethylsulfoxide

and again persists to farther solvation shells (see Fig. 5b). However, the depression of the mutual interaction of water and dioxane is not as far reaching as in the case of aqueous tetrahydrofuran, the minimal $\delta x'_{WS} = -0.052$ at $x_S = 0.55$, and hardly persists ($\delta x'_{WS} > -0.007$) in the third solvation shell due to the existence of two oxygen atoms in the dioxane molecule.

Aqueous Me_2CO

All *Kirkwood-Buff* integrals resemble closely those reported by *Matteoli* and *Lepori* [4] and need not be shown here. The $\delta x'_{WW}$ values shown in Fig. 5c are less pronounced than in the ethers, with water-water interactions reaching the maximal $\delta x'_{WW} = 0.31$ near $x_S = 0.5$ and the mutual water-acetone interactions reaching the minimal $\delta x'_{WS} = -0.045$ at $x_S = 0.35$.

Aqueous Pip

The *Kirkwood-Buff* integrals $G_{WW}(x_S)$ and $G_{WS}(x_S)$ resemble closely those reported by *Matteoli* and *Lepori* [4], but $G_{SS}(x_S)$ is considerably lower (maximally 12 rather than 300 $cm^3\,mol^{-1}$), although of the same shape. The present $G_{WW}(x_S)$ curve extends towards higher x_S than the *Matteoli* and *Lepori* curve and confirms the upturn reported there, leading to a second hump. The calculated positive $\delta x'_{WW}$ and negative $\delta x'_{WS}$ are quite small, however, their absolute values being smaller than 0.1 (see Fig. 5d).

Aqueous Py

The two existing sets of data (*Abe et al* [61] for G^E, *Chan* and *van Hook* [61] for the activity coefficients) are inconsistent as far as the derived *Kirkwood-Buff* integrals are concerned; both sets are shown in Fig. 6b. Those obtained from the former data agree with the curves reported by *Matteoli* and *Lepori* [4], whereas those obtained from latter data do not. The derived $\delta x'_{WW}$ and $\delta x'_{WS}$ values differ too, of course, but only for $x_S < 0.4$, and both sets are shown in Fig. 5e. Those obtained from the *Abe et al.* [61] data are about twice as large (in absolute values) than those calculated from the *Chan* and *van Hook* [60] data, and it is difficult to decide which set resembles closer the actual situation. Both sets lead to small ($\delta x' < 0.07$) self- and mutual preferences at $x_S > 0.4$ (Fig. 5e).

Aqueous NMPy

The *Kirkwood-Buff* integrals $G_{WW}(x_S)$, $G_{WS}(x_S)$, and $G_{SS}(x_S)$ have not been reported previously, and the values obtained for 303.15 K are shown in Fig. 6c. The latter two curves resemble closely the expected ones for ideal mixing (Fig. 1), but $G_{WW}(x_S)$ deviates at $x_S > 0.3$. The calculated $\delta x'_{WW}$ and $\delta x'_{WS}$ values are very small, as shown in Fig. 7a, and at $x_S > 0.2$ are negative for $\delta x'_{WW}$ and positive for $\delta x'_{WS}$, signifying slight preference of mutual over self-association in the solvation shells as in the cases of the ethanolamines.

Aqueous DMSO

The *Kirkwood-Buff* integrals $G_{WW}(x_S)$ and $G_{SS}(x_S)$ differ somewhat from those reported by *Matteoli* and *Lepori* [4], and only $G_{WS}(x_S)$ agrees with it; all three curves are shown in Fig. 6d. They are altogether fairly small and lead accordingly to rather small negative $\delta x'_{WW}$ and positive $\delta x'_{WS}$ values (Fig. 7b). As expected, the self interaction of water in the solvation shells is depressed, the minimal $\delta x'_{WW} =$

-0.058 at $x_S = 0.75$, whereas the mutual interaction, with a maximal $\delta x'_{WS} = 0.006$ near $x_S = 0.45$, is small.

Review of Previously Reported Systems

Although the preferential solvation parameters of many additional aqueous co-solvent mixtures have been previously reported, these had not been corrected for the volume effects, so that nearly ideal systems still apparently showed preferential solvation. The preferential solvation in such mixtures is indeed real, due to the discrepancy in molar volumes, but does not reflect the expected similarity of the self- and mutual interactions, leading to no preferential solvation interactions. That is, such systems have very small values of $|\delta x'_{WW}|$ and $|\delta x'_{SW}|$, considered to be insignificant when ≤ 0.01. It is therefore instructive to include here also additional systems to those recorded in Table 1 for the sake of comparison with the information presented above. The thermodynamic data that had been employed previously [10, 11, 14, 17, 20] were used also for the present calculations. The first solvation shell volume-corrected preferential solvation parameters at 298.15 K are shown in Figs. 7–9.

Aqueous H_2O_2

This system is close to ideal (see Fig. 7c). For the disfavoured water self-interactions, the minimal $\delta x'_{WW} = -0.013$ at $x_S = 0.70$, whereas the favoured mutual interaction has a maximal $\delta x'_{WS} = 0.009$ near $x_S = 0.35$, both extrema being quite small, preferential solvation is hardly significant.

Aqueous MeOH

This solvent system has been dealt with in detail over a wide temperature range [17]. For 298.15 K, the resulting volume-corrected preferential solvation parameters are shown in Fig. 7d. For the water self-interactions, the maximal $\delta x'_{WW} = 0.037$ at $x_S = 0.65$, whereas for the mutual interaction the minimal $\delta x'_{WS} = -0.017$ near $x_S = 0.10$. The maximal $\delta x'_{WW}$ value can be compared with those shown above for aqueous ethanol and *n*-propanol: 0.16 and 0.42, showing the expected gradual change.

Aqueous i-PrOH

The results for this system are shown in Fig. 9a, where for the water self-interactions the maximal $\delta x'_{WW} = 0.165$ at $x_S = 0.50$, whereas for the mutual interaction the minimal $\delta x'_{WS} = -0.031$ near $x_S = 0.30$. The preferential solvation is considerably smaller than for *n*-propanol and *t*-butanol (maximal $\delta x'_{WW} = 0.40$) and is comparable with that of ethanol.

Aqueous EG

This system is, as expected, much more nearly ideal than the monoalcohols and is depicted in Fig. 7e. As for aqueous hydrogen peroxide, for the disfavoured water self-interactions the minimal $\delta x'_{WW} = -0.012$ at $x_S = 0.50$, whereas the favoured

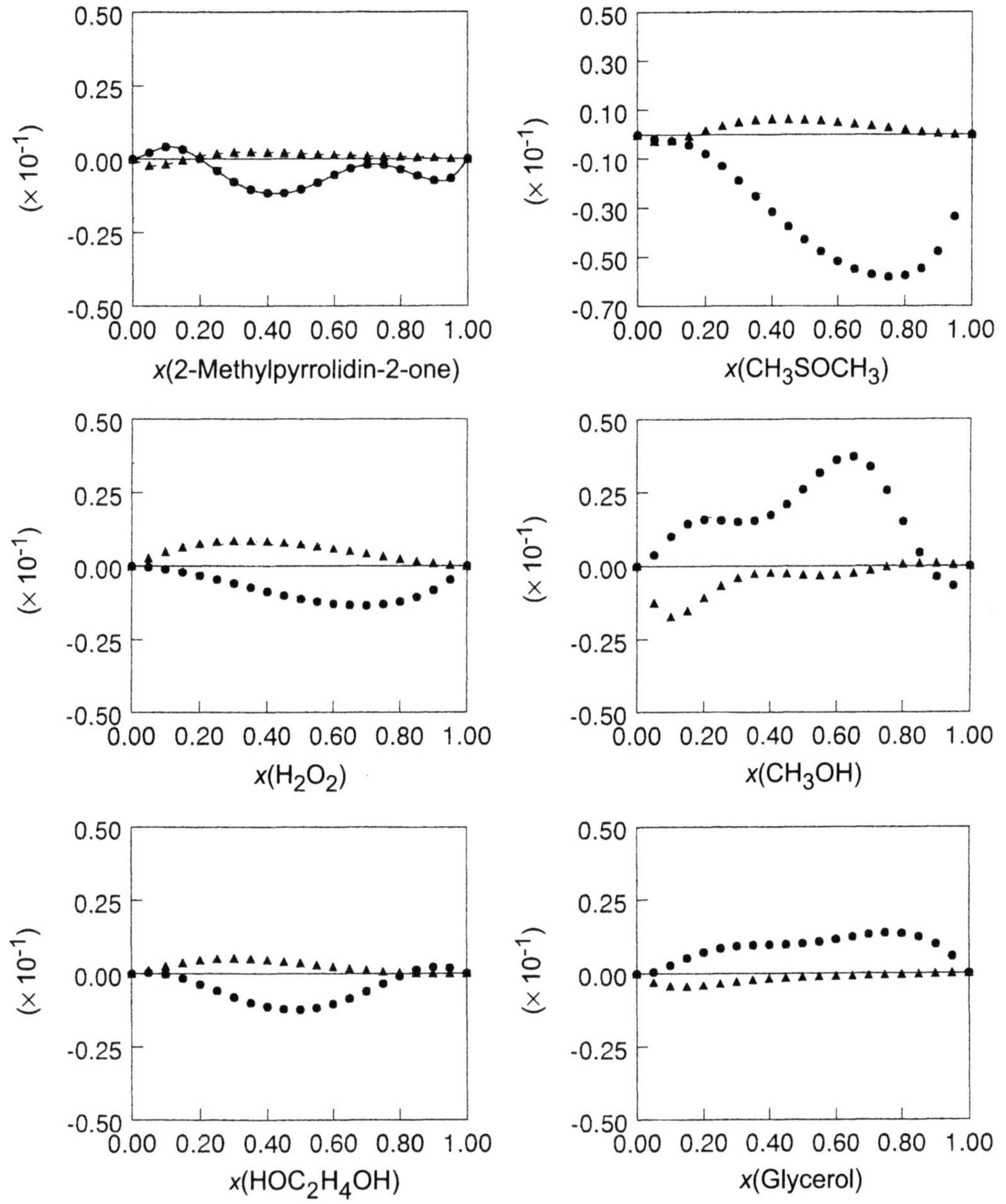

Fig. 7. Volume-corrected preferential solvation parameters $\delta x'_{WW}$ (●) for water-water interactions and $\delta x'_{WS}$ (▲) for water – co-solvent interactions in the first solvation shell at 298.15 K of aqueous a) N-methylpyrrolidin-2-one (at 303.15 K), b) dimethylsulfoxide, c) hydrogen peroxide, d) methanol, e) ethylene glycol, and f) glycerol; note the ordinate scale (the numbers are multiples of 10^{-1})

mutual interaction has a maximal $\delta x'_{WS} = 0.005$ near $x_S = 0.35$, both extrema being quite small and preferential solvation being hardly significant.

Aqueous Gly

This binary solvent system has been dealt with in detail previously [20], and the volume-corrected preferential solvation parameters are shown in Fig. 7f. As for

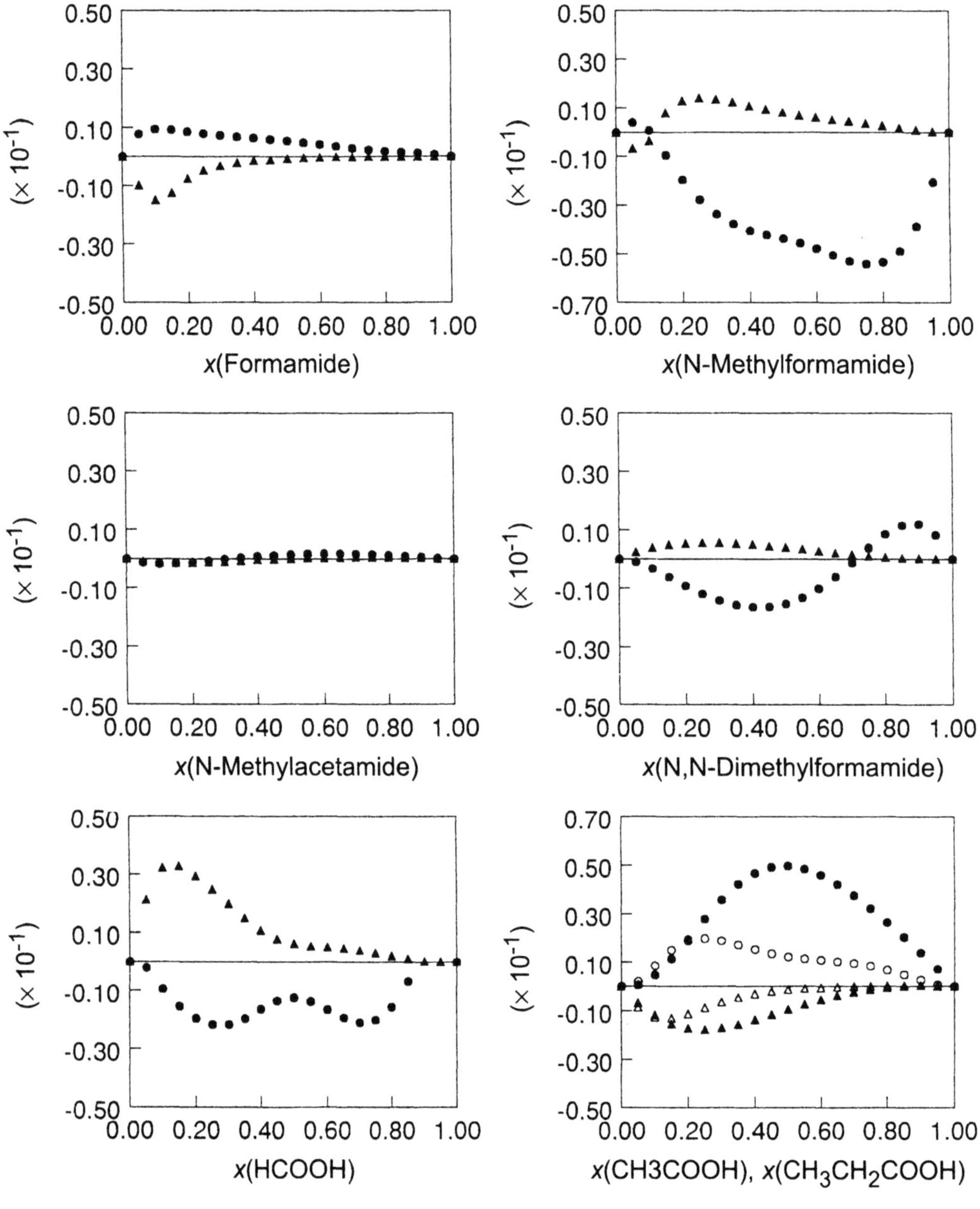

Fig. 8. Volume-corrected preferential solvation parameters $\delta x'_{\mathrm{WW}}$ (●) for water-water interactions and $\delta x'_{\mathrm{WS}}$ (▲) for water – co-solvent interactions in the first solvation shell at 298.15 K of aqueous a) formamide, b) N-methylformamide, c) N-methylacetamide, d) N,N-dimethylformamide, e) formic acid, and f) acetic acid (filled symbols) and propanoic acid (empty symbols); note the ordinate scale (the numbers are multiples of 10^{-1})

aqueous ethylene glycol, the system is close to ideal, but the self interactions of water are now favoured over the mutual interactions. The maximal $\delta x'_{\mathrm{WW}} = 0.014$ at $x_{\mathrm{S}} = 0.75$, whereas the disfavoured mutual interaction has a minimal $\delta x'_{\mathrm{WS}} = -0.005$ near $x_{\mathrm{S}} = 0.15$, both extrema being again rather small.

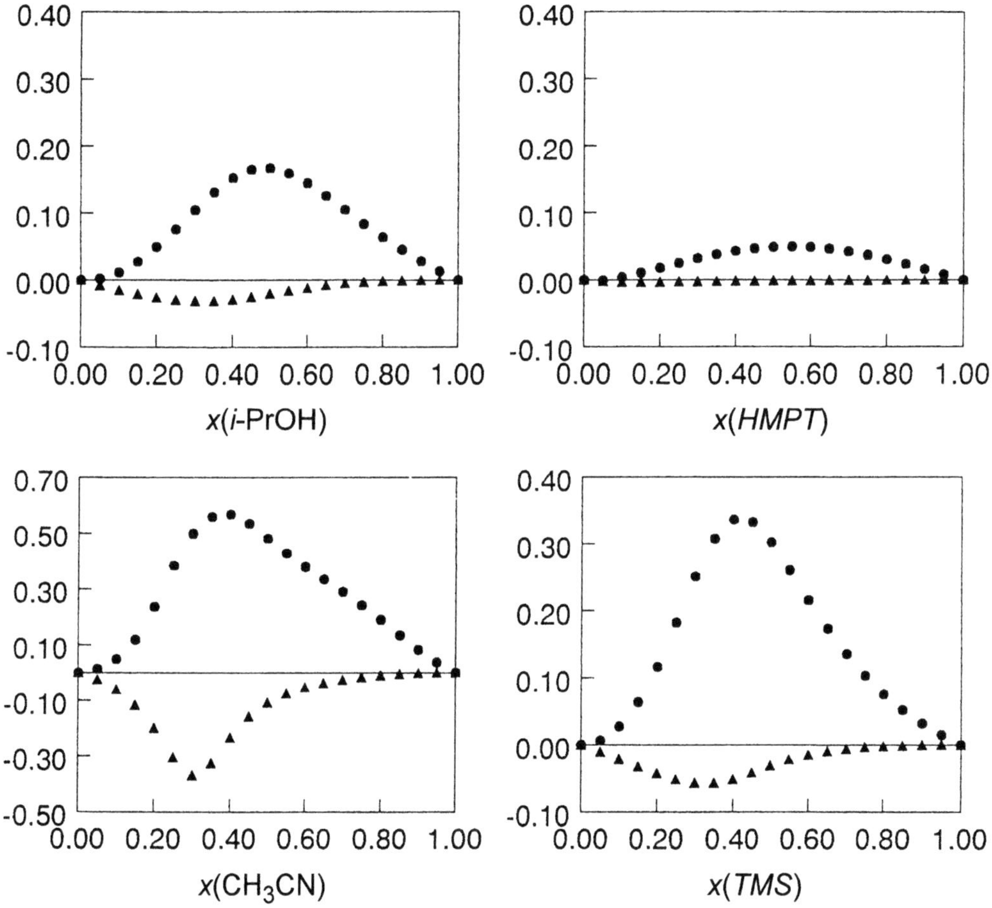

Fig. 9. Volume-corrected preferential solvation parameters $\delta x'_{WW}$ (●) for water-water interactions and $\delta x'_{WS}$ (▲) for water – co-solvent interactions in the first solvation shell at 298.15 K of aqueous: a) 2-propanol, b) hexamethyl phosphoric triamide, c) acetonitrile, and d) tetramethylenesulfone (at 303.15 K)

Aqueous HCOOH

The volume-corrected preferential solvation parameters are shown in Fig. 8e. For this very hydrophilic co-solvent, the water self-interactions are disfavoured, with a minimal $\delta x'_{WW} = -0.022$ at $x_S = 0.30$, whereas the favoured mutual interaction has a maximal $\delta x'_{WS} = 0.033$ near $x_S = 0.15$. These values for the extrema are rather small, showing that the system is not far from ideality.

Aqueous MeCOOH

The volume-corrected preferential solvation parameters are shown in Fig. 8f (filled symbols). This system shows an intermediate magnitude of non-ideality, the water-water interactions being favoured, with maximal $\delta x'_{WW} = 0.050$ at $x_S = 0.50$, whereas the disfavoured mutual water-acid interaction has a minimal $\delta x'_{WS} = -0.018$ at $x_S = 0.25$.

Aqueous EtCOOH

The volume-corrected preferential solvation parameters are shown in Fig. 8f, too (empty symbols). Surprisingly, this system exhibits lower non-ideality than the aqueous acetic acid system, the water-water interactions being favoured, but with maximal $\delta x'_{WW} = 0.020$ only at $x_S = 0.25$, whereas the disfavoured mutual water-acid interaction has a minimal $\delta x'_{WS} = -0.013$ at $x_S = 0.15$. No good explanation for this behaviour is apparent.

Aqueous FA

The volume-corrected preferential solvation parameters are shown in Fig. 8a, indicating that this system is nearly ideal. The extrema are very small, and preferential solvation is hardly significant. The water self-interactions are favoured, but with maximal $\delta x'_{WW} = 0.009$ only at $x_S = 0.15$, and the mutual interactions have a minimal $\delta x'_{WS} = -0.013$ at $x_S = 0.15$.

Aqueous NMF

The volume-corrected preferential solvation parameters are shown in Fig. 8b, the water self interactions being disfavoured in this case, with minimal $\delta x'_{WW} = -0.054$ at $x_S = 0.75$ and the favoured mutual interactions have a maximal $\delta x'_{WS} = 0.014$ at $x_S = 0.25$. The better compatibility of this co-solvent than of formamide with water is unexpected, although the extreme values are fairly small in either case.

Aqueous NMA

The volume-corrected preferential solvation parameters are shown in Fig. 8c. The behavior of this system is practically indistinguishable from ideal behavior: the absolute extreme values of the volume-corrected preferential solvation parameters are below 0.002, so that no preferential solvation takes place.

Aqueous DMF

The volume-corrected preferential solvation parameters are shown in Fig. 8d. The water self-interactions are disfavoured at low co-solvent concentrations, with minimal $\delta x'_{WW} = -0.016$ at $x_S = 0.40$ but somewhat favoured at high co-solvent concentrations, reaching $\delta x'_{WW} = 0.012$ at $x_S = 0.90$. The favoured mutual interactions have a small maximal $\delta x'_{WS} = 0.006$ at $x_S = 0.30$. Little preferential solvation is noted for this system.

The behavior of these four aqueous amide systems can be compared with that of aqueous N-methylpyrrolidin-2-one (Fig. 7a) and with that of the following system (Fig. 9b) which are also fairly close to being ideal.

Aqueous HMPT

The volume-corrected preferential solvation parameters are shown in Fig. 9b. The extrema are not large the water self-interactions being favoured but with maximal

$\delta x'_{WW} = 0.050$ at $x_S = 0.55$; the mutual interactions have a minimal $\delta x'_{WS} = -0.003$ at $x_S = 0.20$. Although the phosphoryl group and the three amidic oxygen atoms are exposed to water, the six methyl groups exert effectively their hydrophobic nature.

Aqueous acetonitrile

This solvent system has been dealt with in detail over a wide temperature range [11]. For 298.15 K, the resulting volume-corrected preferential solvation parameters are shown in Fig. 9c. This system exhibits large deviations from ideal mixing (immiscibility is noted below 272.0 K), and preferential solvation that extends over many solvation shells (microheterogeneity) still occurs at 298.15 K [11]. For the water self-interactions, the maximal $\delta x'_{WW}$ is 0.56 at $x_S = 0.35$, whereas for the mutual interaction the minimal $\delta x'_{WS} = -0.37$ near $x_S = 0.30$. Of the systems considered here, only the aqueous *n*-propanol, *t*-butanol, and tetramethylenesulfone mixtures approach aqueous acetonitrile in exhibiting that large preferential solvation. The temperatures at which the latter two systems are presented here are, however, only slightly higher than the melting points of the neat co-solvents, a fact that ought to affect the behavior of these two systems.

Aqueous TMS

The volume-corrected preferential solvation parameters (for 303.15 K) are shown in Fig. 9d. The water self-interactions are strongly favoured, with maximal $\delta x'_{WW} = 0.34$ at $x_S = 0.40$, and the disfavoured mutual interactions have a minimal $\delta x'_{WS} = -0.06$ at $x_S = 0.30$. The pronounced preferential solvation extends over several solvation shells (not shown) and is similar in magnitude to that noted in aqueous acetone or tetrahydrofuran.

Limiting Self Interaction of the Co-Solvent

A further interesting piece of information that can be derived from the *Kirkwood-Buff* integrals is the limiting (*i.e.* infinite dilution) self-interaction of the co-solvent in aqueous solution, $G^{\infty}_{SS} = \lim(x_S \rightarrow 0)G_{SS}$. The values that have been obtained in the present work and previously [10] are shown in Table 1. It has been previously suggested that large negative values of this quantity result for hydrophilic co-solvents that are good electron-pair donors. Conversely, hydrophobic co-solvents that are poor electron-pair donors should have positive values, *i.e.* should self-aggregate. These expectations are borne out qualitatively by the data in Table 1, although the multivariate expression previously reported [10] agrees with the extended data only in view of the large errors assigned to the coefficients. It must also be recognized that the molar volumes of the co-solvents should play a role in the magnitude of G^{∞}_{SS}, since in an ideal system $G^{\infty,\text{id}}_{SS} = RT\kappa_T - V^{\circ}_S$. Thus, for *HMPT* $G^{\circ}_{SS} = -370\,\text{cm}^3 \cdot \text{mol}^{-1}$ is understandable in view of its large electron-pair donicity and molar volume, as is $G^{\infty}_{SS} = 155\,\text{cm}^3 \cdot \text{mol}^{-1}$ for MeCN in view of its small molar volume and poor donicity. Some of the $G^{\infty}_{SS}/\text{cm}^3 \cdot \text{mol}^{-1}$ entries in Table 1, however, are probably erroneous (*i*-PrOH (+40), 2-BuOEtOH (< −500), and *FA* (>600)), and can be ascribed to inaccuracies in the double derivative of G^E at the limiting concentrations.

References

[1] Ben-Naim A (1977) J Chem Phys **67**: 4884
[2] Donkersloot MCA (1979) J Solution Chem **8**: 293
[3] Patil KJ (1981) J Solution Chem **10**: 315
[4] Matteoli E, Lepori L (1984) J Chem Phys **80**: 2856
[5] Huot J-Y, Battistel E, Lumry R, Villeneuve G, Lavallee J-F, Anusien A, Jolicoeur C (1988) J Solution Chem **17**: 601
[6] Zaitsev AL, Petrenko VE, Kessler YuM (1989) J Solution Chem **18**: 115
[7] Blandamer MJ, Blundell NJ, Burgess J, Cowles HJ, Horn IM (1990) J Chem Soc Faraday Trans **86**: 277
[8] Blandamer MJ, Blundell NJ, Burgess J, Cowles HJ, Horn IM, (1990) J Chem Soc Faraday Trans **86**: 283
[9] Blandamer MJ, Burgess J, Cooney A, Cowles HJ, Horn IM, Martin KJ, Morcom KM, Warrick PW Jr (1990) J Chem Soc Faraday Trans **86**: 2209
[10] Marcus Y (1990) J Chem Soc Faraday Trans **86**: 2215
[11] Marcus Y, Migron Y (1991) J Phys Chem **95**: 400
[12] Cheng Y, Pagé M, Jolicoeur C (1993) J Phys Chem **97**: 7359
[13] Patil KJ, Mehta GR, Dhondge SS (1994) Indian J Chem **33A**: 1069
[14] Marcus Y (1995) J Chem Soc Faraday Trans **91**: 427
[15] Matteoli E (1997) J Phys Chem **B101**: 9800
[16] Zielenkiewicz J (1998) J Chem Soc Faraday Trans **94**: 1713
[17] Marcus Y (1999) PCCP **1**: 2975
[18] Shulgin I, Ruckenstein E (1999) J Phys Chem **B103**: 2496
[19] Zielenkiewicz J (2000) PCCP **2**: 2925
[20] Marcus Y (2000) PCCP **2**: 4891
[21] Kato T (1984) J Phys Chem **88**: 1248
[22] Nishikawa K, Iijima T (1995) J Phys Chem **91**: 10824
[23] Hansen RS, Miller FA (1954) J Phys Chem **58**: 193
[24] Dobson HJE (1925) J Chem Soc 2866
[25] Shaw R, Butler JAV (1930) Proc Roy Soc (London) **A129**: 519
[26] Lenderstrøm-Lang CU, Vaslow F (1968) J Phys Chem **72**: 2645
[27] Benson GC, Kiyohara O (1980) J Solution Chem **9**: 791
[28] D'Arrigo G, Paparelli A (1988) J Chem Phys **88**: 405
[29] Omori G (1988) J Chem Phys **89**: 4325
[30] Matteoli E, Lepori L (1990) J Mol Liq **47**: 89
[31] Butler JAV, Thomson DW, Maclennan WH (1933) J Chem Soc 674
[32] Westmeier S (1976) Chem Techn **28**: 480
[33] Moriyoshi T, Uosaki Y (1984) Zairyo **33**: 127
[34] Westmeier S (1977) Chem Techn **29**: 218
[35] Brown AC, Ives DG (1962) J Chem Soc 1608
[36] Sehgal CM, Porter BR, Greenleaf JF (1986) J Acoust Soc Am **79**: 566
[37] Cooney A, Morcom KW (1988) J Chem Thermodyn **20**: 735
[38] Rochester CH, Symonds JR (1974) J Fluorine Chem **4**: 141
[39] Marcus Y, Hefter GT (1997) J Mol Liq **73/74**: 61
[40] Chiavone-Filho O, Proust P, Rasmussen A (1993) J Chem Eng Data **38**: 128
[41] Pagé M, Huot J-Y, Joliceour C (1993) J Chem Thermodyn **25**: 139

[42] Tamura K, Sabata S, Murakami S (1998) J Chem Thermodyn **30**: 1319
[43] Douheret G, Pal A (1988) J Chem Eng Data **33**: 40
[44] Scatchard G, Wilson GM (1964) J Am Chem Soc **86**: 133
[45] Koga Y (1991) J Phys Chem **95**: 4119
[46] Lara J, Desnoyers JE (1971) J Solution Chem **10**: 465
[47] Touhara H, Okazaki S, Okino F, Tanaka H, Ikari K, Nakanishi K (1982) J Chem Thermodyn **14**: 145
[48] Hawrylak B, Burke SE, Palepu R (2000) J Solution Chem **29**: 575
[49] Treiner C, Bocquet JF, Chemla M (1973) J Chim Phys **70**: 72
[50] Kiyohara O, Benson GC (1977) Can J Chem **55**: 1354
[51] Aminabhavi TM, Gopalakrishna B (1995) J Chem Eng Data **40**: 856
[52] Goats JR, Sullivan RJ (1958) J Phys Chem **62**: 188
[53] Vierk AL (1950) Z Anorg Chem **261**: 283
[54] Bacarella AL, Finch A, Grunwald E (1956) J Phys Chem **60**: 573
[55] Rebollo MB, Buep AH, Baron M (1988) An Asoc Quim Argent **76**: 69
[56] Orye RV, Prausnitz JM (1965) Ind Eng Chem **57(5)**: 18
[57] Villamañan MA, van Ness HC (1984) J Chem Eng Data **29**: 429
[58] Noda K, Ohashi M, Ishida K (1982) J Chem Eng Data **27**: 326
[59] Egorov GI, Gruznov EL, Kolker AM (1996) Zh Fiz Khim **70**: 216
[60] Nakanishi K, Touhara H, Maya K, Abe J-I (1981) Netsokutei **8**: 135
[61] Patil KK (1978) Indian J Pure Appl Phys **16**: 608
[62] Chan TC, Van Hook WA (1976) J Chem Soc Faraday Trans **72**: 583
[63] Abe J-I, Nakanishi K, Touhara H (1978) J Chem Thermodyn **10**: 483
[64] Sharma VK, Singh J (1998) Indian J Chem **37A**: 59
[65] Zielenkiewicz J (1998) J Chem Eng Data **43**: 650
[66] Davis MI (1987) Thermochim Acta **120**: 299
[67] Pal A, Singh YP, Singh W (1994) Indian J Chem **33A**: 59
[68] Lam SY, Benoit RL (1974) Can J Chem **52**: 718
[69] Chan TC, Van Hook WA (1976) J Solution Chem **5**: 107
[70] de Visser C, Heuvelsland WJM, Dunn LA, Somsen G (1978) J Chem Soc Faraday Trans 1 **74**: 1159
[71] Kim J-I (1978) Z Phys Chem N F **113**: 129
[72] Matteoli E (1997) J Phys Chem **B101**: 9800

Received January 10, 2001. Accepted (revised) February 20, 2001

Invited Review

Phase Transitions and Critical Behaviour of Binary Liquid Mixtures

Gerhard Kahl*, **Elisabeth Schöll-Paschinger,** and **Andreas Lang**

Institut für Theoretische Physik und Center for Computational Materials Science, TU Wien, A-1040 Wien, Austria

Summary. Compared to the simple one-component case, the phase behaviour of binary liquid mixtures shows an incredibly rich variety of phenomena. In this contribution we restrict ourselves to so-called binary symmetric mixtures, *i.e.* where like-particle interactions are equal ($\Phi_{11}(r) = \Phi_{22}(r)$), whereas the interactions between unlike fluid particles differ from those of likes ones ($\Phi_{11}(r) \neq \Phi_{12}(r)$). Using both the simple mean spherical approximation and the more sophisticated self-consistent *Ornstein-Zernike* approximation, we have calculated the structural and thermodynamic properties of such a system and determine phase diagrams, paying particular attention to the critical behaviour (critical and tricritical points, critical end points). We then study the thermodynamic properties of the same binary mixture when it is in thermal equilibrium with a disordered porous matrix which we have realized by a frozen configuration of equally sized particles. We observe – in qualitative agreement with experiment – that already a minute matrix density is able to lead to drastic changes in the phase behaviour of the fluid. We systematically investigate the influence of the external system parameters (due to the matrix properties and the fluid–matrix interactions) and of the internal system parameters (due to the fluid properties) on the phase diagram.

Keywords. Binary fluid mixtures; Fluids in porous media; Phase transitions; Phase stability; Critical behaviour.

Introduction

The phase behaviour and the critical phenomena encountered in binary mixtures are considerably more complex than in the case of a pure fluid. Following the *Gibbs* rule, up to four phases can be observed simultaneously, and the way these phases can coexist often leads to rather complex phase diagrams. The phase behaviour is mainly triggered by two mechanisms (and their interrelation): first, there is the size difference of the particles of the two components and their (partial) penetrability; second, there is the chemical influence, expressed *via* the set of the three interatomic potentials, *i.e.* the interaction potentials between particles of the same species and the cross interaction between the unlike particles. The reader might get an impression of the complexity of the phase behaviour of a binary

* Corresponding author. E-mail: gkahl@tph.tuwien.ac.at

mixture from the review article of *Konynenburg* and *Scott* [1]: using a simple *van der Waals* model (which incorporates all relevant aspects of a binary mixture), these authors have classified several types of phase diagrams that can be encountered for such a system.

In this contribution we consider only a simplified version of a binary fluid mixture, which is usually called symmetric binary mixture in the literature. It is characterized by the fact that all particles are of the same size (ignoring thus the influence of particle size on the phase behaviour mentioned above). Further, the interactions between like particles are equal, only the unlike interaction (between particles of different species) is different. If we denote the fluid particles of the two species by indices '1' and '2', the set of interactions for a symmetric binary mixture are thus given by $\Phi_{11}(r) = \Phi_{22}(r) \neq \Phi_{12}(r)$. To further simplify the model, we assume the same functional form for all interatomic potentials; hence, $\Phi_{12}(r) = \alpha\Phi_{11}(r)$, α being different from 1. We additionally demand that our system be homogeneous and isotropic; therefore, potentials and the correlation functions (introduced later) are functions of the distance only. For such a system we will expect – apart from solid phases, which are not considered in this contribution – three different phases: the homogeneous equimolar gas phase (G), the homogeneous equimolar liquid (L), and the nonequimolar (demixed) fluid (DF), which consists itself of two coexisting phases, a 1-rich and a 2-rich fluid. Due to the symmetry, these two fluids have the same density. For $\alpha > 1$, where the unlike particle attraction dominates the like one, it is energetically more favourable for the mixture to exist in the equimolar phase; here we only expect a G-L transition (similar to that encountered in the one-component case), and no DF will be observed. In this contribution we have considered the more interesting case, *i.e.* $\alpha < 1$, where we expect both a G-L transition (as in the one-component case) and a demixing transition as well as the interplay of these two transitions.

The symmetry of our system leads to a reduction of types of phase diagrams that we encounter [2]: we observe three different archetypes which are distinguished by the location where the so-called λ-line (*i.e.* the critical line of the fluid demixing transition) intersects the first order G-L coexistence curve. In type I, where the λ-line meets the G-L coexistence curve well below the G-L critical point, we observe a critical end point (CEP) behaviour. In type II we observe a triple point (coexistence of G-, L-, and DF-phase) and a tricritical point (where an equimolar liquid, a 1-rich, and a 2-rich fluid become simultaneously critical); this point is characterized by two vanishing order parameters, the difference in the densities of the equimolar liquid and the DF-phase, and the difference in the concentrations of the two coexisting DF-phases. Type III is characterized again by a tricritical point; however, the G-L transition is suppressed. The variation I $\rightarrow$ II $\rightarrow$ III turns out to be induced by decreasing α.

In the second part of the present contribution we investigate this system when it is in thermal equilibrium with a disordered porous matrix of immobile particles, realized in our case by a frozen liquid configuration of particles of the same size as the fluid particles. Investigations of fluids that are in equilibrium with a disordered matrix have become a very challenging field in liquid-state physics during the past years [3]. Experimental and theoretical studies revealed that a porous matrix (even if it occupies only a minute fraction of the volume) can have a substantial influence on the phase behavior of the liquid: ^{4}He and N_2 in high-porosity aerogel [4] are two examples where the near-critical liquid–vapor (LV) curve is narrowed drastically under the influence of a matrix. The effects become of course more interesting in the case that the fluid is a binary mixture; for example, experiments on a ^{3}He-^{4}He mixture inside a highly porous silica-gel or a porous gold matrix have shown a drastic modification of the superfluid transition [5]. A deeper understanding of these obviously very complex effects is all the more desirable as it might help to predict properties of materials of technological relevance (with widespread applications in catalysis, adsorption, *etc.*); for an overview, see Refs. [3, 6, 7].

Our simple model mimics the characteristic features of the matrix–fluid and fluid–fluid interaction of realistic systems and hence is able to give (at least qualitative) information of how a variation of the system parameters influences the phase behaviour. We present results of systematic variations of the system parameters and their influence on the phase behaviour of these complex systems. At this occasion one should point out that similar archetypes of phase diagrams (and sequences of these) are encountered in systems with completely different interatomic potentials: as examples we list here the *Heisenberg* fluid [8] (a fluid where the particles interact *via* hard-core and a *Heisenberg*-type interaction of their magnetic moments) and the *Stockmayer* fluid [9] (a fluid where the particles carry dipolar moments and interact – in addition – *via Lennard-Jones* potentials). In contrast to the present system, however, the changes between the three types of phase diagrams in these two examples are triggered by only one parameter.

In our calculations we have used liquid-state theories that are based on statistical mechanics for both systems. The central equations are the *Ornstein-Zernike* (OZ) equations [10] that relate the correlation functions of the particles of different species. For the binary case, the OZ-equations are three coupled integral-equations for the sets of correlation functions, each of them consisting of three functions. These equations are supplemented with a so-called closure relation that relates these functions with the set of pair potentials. In this contribution we have used the simple (and thermodynamically inconsistent) mean spherical approximation (MSA) [11] and the rather sophisticated (and thermodynamically self-consistent) self-consistent *Ornstein-Zernike* approximation (SCOZA) [12]. Despite its simplicity, the MSA model is able to give qualitative results for the phase diagram; thus, systematic trends can be worked out on a qualitative basis. However, this approach fails near the critical region: it is neither able to locate the critical point, nor is it able to give the correct critical exponents. The more refined SCOZA technique, on the other hand, gives highly accurate results for the phase boundaries and remains accurate as one approaches the critical region; its application is, however, restricted to the case of the bulk fluid as a consequence of the complexity of its concepts.

We have applied both methods to the case of the bulk fluid. We have investigated the influence of the interaction parameter α on the phase behaviour employing the MSA method, showing that the transition between the types of phase diagrams can be triggered by this parameter. Hence, in this investigation we refine in a quantitative way a study that was based on a simple mean field approach [2]. The MSA permits in addition, to identify metastable transitions that might be of relevance when studying dynamic processes. Using the SCOZA we have studied the stability of this fluid both with respect to mechanical and to chemical aspects.

The situation is much more delicate for the system where the fluid is in thermal equilibrium with the porous matrix. Here, a special approach *via* the so-called replica trick (explained in more detail below) has to be used: the concept (introduced to liquid-state physics by *Given* and *Stell* [13]) allows to map the system 'fluid + matrix' (which represents a very special three-component system) onto the limiting case of a fully equilibrated fluid mixture built up by the matrix and s identical and non-interacting copies of the fluid; thermodynamic and structural properties of the partly quenched original system ('fluid + matrix') are then obtained by taking the ($s \to 0$)-limit of the corresponding quantities of the fully equilibrated replicated system. The formalism arising from this route is rather complex even for a simple closure relation like the MSA (for more details, cf. Ref. [14]). Hence, we present only MSA results.

In our investigations we have studied the trends in the phase diagrams of such a system. Distinguishing between internal parameters (*i.e.* parameters that characterize the fluid–fluid interactions) and external parameters (*i.e.* parameters that characterize the matrix and the matrix–fluid interactions) we found that, similar as in the case of the bulk fluid, one can produce – for a given matrix density – trends in the types of phase diagram by varying the internal parameters. However, it is surprising that such trends can also be induced by varying the external parameters, keeping the internal parameters fixed.

The System and the Theoretical Methods

The System

The liquid chosen for this study is a symmetric binary mixture: the like-particle potentials (the components are denoted by indices '1' and '2') are identical ($\Phi_{11}(r) = \Phi_{22}(r)$), whereas the interaction between unlike particles, $\Phi_{12}(r)$, is different and is related to $\Phi_{ii}(r)$ *via* $\Phi_{12}(r) = \alpha\Phi_{ii}(r)$. This choice restricts the number of system parameters of a binary mixture drastically and simplifies systematic investigations. Nevertheless, a rich phase behaviour is to be expected. For $\Phi_{ij}(r)$ we have chosen hard-core *Yukawa* (HCY) potentials:

$$\beta\Phi_{ij}(r) = \begin{cases} \infty & \forall\, r \leq \sigma \\ -\frac{K_{ij}}{r}\exp(-z(r-\sigma)) & \forall\, r > \sigma \end{cases} \tag{1}$$

The screening length z and the hard-sphere diameter σ are assumed to be equal for all potentials, whereas the interaction strengths are given by $K_{11} = K_{22}$ and $K_{12} = \alpha K_{11}$. Further system parameters are the concentrations of the two species of the fluid, $x_1 \equiv x$ and $x_2 = 1 - x_1$, and the number density of the fluid $\rho_f = \rho_1 + \rho_2$, ρ_i being the partial densities.

We then bring this binary fluid in contact with a disordered porous matrix. This matrix is assumed to be obtained by rapidly quenching a liquid configuration of matrix particles (index "0"), which are then immobile and not affected by the fluid particles; they interact with the fluid particles *via* potentials $\Phi_{0i}(r)$, $i = 1, 2$. For simplicity, the matrix particles have the same size as the fluid particles, and again a HCY potential is assumed for $\Phi_{0i}(r)$: the contact values K_{0i} are related to the fluid–fluid contact values *via* $K_{01} = K_{02} = yK_{11}$, and the number density of the matrix will be denoted by ρ_0. Further, we assume a pure hard-sphere matrix, *i.e.* $K_{00} = 0$.

We define a reduced temperature *via* $T^\star = \sigma/K_{11}$ (in the following, $T^\star$ is symbolized by T), and we put σ to unity. The reduced density is hence given by $\rho_f^\star = \sigma^3 \rho_f = \rho_f$.

Theoretical Methods

Liquid state theory for the bulk fluid

The structural and thermodynamic properties of the system have been calculated with methods based on statistical mechanics. The systems are assumed to be homogeneous and isotropic; hence, the correlation functions that describe the structure of the system depend on the distance of the particles only. The central correlation function in our approach is the so-called pair distribution function (PDF), $g(r)$. It is a function of the distance r between two particles and depends on temperature, density, and the interatomic potential. The PDF measures the extend to which the structure of a fluid deviates from complete randomness. Let us assume that we pick out one particle of the fluid and move with this particle through the volume that is occupied by the fluid. Counting at each time-step the number of particles separated by a distance r from this particle, we find that $4\pi\rho g(r)r^2dr$ is the mean number of particles in a spherical shell of thickness dr and radius r. Since the fluid particles become uncorrelated for large distances, $\lim_{r\to\infty} g(r) = 1$. One then introduces the total correlation function *via* $h(r) = g(r) - 1$. The direct correlation function, $c(r)$, is defined *via* the *Ornstein-Zernike* (OZ) equation [10] *via*

$$h(r) = c(r) + \rho \int c(r')h(|\mathbf{r} - \mathbf{r}'|)d^3r' \tag{2}$$

which expresses that the total correlation between two particles separated by a distance r is given by the direct correlation between these particles plus the indirect correlation, mediated *via* an arbitrary number of other particles in the fluid and expressed by the convolution integral in Eq. (2). For a broader overview over correlation functions we refer to [11].

In the binary cases these correlation functions are generalized to sets of correlation functions, introducing pairs of indices (ij) with $i, j = 1, 2$ and indicating correlations between particles of species i and j. These functions are denoted by $g_{ij}(r)$, $h_{ij}(r)$, and $c_{ij}(r)$. The OZ equations (now a set of three coupled integral-equations) read

$$h_{ij}(r) = c_{ij}(r) + \sum_{k=1,2} \rho_k \int c_{ik}(r')h_{kj}(|\mathbf{r} - \mathbf{r}'|)d^3r' \quad i,j = 1,2. \tag{3}$$

For a given set of $h_{ij}(r)$, the OZ equations can be viewed as defining relations for the direct correlation functions $c_{ij}(r)$. However, in general, both sets of correlation functions are unknown. In order to make a solution of the OZ equations possible, a further relation between these sets of functions and the set of pair potentials, $\Phi_{ij}(r)$, has to be provided; such a (functional) relation is called a closure relation and can formally be written as $F(\{c_{ij}\}, \{h_{ij}\}, \{\Phi_{ij}\}) = 0$. Closure relations are derived from exact expressions of statistical mechanics using simplifying assumptions. Since the pair distribution functions $g_{ij}(r)$ obtained with these (approximate) methods are not exact for a given set of pair potentials $\Phi_{ij}(r)$, this leads to an effect that is known in the literature as thermodynamic inconsistency which will be discussed later. A considerable number of such closure relations have been used in the literature during the past years; for an overview, see Refs. [11, 15]. The closure relations employed here are the mean spherical approximation (MSA) [11, 15] and the self-consistent *Ornstein-Zernike* approximation (SCOZA) [12]; they will be discussed in the subsequent subsections.

Once the structure is known in terms of the correlation functions, the thermodynamic properties can be calculated in a straightforward way. To this end, statistical mechanics provides different routes, such as the energy, the virial, or the compressibility route. In the energy route, the excess (over ideal gas) internal energy U^{ex} is given by

$$\frac{U^{\text{ex}}}{V} = u = 2\pi \sum_{ij} \rho_i \rho_j \int \Phi_{ij}(r) g_{ij}(r) r^2 dr \tag{4}$$

where V is the volume of the system. In the virial route, the pressure P is calculated *via*

$$\beta P = \rho - \frac{2\pi}{3} \beta \sum_{ij} \rho_i \rho_j \int \Phi'_{ij}(r) g_{ij}(r) r^3 dr \tag{5}$$

(the prime stands for the derivative of the potentials with respect to the distance). Finally, the compressibility route relates the isothermal compressibility, $\chi_T = (-1/V)(\partial V/\partial P)_T$, to the zero-$q$ limit of the direct correlation functions *via*

$$\frac{\beta}{\rho \chi_T} = \frac{1}{\chi_T^{\text{red}}} = 1 - \frac{1}{\rho} \sum_{ij} \rho_i \rho_j \tilde{c}_{ij}(q=0) \tag{6}$$

where the tilde represents a three-dimensional *Fourier* transform. Further thermodynamic quantities relevant for the determination of phase equilibria (such as the free energy A or the chemical potentials μ_i, $i = 1, 2$) can be derived from the above quantities *via* thermodynamic integration or differentiation (using an appropriately chosen *Maxwell* relation).

If the structure functions were exact for the given system, one should obtain the same results for the thermodynamic properties *via* each of these three routes; the system is then called thermodynamically self-consistent. Computer simulations are one of the few methods that – apart from size-effects – provide self-consistent results. If approximations are introduced in the calculation of the correlation functions (and this is certainly the case when using closure relations), the different routes will lead to different results for the thermodynamic properties; the method is then called thermodynamically inconsistent.

Given the thermodynamic properties, one can then determine the coexistence parameters of two (or more) coexisting phases (denoted by Greek indices). These parameters are calculated from the coexistence relations which, for a given temperature T, read

$$P^{\alpha}(\rho^{\alpha}, x_1^{\alpha}) = P^{\beta}(\rho^{\beta}, x_1^{\beta}) \tag{7}$$

and

$$\mu_i^\alpha(\rho^\alpha, x_1^\alpha) = \mu_i^\beta(\rho^\beta, x_1^\beta) \quad (i = 1, 2). \tag{8}$$

For a general binary mixture, up to four phases can coexist.

Liquid state theory for a fluid in contact with a porous matrix

The route outlined above for a bulk fluid and leading from the set of interactions to the thermodynamic properties of a system is considerably more complex in the case where the fluid is in contact with a porous matrix. In our investigations the matrix is obtained by a thermal quench of an equilibrated liquid configuration (matrix particles carry the index '0'). Hence, our system represents a very special three-component fluid: the particles of the matrix are assumed to be fixed in place and are not affected by the mobile particles of the fluid, whereas the fluid particles are allowed to equilibrate in the rigid matrix structure. Physical quantities are therefore obtained by two successive averages: a first average is taken over the degrees of freedom of the fluid particles (keeping the positions of the matrix particles fixed), and a second average is taken over all possible degrees of freedom of the matrix particles. Perhaps the most promising approach to solve this problem has been proposed by *Given* and *Stell* [13] who have introduced the replica trick [16] in this field: it exploits an isomorphism between the partly quenched and a fully equilibrated fluid mixture (the so-called replicated system) which consists of the now mobile matrix particles and of s noninteracting identical copies of the liquid. The properties of the partly quenched system are obtained by considering the limit $s \to 0$ of the properties of the equilibrium system which, in turn, can be treated by standard liquid state theories (see preceding subsection). Applying this framework for our system, nine OZ-type integral equations [14] (so-called replica *Ornstein-Zernike* equations, ROZ) can be derived that relate the direct and total correlation functions of the fluid and of the matrix particles. One of these equations describes the matrix correlation functions and is decoupled from the other equations; the remaining eight equations are coupled. Again, they have to be solved along with a closure relation.

In a similar way one can calculate the thermodynamic properties of the replicated system (using the standard relations mentioned in the preceding section) and then apply the limit $s \to 0$. In this manner, thermodynamic relations corresponding to Eqs. (4)–(6) are obtained which are considerably more complicated and are therefore not reproduced here; for details, see Ref. [14]. The phase equilibrium between coexisting phases is again calculated by equating the pressure and the chemical potentials for a given temperature T.

The mean spherical approximation

The closure relation of the mean spherical approximation (MSA) is given by [11]

$$g_{ij}(r) = 0 \quad (r < \sigma = 1) \tag{9}$$

and

$$c_{ij}(r) = -\beta\Phi_{ij}(r) \quad (r > \sigma = 1). \tag{10}$$

Whereas the first relation is – due to the impenetrability of the particles – exact, the second relation represents an approximation: one assumes that the long-distance behaviour of the direct correlation functions (*i.e.* $c_{ij}(r) \rightarrow -\beta\Phi_{ij}(r)$ for $r \rightarrow \infty$) is valid for all distances. As a consequence, the MSA leads to thermodynamically inconsistent data for the thermodynamic properties. Nevertheless, as will be shown in subsequent sections, it provides qualitatively good results in comparison with computer simulation data for systems not too close to phase boundaries or to a critical region.

The self-consistent Ornstein-Zernike approximation

The self-consistent *Ornstein-Zernike* approximation (SCOZA) was introduced in its original version [12] by *Høye* and *Stell* and has been refined ever since: meanwhile, it exists in several versions of different levels of sophistication (for an overview, Ref. [17]). The aims of the SCOZA are two-fold: first, it overcomes the lack of thermodynamic inconsistency of the MSA by imposing thermodynamic consistency between the energy and the compressibility route; second, the SCOZA is well-suited to treat fluids near phase transitions and near criticality accurately.

We present here the formulation of the SCOZA in its simplest version (for the more sophisticated versions, cf. Ref. [17]); the closure relations now read

$$g_{ij}(r) = 0 \quad (r < \sigma = 1) \tag{11}$$

and

$$c_{ij}(r) = K_{ij}(\rho, \beta, x)\Phi_{ij}(r) \quad (r > \sigma = 1). \tag{12}$$

A comparison with the MSA closure relations shows that the state-dependent functions $K_{ij}(\rho, \beta)$ replace the prefactor $-\beta$ in Eq. (10). We use the simplifying assumption that $K_{ij}(\rho, \beta, x) = K(\rho, \beta, x)$. In contrast to the MSA, where $K(\rho, \beta, x) = -\beta$, this function is not known *a priori* but is determined in the SCOZA by imposing consistency between the energy and the compressibility route, *i.e.* u (determined *via* the energy route) and χ_T^{red} (calculated *via* the compressibility route) have to fulfill the following partial differential equation (PDE):

$$\rho \frac{\partial^2 u}{\partial \rho^2} = \frac{\partial}{\partial \beta}\left(\frac{1}{\chi_T^{\mathrm{red}}}\right). \tag{13}$$

Equation (13) supplemented with the OZ equation (Eq. (3)), the closure relations (Eqs. (11) and (12)), and Eqs. (4) and (6) represents a partial differential equation for $K(\rho, \beta, x)$, which is transformed (*via* a rather complex and heavy formalism) into a (diffusion-type) PDE for u with a known function $B(\rho, u)$:

$$B(\rho, u)\frac{\partial u}{\partial \beta} = \rho \frac{\partial^2 u}{\partial \rho^2} \tag{14}$$

The steps from Eq. (13) to Eq. (14) are based on the fact that the OZ-equations supplemented by the closure relations *vide supra* can be solved *semi*-analytically for a HCY-mixture [18–20]. Using these results, the formulation of the SCOZA can be derived in a rather straightforward (but cumbersome) way. The numerical implementation is very demanding.

This simplifying fact represents a considerable advantage for the system investigated here. However, it also points out the main drawback of the SCOZA: its application is restricted (at least up to now) to such systems where a basic analytic concept is available. Hence, it is not surprising, that the SCOZA has been solved up to now only for HCY-systems and for the lattice gas model [21]. In a more general case, where such a (*semi*-)analytic help is not available, all calculations have to be performed numerically.

Results

Bulk Fluid

Phase diagrams

In the first part of this study we have investigated the phase diagram of the pure binary symmetric mixture defined in the preceding section. For the interaction parameter α, which characterizes the ratio between the unlike and the like interactions, we have chosen values smaller than one. This guarantees, that, apart from the vapour–liquid transition, we shall also encounter a fluid–fluid decomposition for a sufficiently small α. We expect the following phases: a homogeneous (equimolar) gas (G), a homogeneous (equimolar) liquid (L), and a nonequimolar (demixed) fluid (DF). The solid phases have not been considered here.

In some of the figures we shall present the MSA data; several of these are supplemented by results from grand canonical Monte-Carlo simulations performed by *D. Levesque* and *J.-J. Weis* [22] (for details of the simulation, see references in this paper). The phase diagrams presented are projections of the three dimensional (T, ρ, x) phase diagram onto the $(x = 1/2)$ plane. In Fig. 1, such a three-dimensional presentation is shown; the phase diagram is of type II (see below).

In a first step we have applied the (simple and thermodynamically inconsistent) MSA (Eqs. (9) and (10)). The results are shown in Fig. 2 for five different α-values. One can clearly follow the strong influence of the parameter α: for $\alpha = 0.9$ we observe a coexistence between the G and the L phases. There is probably no fluid-fluid phase separation: the existence of a DF phase and the corresponding transition is characterized in these figures by the so-called λ-line, the critical line of the fluid–fluid demixing transition: for a given temperature T, the critical density for this demixing transition is given by $\rho_c(T)$; these values of $\rho_c(T)$ represent the λ-line as a function of T. For densities larger than $\rho_c(T)$ we observe the DF phase, whereas for densities smaller than $\rho_c(T)$ we encounter the equimolar phase. As we decrease α, this λ-line approaches the G-L coexistence line from the high-density side and intersects this first order transition in a so-called critical end point (CEP); here, a non-critical vapour coexists with a critical nonequimolar (demixed) fluid (DF). Such a phase diagram will be called a type I phase diagram. Below the CEP temperature one observes a triple line where a vapour coexists with a 1-rich and a 2-rich fluid. By further decreasing α, the λ-line intersects the G-L coexistence curve at temperatures close to the G-L critical point, leading to a so-called

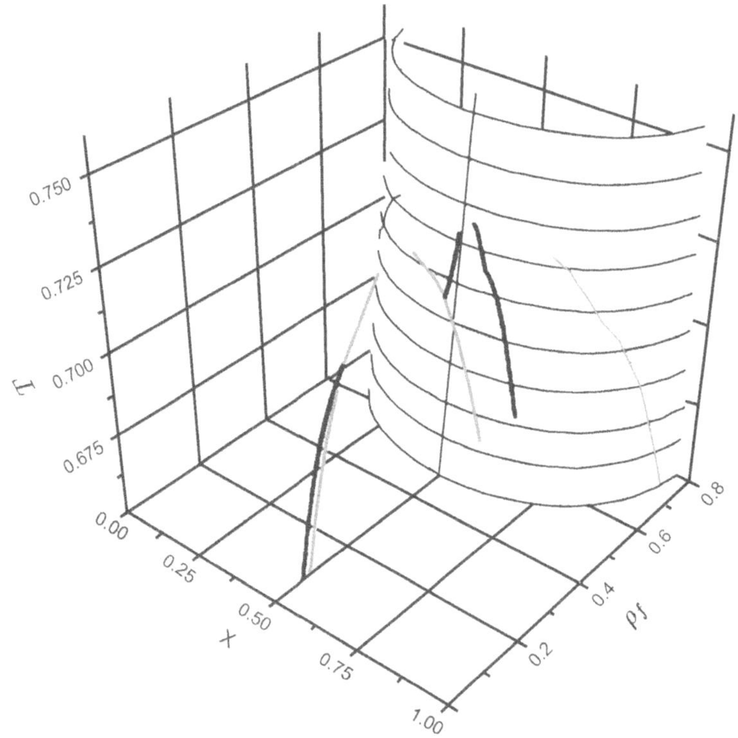

Fig. 1. Phase diagram of a binary bulk mixture in (T, ρ_f, x)-space; the curves in the ($x = 1/2$) plane represent the phase diagram that will be presented in the following figures (thick black and grey lines); demixing transitions are depicted for several isothermal planes by the thin grey lines in the high density regime; the critical points of this demixing transition are connected by the λ-line; from Ref. [23]

tricritical point: here L and DF become critical simultaneously, characterized by two vanishing order parameters; such a phase diagram is denoted to be of type II. We also observe a point where four phases (L, G, 1-rich DF, 2-rich DF) coexist. In addition, we encounter in phase diagrams of type I and II the usual critical point of the G-L phase coexistence. Finally, upon further decreasing α we observe the type III phase diagram, where the G-L transition is completely suppressed. Again we obtain a tricritical point where the L and the DF phase become simultaneously critical. Apart from the stable phase transitions we have also depicted in these figures the metastable phase equilibria; they are of relevance for the dynamic properties of the system (for a more detailed discussion, see Refs. [2, 22]). A comparison with computer simulations shows that the MSA gives reasonably good results for the bulk case.

Stability

The *semi*-analytic character of the SCOZA formalism offers the possibility to study the stability of our system in detail. As shown in Ref. [20], the mechanical stability (*i.e.* with respect to L-G separation) and the chemical stability (*i.e.* with respect to the demixing transition) can be studied using the *semi*-analytical expressions of the

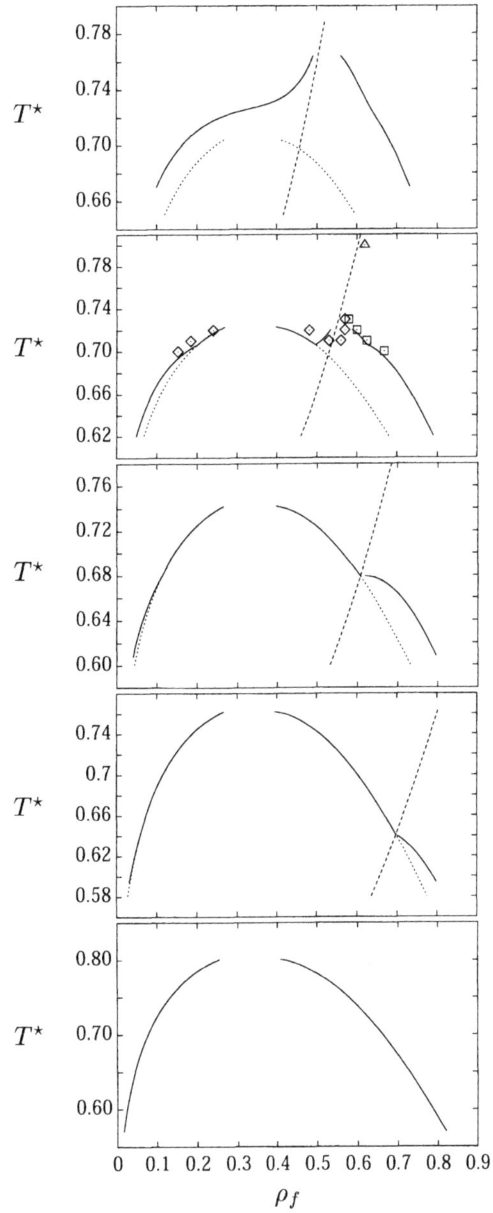

Fig. 2. Variation of α of the phase diagram of the binary symmetric bulk mixture from MSA: $\alpha = 0.65$, 0.70, 0.75, 0.80, and 0.90 (from top to bottom); full line: G-L, G-DF, or L-DF coexistence curves; dotted line: metastable G-L transition; dashed line: λ-line; symbols: grand canonical Monte Carlo simulations from Ref. [22]

MSA formalism. In Fig. 3 we depict this border, which in (T, ρ, x)-space is a surface: above this surface, the system is stable, whereas below it is unstable. The mechanical instability is expressed by a divergence of $\chi_T^{\rm red}$: it is encountered only along the $(x = 1/2)$ line on this surface. The onset of the chemical instability, on the other hand, is characterized by a change in the sign of $(\partial^2 G/\partial x^2)$, where G is the *Gibbs* free energy [15].

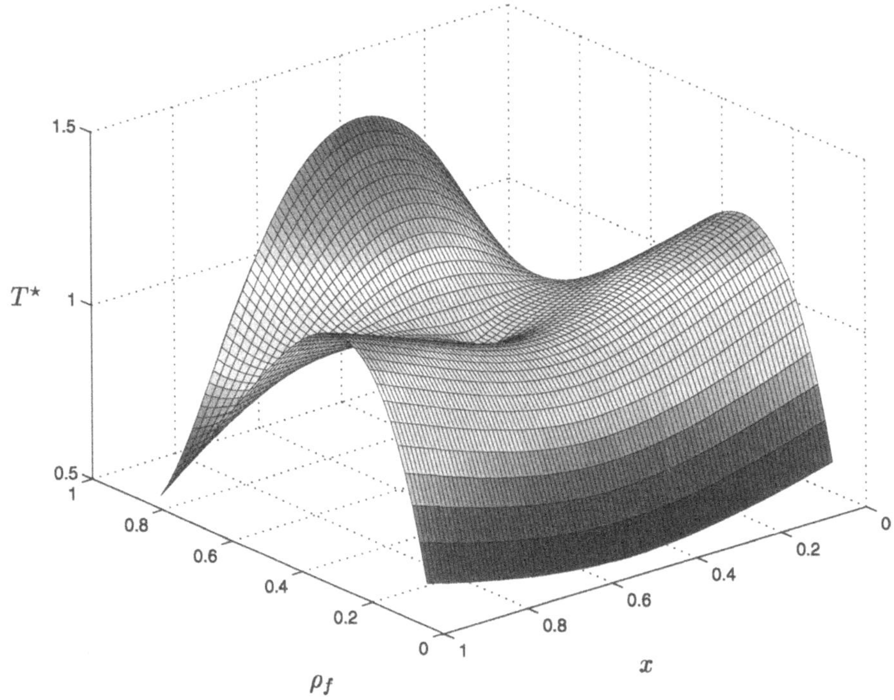

Fig. 3. Surface separating the stable (above) from the unstable (below) states of a symmetric binary bulk fluid in (T, ρ_f, x) – space ($z\sigma = 1.8$, $\alpha = 0.65$); see text

Fluid in a Porous Matrix

We have now brought our bulk fluid in thermal equilibrium with a porous matrix. Four parameters can be varied: (*i*) α, the ratio between unlike and like fluid particle interactions, (*ii*) y, the ratio between the matrix–fluid interactions and the potential of the like fluid particles: positive and negative values of y correspond to attractive and repulsive matrix–fluid interactions, (*iii*) the matrix density ρ_0 and (*iv*) the screening length z of the potentials. In an effort to enable a comparison with the computer simulations we have truncated our HCY-potentials (unless otherwise stated) at a cut-off radius of $r_c = 2.5\ \sigma$.

Comparison with simulations

We have first compared the MSA data with simulation results. The computer simulations were carried out for four different matrix densities $\rho_0 = 0$, 0.05, 0.15, and 0.3 at $\alpha = 0.7$, $y = 1$, $z\sigma = 2.5$, and $r_c = 2.5\ \sigma$ (for technical details, see Ref. [22] and references quoted therein). The results are shown in Fig. 4. For $\rho_0 = 0$ one observes a type II phase diagram: a first order G-L transition with a critical temperature of $T_c \approx 0.72$–0.73 and a critical density $\rho_c \approx 0.35$ and a line of second order demixing transitions terminating at a tricritical point with a temperature of $T_{tc} \approx 0.73$, slightly higher than the critical temperature, and a density of $\rho_{tc} \approx 0.57$. As we increase the matrix density, the temperature range within which the equimolar liquid exists decreases, and the phase diagram becomes a type I

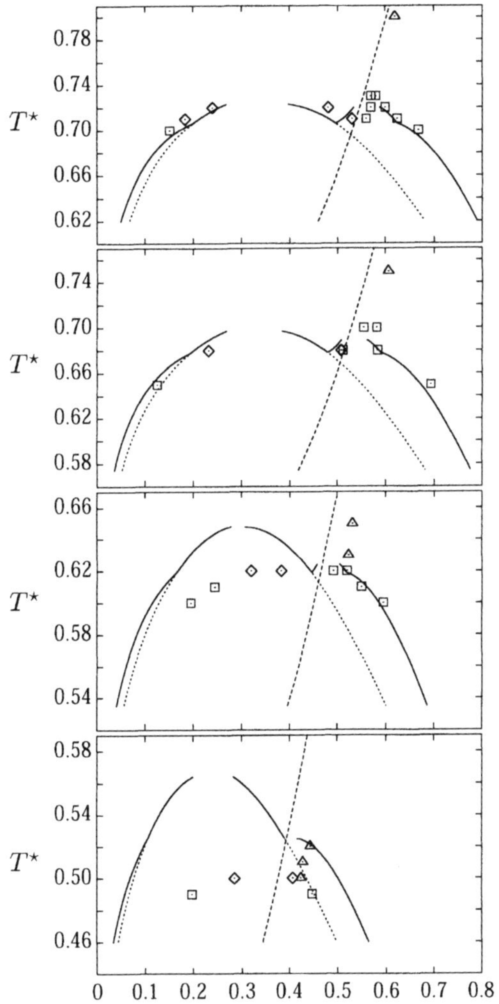

Fig. 4. Variation of the phase diagram of a binary fluid mixture in contact with a porous matrix with matrix density ρ_0 ($\alpha = 0.7$, $y = 1$, $z\sigma = 2.5$): comparison between grand canonical Monte Carlo (GCMC) and MSA results; $\rho_0 = 0$, 0.05, 0.15, and 0.30 (from top to bottom); symbols: GCMC simulations (diamonds: G-L equilibrium; squares: G-DF or L-DF equilibrium; triangles: λ-line); lines: MSA results (full line: G-L, G-DF or L-DF coexistence curve; dotted line: metastable G-L transitions; dashed line: λ-line); from Ref. [22]

diagram. At $\rho_0 = 0.3$ the phase diagram reveals (within accuracy of the simulation results) a tricritical point in the temperature range 0.49–0.52, or possibly a critical end point.

Although the trends in the types of phase diagrams are reproduced qualitatively in the simulations, it becomes obvious from Fig. 4 that the quantitative agreement between MSA data and simulation results deteriorates as we increase the matrix density: whereas for the pure bulk fluid ($\rho_0 = 0$) we have excellent agreement (both for the two first-order transitions and for the position of the λ-line), marked differences between these two sets of data appear as the matrix density increases.

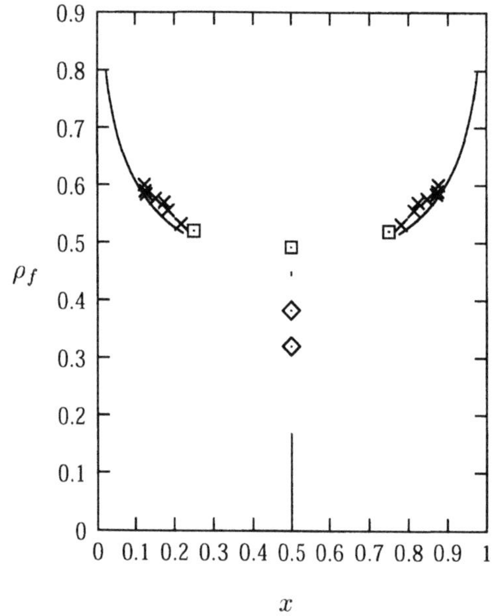

Fig. 5. Demixing transition of a binary fluid in contact with a porous matrix of density $\rho_0 = 0.15$ ($\alpha = 0.7$, $y = 1$, $z\sigma = 2.5$): ρ_f as a function of concentration x for $T = 0.62$; the symbols denote GCMC simulations (diamonds: G-L equilibrium; squares: L-DF equilibrium; crosses: first order demixing transition), the lines MSA results; from Ref. [17]

The critical temperatures obtained in the simulations are considerably lower than for the MSA data, whereas the critical densities are shifted to higher values. In addition, Fig. 5 shows the demixing transition for one of the systems: these phase diagrams are horizontal (isothermal) cuts through the three-dimensional phase diagram of Fig. 1. It is surprising, that – in contrast to the phase diagrams presented in Fig. 4 (projections onto the ($x = 1/2$) plane) – the concentrations of the demixed phase agree very nicely with simulation data even at high matrix densities. It is obvious that these differences are due to deficiencies of the MSA.

Variation of α

The variation of the phase diagram with α (ratio of the interaction strengths between unlike and like particles) is shown in Fig. 6 for a matrix density of $\rho_0 = 0.1$. $z\sigma$ was chosen to be 2.5; variation of this parameter for the bulk fluid was already discussed before. In an effort to study in addition the influence of the matrix–fluid interaction we have considered a pure hard sphere ($y = 0$) and a HCY ($y = 1$) matrix–fluid interaction. Although qualitatively the sequence of types of phase diagrams is the same in the two series (type III $\rightarrow$ type II $\rightarrow$ type I), the character of the matrix–fluid interaction leads to quantitative changes: keeping the matrix density fixed to $\rho_0 = 0.1$, the change of y from 1 to 0 (*i.e.* from an attractive HCY interaction to a pure hard sphere matrix) lowers T_c by *ca.*10%, shifts ρ_c from ~0.32 to ~0.26, and delays the appearance of the CEP as one increases α.

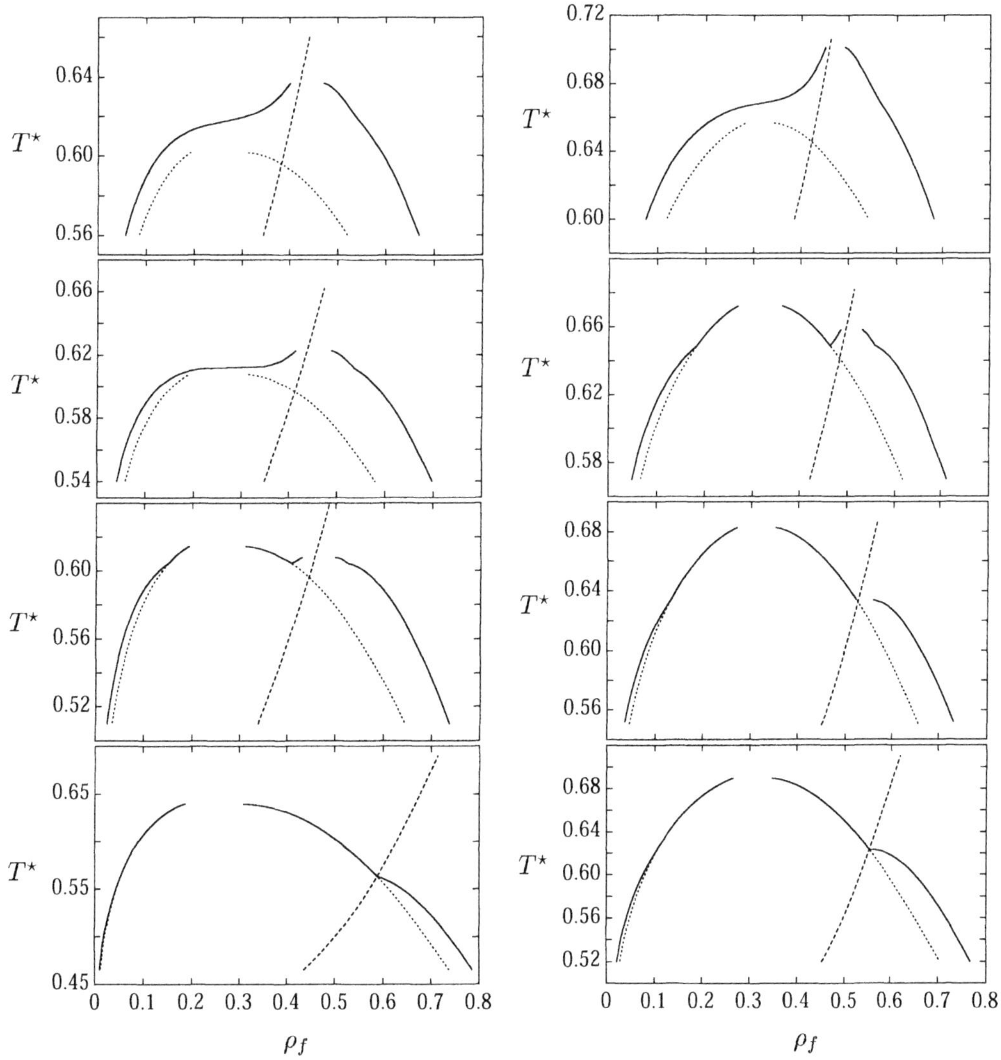

Fig. 6. Variation of the phase diagram of the binary fluid mixture in thermal equilibrium with a porous matrix from MSA with α; left column ($\rho_0 = 0.1$, $y = 0$): $\alpha = 0.68$, 0.70, 0.72, and 0.80 (from top to bottom); right column ($\rho_0 = 0.1$, $y = 1$): $\alpha = 0.65$, 0.70, 0.73, and 0.75 (from top to bottom); full line: G-L, G-DF or L-DF coexistence curves; dotted line: metastable G-L transition; dashed line: λ-line; from Ref. [22]

Variation of y

The parameter y is the ratio between the fluid–fluid and the matrix–fluid interactions; $y > 0$ represents an attraction between matrix and fluid particles, whereas $y < 0$ represents a repulsion. The influence of y on the phase behaviour is shown in Fig. 7 for two different pairs of (ρ_0, α) with $z\sigma = 2.5$: (*i*) $\rho_0 = 0.05$, $\alpha = 0.7$; (*ii*) $\rho_0 = 0.1$, $\alpha = 0.73$. For the lower matrix density, $\rho_0 = 0.05$, the sequence of phase diagrams is found to be type III $\rightarrow$ type II $\rightarrow$ type III when y decreases from positive to negative values. A G-L transition appears near $y \sim 2$ (a precise location cannot be found due to numerical problems in the critical region as

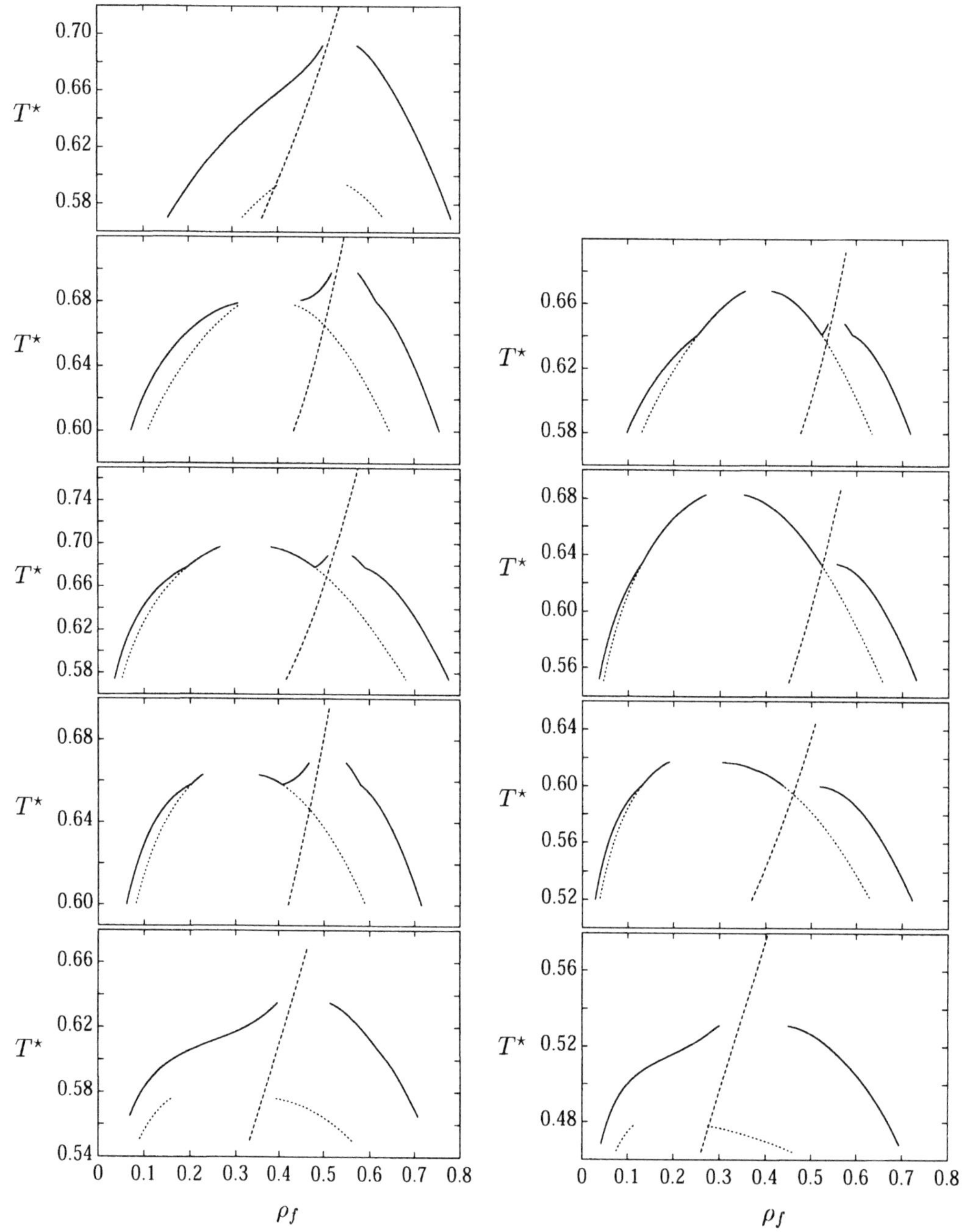

Fig. 7. Variation of the phase diagram of the binary fluid mixture in thermal equilibrium with a porous matrix from MSA with y; left column ($\rho_0 = 0.05$, $\alpha = 0.7$): $y = 3.5$, 2, 1, 0, -1 (from top to bottom); right column ($\rho_0 = 0.10$, $\alpha = 0.73$): $y = 2$, 1, 0, -1 (from top to bottom); full line: G-L, G-DF, or L-DF coexistence curves; dotted line: metastable G-L transition; dashed line: λ-line; from Ref. [22]

mentioned above) and exists only in a small range of y-values, extending roughly from 2 to -0.5. The phase diagram is again of type III for the more strongly repulsive matrix–fluid interaction $y = -1$. The matrix density has only a minor influence on the qualitative sequence of the types of phase diagrams: for the larger matrix density $\rho_0 = 0.1$, the type II phase behaviour occurs at least for $0 < y < 2$.

Variation of ρ_0

The influence of the matrix density ρ_0 on the phase diagram of the mixture is shown in Fig. 8 for two values of the parameter y ($y = 0$ and 1) and $\alpha = 0.7$; we assume $z\sigma = 2.5$. As discussed in the comparison of simulation and theoretical results, in both cases we have a type II diagram characterized by a tricritical point

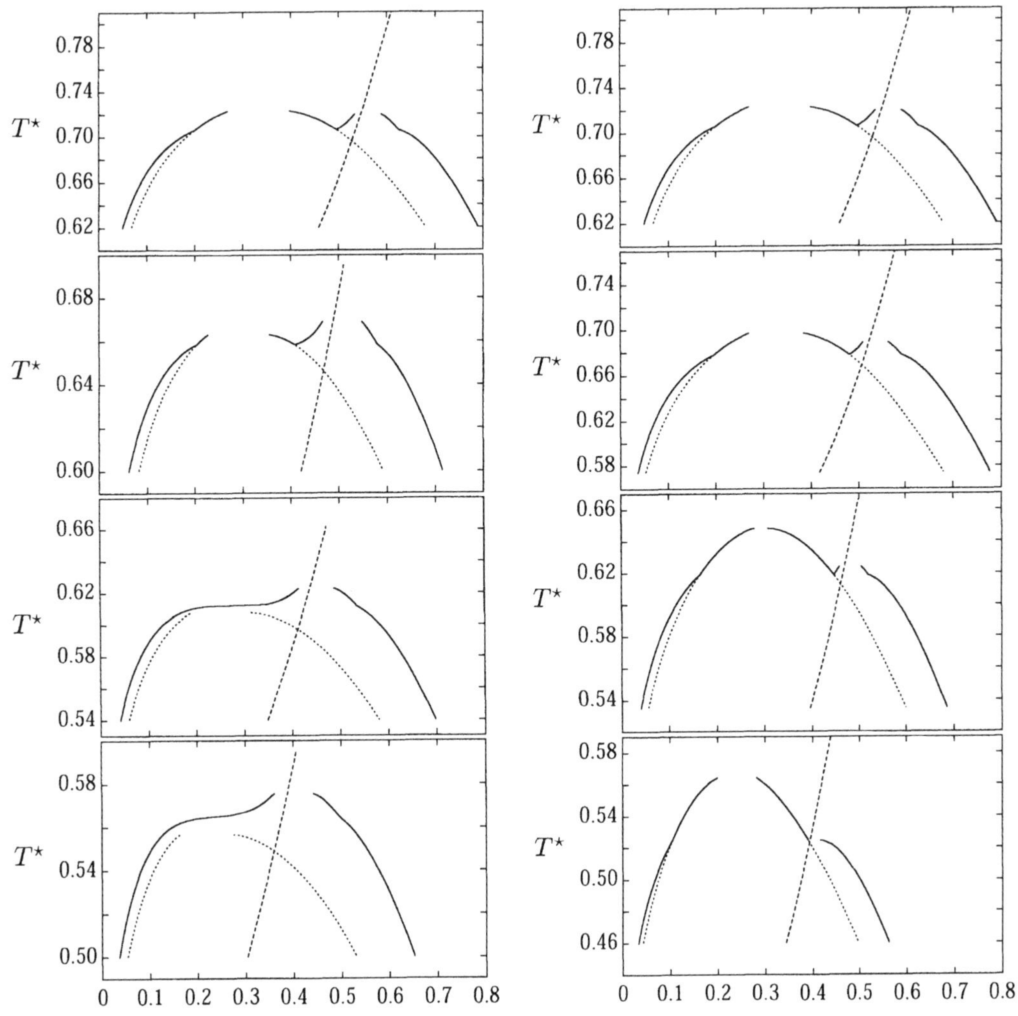

Fig. 8. Variation of the phase diagram of the binary fluid mixture in thermal equilibrium with a porous matrix from MSA with matrix density ρ_0; left column ($y = 0$, $\alpha = 0.7$): $\rho_0 = 0$, 0.05, 0.10, and 0.15 (from top to bottom); right column ($y = 1$, $\alpha = 0.7$): $\rho_0 = 0$, 0.05, 0.15, and 0.30 (from top to bottom); full line: G-L, G-DF, or L-DF coexistence curves; dotted line: metastable G-L transition; dashed line: λ-line; from Ref. [22]

at $\rho_0 = 0$, where the λ-line of the second order demixing transition terminates, as well as a triple point where G, L, and DF coexist. As we increase ρ_0, at $y = 1$ (attractive tail in the matrix–fluid interaction) the tricritical temperature T_{tc} decreases; at $\rho_0 \sim 0.3$ the first order transition between L and DF has vanished, giving rise to a CEP at T_{cep} (type I phase diagram). It can be noted that the existence of a CEP leads to a kink in the G-L curve, clearly visible in the MSA data, a phenomenon which has been discussed in a simulation study of the pure mixture [24]. It is rather surprising that for the case of a pure hard sphere matrix–fluid interaction ($y = 0$) a different scenario is observed: increasing ρ_0 now leads to a type III phase diagram; for $\rho_0 \sim 0.1$ the G-L transition becomes metastable and hidden below the G-DF transition (type III phase diagram). This points out very nicely the strong influence of the matrix (external system parameter) on the phase behaviour of the fluid.

Variation of z

Finally, we have varied the screening length z of the potential. Since its influence is less dramatic than any of the other system parameters, we do not present a figure but rather summarize the results: variation of the screening length $z\sigma$ from 2 to 3 lowers the tricritical temperature and the critical temperature of the metastable G-L transition; the type of the phase diagram is unaffected.

Conclusions

In this contribution we have discussed the phase behaviour and the criticality of a binary fluid mixture, restricting ourselves to a symmetric binary fluid. Here the potentials between the like particles are equal, but differ, on the other hand, from the unlike interactions; the attraction between particles of different species is assumed to be weaker than the like particle interaction. We have first considered the case of the binary bulk fluid and have calculated structural and thermodynamic properties *via* the simple (and thermodynamically inconsistent) MSA and the sophisticated (and thermodynamically self-consistent) SCOZA models. We have seen – leaving the solid phases aside – three coexisting phases: the equimolar gas (G), the equimolar liquid (L), and the nonequimolar (demixed) fluid (DF). As a consequence of the symmetry of the system we encounter three types of phase diagrams that are characterized by the different loci where the λ-line (*i.e.* the line of critical points of the DF transition) intersects the second order G-L transition. We show by using the simple MSA that the three types of phase diagram are induced by the parameter that measures the strength of the unlike interaction. The SCOZA helps us to get more quantitative results: we have calculated critical points and have given a stability analysis (both with respect to mechanical as well as to chemical stability). We then immerse this symmetric binary mixture in a porous matrix, realized by a frozen configuration of a liquid. Structural and thermodynamic quantities are again calculated from liquid state theory; however, one has to apply a special trick (replica trick) to take into account the fact that in this special ternary mixture (binary liquid + matrix) the matrix particles interact with the liquid particles, but are assumed to be immobile: there is no structural response of the fluid to the matrix. Here we have used only MSA to calculate the properties of the fluid. By systematically varying the system parameters we have shown that the trends between the three types of phase diagrams are not only triggered by the internals system parameters (*i.e.* the parameters that characterize the fluid), but also by varying the external parameters (*i.e.* the matrix properties and the interaction fluid–matrix). This might be of relevance for technological applications (catalysis, adsorption, *etc.*) where the phase behaviour of the fluid can be influenced by varying the (chemical) properties of the matrix. Extension of this concept by considering fluids with orientational interactions (dipolar and molecular fluids) and charged particles is currently in progress and will be presented in due course.

Acknowledgments

This work was supported by the *Österreichische Forschungsfonds* under Project Nos. P13062-TPH and P14371-TPH and the *Wiener Handelskammer. E. Schöll-Paschinger* is a student of the Graduate Program "Computational Material Science" of the *Österreichische Forschungsfonds* (Proj. No. W004). The authors are indebted to Profs. *D. Levesque* and *J.-J. Weis* at the Laboratoire de Physique Théorique (Université Paris-Sud, Orsay) for providing computer simulation data and for many stimulating discussions.

References

[1] van Konynenburg HP, Scott RL (1980) Philos Trans R Soc Ser A **298**: 495

[2] Wilding NB, Schmid F, Nielaba P (1998) Phys Rev E **58**: 2201.

[3] Rosinberg M-L (1999) In: Caccamo C, Hansen J-P, Stell G (eds) New Approaches to Problems in Liquid State Theory. NATO Science Series, Kluwer, Dordrecht

[4] Wong APY, Chan MHW (1990) Phys Rev Lett **65**: 2569; Wong APY, Kim SB, Goldburg WI, Chan MHW (1993) Phys Rev Lett **70**: 954; Chan M, Mulders N, Reppy J (1996) Physics Today **49**: 30

[5] Kim SB, Ma J, Chan MHW (1993) Phys Rev Lett **71**: 2268; Tulimieri DJ, Yoon J, Chan MHW (1999) Phys Rev Lett **82**: 121 and references therein.

[6] See e.g. Frisken BJ, Ferri F, Cannell DS (1995) Phys Rev E **51**: 5922 and references therein; Zhuang Z, Casielles AG, Cannell, DS (1996) Phys Rev Lett **77**: 2969; Maher JV, Goldburg WI, Pohl DW, Lanz M (1984) Phys Rev Lett **53**: 60

[7] See e.g. Goh MC, Goldburg WI, Knobler CM (1987) Phys Rev Lett **58**: 1008; Dierker SB, Wiltzius P (1987) Phys Rev Lett **58**: 1865; Wiltzius P, Dierker SB, Dennis BS (1989) Phys Rev Lett **62**: 804; Dierker SB, Wiltzius P (1991) Phys Rev Lett **66**: 1185; Aliev F, Goldburg WI, Wu X-l (1993) Phys Rev E **47**: 3834 Lin MY, Sinha SK, Drake JM, Wu X-l, Thiyagarajan P, Stanley HB (1994) Phys Rev Lett **72**: 2207; Lacelle S, Tremblay L, Bussière Y, Cau F, Fry CG (1995) Phys Rev Lett **74**: 5228; Tulimieri DJ, Yoon J, Chan MHW (1999) Phys Rev Lett **82**: 121; Formisano F, Teixeira J (2000) Eur Phys J E **1**: 1

[8] Weis J-J, Nijmeijer MJP, Tavares JM, Telo da Gama MM (1997) Phys Rev E **55**: 436; Tavares JM, Telo da Gama MM, Teixeira PIC, Weis J-J, Nijmeijer MJP (1995) Phys Rev E **52**: 1915.

[9] Groh B, Dietrich S (1994) Phys Rev Lett **72**: 2422; Groh B, Dietrich S (1999) In: Caccamo C, Hansen J-P, Stell G (eds) New Approaches to Problems in Liquid State Theory. NATO Science Series, Kluwer, Dordrecht

[10] Ornstein LS, Zernike F (1914) Proc Acad Sci (Amsterdam) **17**: 793

[11] Hansen J-P, McDonald IR (1986) *Theory of Simple Liquids*, 2nd edn. New York, Academic Press

[12] Høye JS, Stell G (1977) J Chem Phys **67**: 439; Høye JS, Stell G (1984) Mol Phys **52**: 1071; for an overview see: Pini D, Stell G, Wilding NG (1998) Mol Phys **95**: 483

[13] Given JA (1992) Phys Rev A **45**: 816; Given JA, (1992) J Chem Phys **96**: 2287; Given JA, Stell G (1994) Physica A **209**: 495
[14] Paschinger E, Kahl G (2000) Phys Rev E **61**: 5330
[15] Caccamo C (1996) Phys Rep **274**: 1
[16] Edwards SF, Anderson PW (1975) J Phys F **5**: 965; Edwards SF, Jones C (1976) J Phys A **9**: 1595
[17] Schöll-Paschinger E (2001) PhD Thesis, Technische Universität Wien
[18] Høye JS, Blum L (1977) J Stat Phys **16**: 399
[19] Arrieta E, Jedrzejek C, Marsh KN (1987) J Chem Phys **86**: 3607
[20] Arrieta E, Jedrzejek C, Marsh KN (1991) J Chem Phys **95**: 6806
[21] Dickman R, Stell G (1996) Phys Rev Lett **77**: 996 Pini D, Stell G, Dickman R (1998) Phys Rev E **57**: 2862
[22] Schöll-Paschinger E, Levesque D, Weis J-J, Kahl G (2001) Phys Rev E **64**: 11502
[23] Lang A, PhD Thesis, Technische Universität Wien
[24] Wilding NB (1997) Phys Rev Lett **78**: 1488; (1997) Phys Rev E **55**: 6624

Received June 27, 2001. Accepted July 2, 2001

Extraction of Unprotected Amino Acids by Mixed-Ligand Nickel(II) and Copper(II) Chelates

Yoshinori Ihara*, **Shinji Kurose,** and **Takashi Koyama**

Department of Chemistry, Faculty of Education, Kanazawa University, Kanazawa 920-1192, Japan

Summary. Six mixed-ligand Nickel(II) and Copper(II) chelates with square-planar geometry of the formula [Ni/Cu(O-O)(*S-tmpn*)]$B(C_6H_5)_4$ were prepared, where O-O represents acetylacetonate, tropolonate, or hinokitiolate and *S-tmpn* is (*S*)-tetramethyl-1,2-propanediamine. The compounds were investigated with respect to their function as receptor for unprotected amino acids, taking advantage of their high solubilities in non-polar organic solvents. In liquid-liquid extraction experiments between a 1,2-dichloroethane phase containing the metal chelates and an aqueous phase containing amino acids (*rac*-phenylglycine, *rac*-phenylalanine, or *rac*-tryptophan), the nickel(II) chelates effectively extracted amino acids from the aqueous phases under neutral conditions, forming octahedral ternary chelates.

Keywords. Mixed-ligand chelate; Extraction of amino acids; Solvatochromism.

Introduction

Mixed-ligand nickel(II) and copper(II) chelates consisting of β-diketonates (*β-dike*) and N- or N,N′-alkylated ethylenediamines (R_nen), [Ni/Cu(*β-dike*)(R_nen)]*X* (*X*: univalent anion such as ClO_4^-), are soluble in polar and non-polar solvents, exhibiting pronounced solvatochromic behavior [1–4]. The metal chelates which show high solubilities in non-polar solvents may serve as receptors to extract guest-ligand molecules from their aqueous solutions, and also to recognize chirality of the guest molecules in liquid(polar)-liquid(non-polar) extraction systems. In fact, *Tsukube et al.* have recently reported that some lanthanide complexes act as useful receptors of amino acids in liquid-liquid extraction experiments between non-polar organic phases containing chiral lanthanide(III) *tris*-(β-diketonate) complexes and aqueous phases containing unprotected amino acids; especially the ytterbium complex offered excellent enantioselectivity in the extraction process [5]. It appears that an attempt to utilize metal complexes as extraction reagents for chiral recognition of amino acids and other substrates is unique and interesting, because the stereochemistry of metal complexes can be designed to control steric effects, and chiral recognition owing to electronic effects on the basis of symmetry of metal orbitals is also anticipated [6–9].

* Corresponding author. E-mail: ihara@ed.kanazawa-u.ac.jp

The investigations on the possibilities of extraction of amino acids and of chiral recognition were performed using transition metal complexes with high solubilities in non-polar solvents. We prepared square-planar nickel(II) and copper(II) mixed-ligand chelates containing (*S*)-tetramethyl-1,2-propanediamine (*S-tmpn*), acetylacetonate (*acac*), tropolonate (*trop*), or hinokitiolate (*hino*) as ligands, which are simple chiral model compounds, and tested whether or not they can extract unprotected amino acids as a preliminary examination directed toward chiral recognition.

Results and Discussion

Extraction of amino acids by the metal chelates

Table 1 summarizes solid state properties and analytical data of the chelates prepared for the extraction of amino acids. The presence of two molecules of crystal water in the Cu-*trop* and Cu-*hino* chelates was confirmed by thermogravimetric analyses. All the chelates have square-planar geometry in the solid state and are in most cases, slightly soluble in water but fairly soluble in most organic solvents as expected, exhibiting pronounced solvatochromic behavior. 1,2-Dichloroethane was selected as non-polar solvent in this study.

Extraction experiments were carried out for all metal chelates except for Cu-*acac* because the aqueous solutions of amino acids separated from dichloroethane phases initially containing the Cu-*acac* were colored after extraction, indicating the entering of Cu(II) ions into the aqueous phases. Table 2 shows the results of extraction for aqueous solutions of three aliphatic amino acids, *i.e.* *rac*-phenylglycine (*Phe-Gly*), *rac*-phenylalanine (*Phe-Ala*), and *rac*-tryptophan (*Trp*), using the above five metal chelates dissolved in dichloroethane.

The nickel(II) chelates displayed satisfactory efficiency as receptors of the amino acids; particularly in the Ni-*acac* systems high extractabilities were observed, and *Phe-Ala* was quantitatively extracted. Equilibrium was attained within 2 hours (Table 2); the *pH* values of the aqueous solutions of amino acids before and after the reactions were nearly constant around neutrality (6.6–7.4), suggesting that the amino acids were extracted in the zwitterionic form.

Table 1. Solid-state properties and analytical data of the chelates

Chelate	Color	μ_{eff}/B.M.	$\tilde{\nu}_{max}/10^3\ cm^{-1}$ [a]	Analytical data[b]		
				C (%)	H (%)	N (%)
[Ni(*acac*)(*S-tmpn*)]$B(C_6H_5)_4$	red-orange	diamag	20.4	70.94 (71.21)	7.50 (7.47)	4.67 (4.61)
[Ni(*trop*)(*S-tmpn*)]$B(C_6H_5)_4$	orange	diamag	20.5	71.99 (72.53)	6.75 (6.89)	4.35 (4.45)
[Ni(*hino*)(*S-tmpn*)]$B(C_6H_5)_4$	orange	diamag	20.7	73.62 (73.35)	7.46 (7.36)	4.25 (4.17)
[Cu(*acac*)(*S-tmpn*)]$B(C_6H_5)_4$	blue-violet	1.88	19.2	70.06 (70.64)	7.46 (7.41)	4.53 (4.58)
[Cu(*trop*)(*S-tmpn*)]$B(C_6H_5)_4 \cdot 2H_2O$	violet	1.96	20.1	68.51 (68.11)	6.68 (7.07)	4.13 (4.18)
[Cu(*hino*)(*S-tmpn*)]$B(C_6H_5)_4 \cdot 2H_2O$	brown-violet	1.84	20.5	69.23 (69.14)	7.11 (7.50)	3.86 (3.93)

[a] Wave number of the d-d band in the reflectance spectrum; [b]calculated values in parentheses

Table 2. Extraction of amino acids by the metal chelates (2 h → 4 h[a])

	Extraction (%)		
Chelate	*Phe-Gly*	*Phe-Ala*	*Trp*
[Ni(*acac*)(*S-tmpn*)]$B(C_6H_5)_4$	40→38	86→90	50→52
[Ni(*trop*)(*S-tmpn*)]$B(C_6H_5)_4$	10→9	28→30	24→28
[Ni(*hino*)(*S-tmpn*)]$B(C_6H_5)_4$	3	20→19	16→20
[Cu(*trop*)(*S-tmpn*)]$B(C_6H_5)_4$	4	0	0
[Cu(*hino*)(*S-tmpn*)]$B(C_6H_5)_4$	6→7	0	0

[a] Reaction times

Extraction profiles of amino acids

The Ni-*trop* chelate is considered to form a 1:1 adduct with the amino acids on the basis of the following experiments. When the total concentration of the chelate in the dichloroethane phase and *Phe-Ala* in the aqueous phase was fixed at 10 m*M*, the extracted amounts of *Phe-Ala* showed a bell-shaped dependence on the mole fraction of *Phe-Ala*, and the maximum was observed at 1:1 stoichiometry (Fig. 1). In addition, when the concentration of the chelate in the dichloroethane phase was fixed at 2.5 m*M*, the extracted amount of *Phe-Ala* increased with its concentration in the aqueous phase (0–15 m*M*); the saturated amount of *Phe-Ala* extracted is anticipated to approximate unity which also indicates the formation of a 1:1 adduct in the dichloroethane phase (Fig. 2).

On the other hand, a similar plot for Ni-*acac* and *Phe-Ala* (Fig. 3) suggests the formation of 1:2 and/or 1:3 adducts, in which the chelate accepts 2 and/or 3 *Phe-Ala*

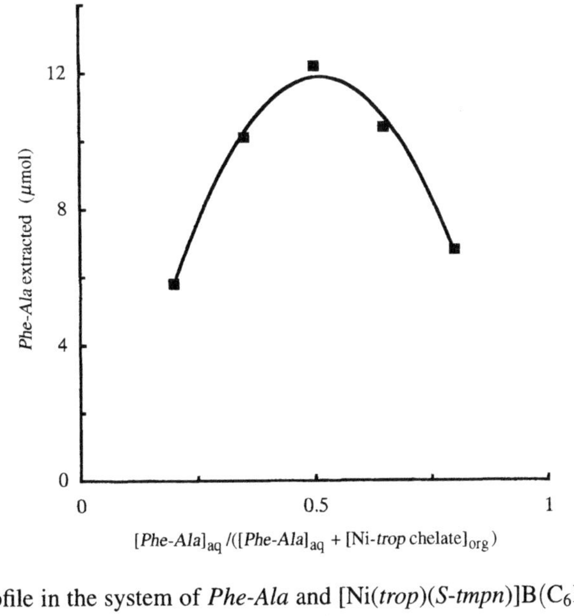

Fig. 1. Extraction profile in the system of *Phe-Ala* and [Ni(*trop*)(*S-tmpn*)]$B(C_6H_5)_4$ at constant total concentration

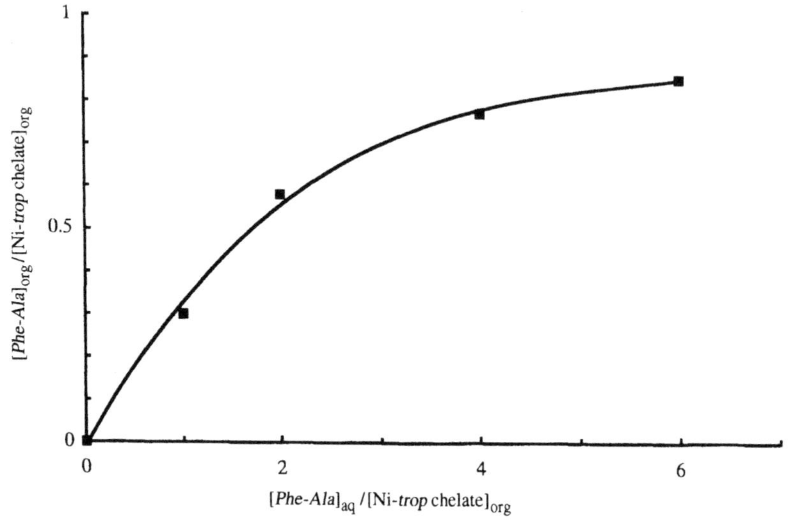

Fig. 2. Extraction profile in the system of *Phe-Ala* and [Ni(*trop*)(*S-tmpn*)]$B(C_6H_5)_4$ with increasing concentration of *Phe-Ala*

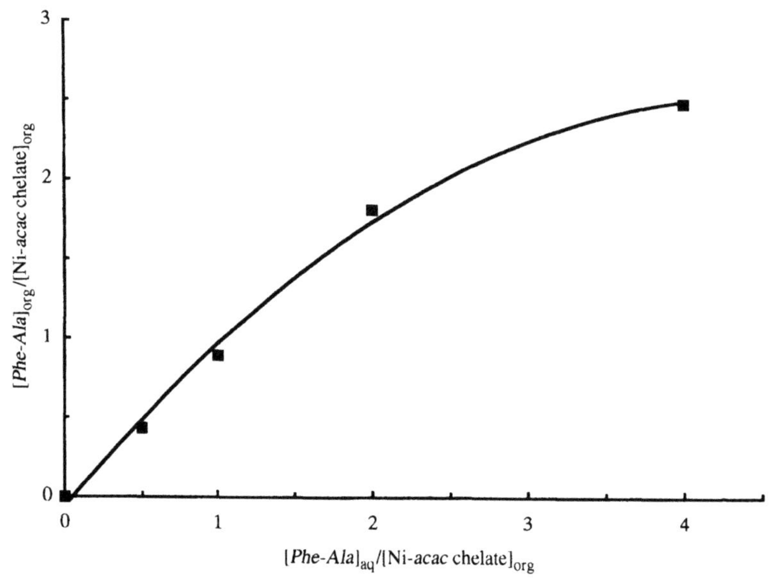

Fig. 3. Extraction profile in the system of *Phe-Ala* and [Ni(*acac*)(*S-tmpn*)]$B(C_6H_5)_4$ with increasing concentration of *Phe-Ala*

molecules per chelate molecule. This result implies that the reactions of Ni-*acac* proceed by a mechanism different from Ni-*trop*.

Spectroscopic changes of the dichloroethane phases

When reacting nickel(II) chelates with amino acids, the dichloroethane phases containing chelates with an effective extraction ability turned into weak green

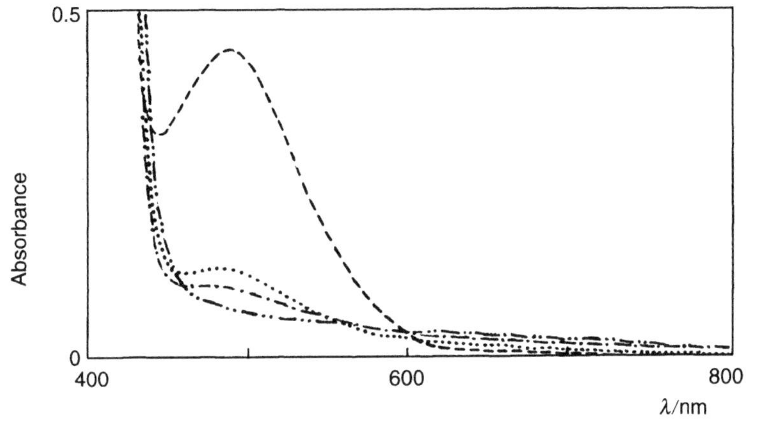

Fig. 4. Spectroscopic changes of the dichloroethane phases in the system of *Phe-Ala* and [Ni(*trop*)(*S-tmpn*)]B$(C_6H_5)_4$ with increasing concentration of *Phe-Ala*: $[\text{chelate}]_{org} = 2.5\,\text{m}M$ (- - -); $[Phe\text{-}Ala]_{aq}/[\text{chelate}]_{org} = 2.5\,\text{m}M/2.5\,\text{m}M$ (· · ·); $[Phe\text{-}Ala]_{aq}/[\text{chelate}]_{org} = 5.0\,\text{m}M/2.5\,\text{m}M$ (– · –); $[Phe\text{-}Ala]_{aq}/[\text{chelate}]_{org} = 10\,\text{m}M/2.5\,\text{m}M$ (– · · –)

characteristic for octahedral species. Figure 4 shows the UV/Vis spectroscopic changes of the dichloroethane phases in the system Ni-*trop*/*Phe-Ala* as a function of increasing concentration of *Phe-Ala* in the aqueous phase at constant concentration of the chelate in the dichloroethane phase. It can be seen that a strong absorption band at about 530 nm, which is observed in the absence of *Phe-Ala* and is evidently due to the square-planar chelate, decays gradually, and a very weak absorption appears in the red region. The total amount of *Phe-Ala* extracted increases with increasing relative concentration of *Phe-Ala* (Fig. 2); the spectroscopic changes represented by a drop of intensity at about 530 nm must therefore result from an increase of the amount of octahedral species newly formed in the dichloroethane phase. Accordingly, it seems reasonable to assume that the spectroscopic changes correspond mainly to structural conversion of the chelate from square-planar to octahedral. In addition, the fact that the formation of the adduct took place at a molar ratio of 1:1 for Ni-*trop* and *Phe-Ala* (Figs. 1 and 2) indicates that [Ni(*trop*)(S-*tmpn*)]$^+$ incorporates an amino acid in its coordination sites to produce the ternary octahedral complex [Ni(*Phe-Ala*)(*trop*)(*S-tmpn*)] in the extraction process.

The system of Ni-*acac* and *Phe-Ala* exhibits the same features with respect to the spectroscopic changes of the dichloroethane phases upon an increase of the relative concentration of *Phe-Ala* in the aqueous phase. From a higher extractability and the extraction profile (Fig. 3) it is supposed that the octahedral species consist of not only the ternary complex similar to that of the Ni-*trop* chelate, but other adducts such as complexes in which amino acid is combined with the *acac* ion by ligand reaction are also formed.

Further investigations will be directed towards a discussion of the extraction mechanism, the effects of different amino acids, and the possibility of enantio-selective recognition.

Experimental

Materials

(*S*)-Tetramethyl-1,2-propanediamine (*S-tmpn*) was obtained by N-methylation of (*S*)-1,2-propanediamine [10]. Tropolone (*trop*) and hinokitiol (*hino*) were supplied by Tokyo Chem. Ind. Co. The mixed metal chelates were prepared as follows: Copper(II) chloride dihydrate (2 mmol) was dissolved in MeOH (5 cm^3), and a methanolic solution (5 cm^3) containing (*S*)-*tmpn* (4 mmol) and *acac*, *trop*, or *hino* (2 mmol) was added with stirring. After Na_2CO_3 (1 mmol) was added to the solution, it was gently heated for 1 h and then filtered. The crude crystals obtained upon addition of sodium tetraphenylborate (3 mmol) to the filtrate were recrystallized from dichloroethane.

Nickel(II) chloride hexahydrate (2 mmol) was dissolved in MeOH (5 cm^3), and a methanolic solution (5 cm^3) containing *S-tmpn* (4 mmol) and *acac*, *trop*, or *hino* (2 mmol) was added with stirring. After Na_2CO_3 (1 mmol) was added to the solution, it was gently heated to dryness. The resulting oily solid was treated with dichloroethane, and sodium tetraphenylborate (3 mmol) was added. The insoluble residue was filtered off, and the crude crystals precipitated from the filtrate were recrystallized from dichloroethane.

Extraction experiments

A dichloroethane solution of metal chelate (5 m*M*, 5 cm^3) was mixed with an aqueous solution of racemic amino acid (2.5 m*M*, 5 cm^3). After the mixture had been stirred for 2 h at room temperature, the aqueous phase was separated and characterized. The extraction percentage of amino acid was determined on the basis of the estimation of the residual amino acid in the aqueous phase by HPLC analysis (Kanto Chem. Co. Inc., Mightysil RP-18). Extraction values in Table 2 (%) were calculated as follows: extraction (%) = (((amino acid extracted in the presence of metal chelate) – (amino acid extracted in the absence of metal chelate))/(amino acid intially contained in the aqueous solution)) × 100.

Measurements

The electronic spectra were recorded with a Jasco V-570 spectrophotometer. The effective magnetic moments were evaluated from the susceptibilities measured by a Sherwood magnetic balance MSB-MKI. Thermogravimetry was carried out with a Seiko SSC/580 TG/DTA-30 apparatus.

References

[1] Sone K, Fukuda Y (1987) Inorganic Thermochromism. Springer, Heidelberg
[2] Tsiamis C, Alkam HH, Hadjikostas CC (1994) Inorg Chim Acta **221**: 17
[3] Ihara Y, Yoshizakiya M, Yoshiyama R, Sone K (1996) Polyhedron **15**: 3643
[4] Ihara Y, Teranishi K, Hirose N, Sone K (1998) Polyhedron **17**: 3247
[5] Tsukube H, Shinoda S, Uenishi J, Kanatani T, Itoh H, Shiode M, Iwachido T, Yonemitsu O (1998) Inorg Chem **37**: 1585
[6] Tsukube H (1982) Angew Chem Int Ed Engl **21**: 304
[7] Aoyama Y, Asakawa M, Yamagishi A, Toi H, Ogoshi H (1990) J Am Chem Soc **112**: 3145
[8] Beer PD, Chen Z, Goulden AJ, Grieve A, Hesek D, Szemes F, Wear T (1994) Chem Commun **1994**: 1269
[9] Campagna T, Grasso G, Rizzarelli E, Vecchio G (1998) Inorg Chim Acta **275–276**: 395
[10] Moshier RW, Spialter L (1956) J Org Chem **21**: 1050

Received February 19, 2001. Accepted April 2, 2001

Solvent Effects on Ion-Pair Distribution and Dimerization of Tetraalkylammonium Salts

Kiyoshi Sawada*, **Eiji Takahashi**, **Tomokazu Horie**, and **Keiichi Satoh**

Department of Chemistry, Faculty of Science, Niigata University, Niigata 950-2181, Japan

Summary. The distribution of tetraalkylammonium ions $(C_nH_{2n+1})_4N^+$ (R^+, $TAAn^+$, $n=4$–7) with picrate ion (pic^-) and inorganic anions X^-, (Cl^-, Br^-, ClO_4^-), into various inert organic solvents was studied at 25.0°C. The distribution data were analyzed by taking into consideration the distribution of ion pairs, $R^+ \cdot X^-$, and the dimerization of the ion pairs, $(R^+ \cdot X^-)_2$, in the organic phase. The ion-pair, distribution constant, K_{dist}, increases with increasing chain length of the tetraalkylammonium ion and with increasing ionic radius of the anion. The values of K_{dist} show a good correlation with the E_T value of solvent, *i.e.* the solvation ability with respect to the anion, and smoothly increase with increasing E_T. The effect of the solvent on the dimerization constants, K_{dim}, is markedly different between the ion pairs of picrate ion and inorganic anions. In the case of picrate, K_{dim} significantly decreases with decreasing length of the alkyl chain of the tetraalkylammonium ion, but hardly changes by changing the solvent. On the other hand, in the case of ion pairs of inorganic anions the value of K_{dim} decreases with decreasing E_T and is almost constant for all anions. These results were reasonably explained by the difference of the solvation of the anion moieties of the monomeric and dimeric ion pairs.

Keywords. Anions; Cations; Electrolytes; Equilibrium; Solvent effect.

Introduction

Tetraalkylammonium ($TAAn^+$) salts with short alkyl chains, such as methyl and ethyl groups, are the most common electrolytes for non-aqueous solvents, and their thermodynamic properties in a variety of solvents have been studied. $TAAn^+$ salts with long alkyl chains are quite popular as cationic surfactants and ion-pair extraction reagents. Various kinds of $TAAn^+$ salts have been used for the separation and determination of anionic compounds and complexes [1], and the fundamental data of the distribution equilibria and the thermodynamic properties of these ion pairs have extensively been collected [2]. It has been reported that the logarithmic distribution constants of ion pairs, $\log K_{dist}$, is approximately given by the sum of the contributions of cation and anion and that $\log K_{dist}$ linearly correlates with the number of the methylene groups of the tetraalkylammonium ion [3–5]. The mechanism of ion-pair distribution, however, is not so simple. Although the electrolytes are present in a form of ion pair in low permittivity (ε) solvents, the

* Corresponding author. E-mail: sawada@sc.niigata-u.ac.jp

dissociation of the ion pair must be taken into consideration even in relatively low permittivity solvents such as 1,2-dichloroethane ($\varepsilon = 10.4$) [6–11]. Thus, the distribution of ion pairs has been reasonably rationalized by two elemental steps: the distribution of each ion from aqueous phase to organic phase and the ion-pair formation in the organic phase [9, 10].

In addition to ion-pair dissociation, the formation of higher aggregates of ions such as triple and quadruple ions is important in low ε solvents [6, 7, 10]. In a previous paper, we have studied the dimerization of ion pairs (quadruple ions) of a series of tetraalkylammonium salts in xylene and reported the effects of the chain length of $TAAn^+$ and of the ion size of counter anions [10]. Thus, the elucidation of side reactions of electrolytes in organic solvents, such as ion-pair dissociation and aggregation, is quite important not only from the viewpoint of solution thermodynamics but also for the exact treatment of the distribution data.

In the present investigation, the distribution of tetraalkylammonium salts into various organic solvents was studied, and the effects of the solvent and of the structure of the electrolyte on the distribution and dimerization equilibria of ion pairs are discussed.

Results and Discussion

Distribution equilibria

The distribution equilibrium of monomeric ion pairs of tetraalkylammonium ion $TAAn^+$ (R^+) with counter anions (X^-) is given by

$$R^+ + X^- \overset{K_{\mathrm{dist}}}{\rightleftharpoons} R^+ \cdot X^-_{,\mathrm{org}} \tag{1}$$

where the subscript 'org' refers to the organic phase. The distribution constant, K_{dist}, of the ion pair $R^+ \cdot X^-$ is defined by

$$K_{\mathrm{dist}} = \frac{[R^+ \cdot X^-]_{\mathrm{org}}}{[R^+][X^-]} \tag{2}$$

where the square brackets denote the equilibrium concentration in solution. The distribution ratio of R^+, D, is obtained from the total concentrations of R^+ in the organic phase, $C_{R^+,\mathrm{org}}$ and in the aqueous phase, C_{R^+}, as $D = C_{R^+,\mathrm{org}}/C_{R^+}$. As the side reactions for R^+ and X^- in the aqueous phase are negligible, the equilibrium concentrations $[R^+]$ and $[X^-]$ are given by the total concentrations C_{R^+} and C_{X^-}, respectively. If we assume the distribution of a monomeric species $R^+ \cdot X^-$ into the organic phase, the equilibrium concentration of the ion pair, $[R^+ \cdot X^-]_{\mathrm{org}}$, is equal to the total concentration of R^+ in the organic phase, $C_{R^+,\mathrm{org}}$. Thus,

$$\log D - \log C_{X^-} = \log K_{\mathrm{dist}} \tag{3}$$

The distribution ratio of the tetraalkylammonium ion $(C_nH_{2n+1})_4N^+$ (R^+, $TAAn^+$, $n = 4$–7) with an anion X^- (picrate ion (pic^-) or inorganic anion (Cl^-, Br^-, ClO_4^-)) into an inert organic solvent (chloroform, bromobenzene, chlorobenzene, 1,1,1-trichloroethane, benzene, toluene, xylene, carbon tetrachloride, *n*-hexane) was measured as a function of $TAAn^+ \cdot X^-$ concentration. Results (plots of $\log D - \log C_{X^-}$ *vs.* $\log C_{R^+,\mathrm{org}}$) are shown in Fig. 1 (distribution of various kinds of

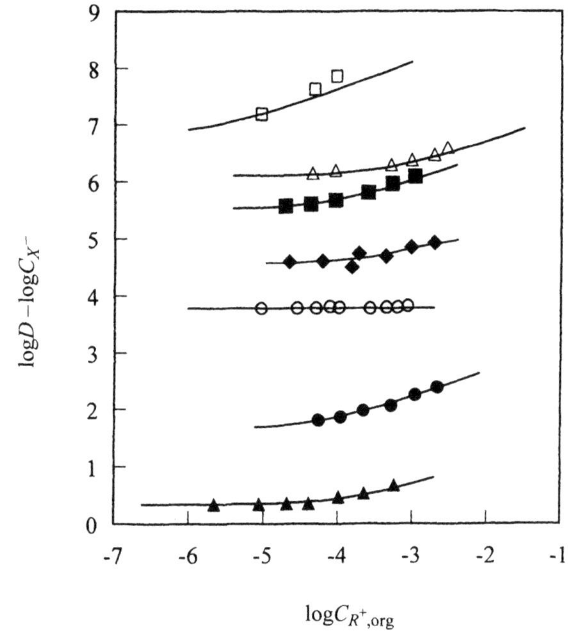

Fig. 1. Plot of $\log D - \log C_{X^-}$ as a function of $\log C_{R^+,org}$ in benzene; $R^+ \cdot X^-$: □, *TAA*6$^+ \cdot$ *pic*$^-$; △, *TAA*5$^+ \cdot$ *pic*$^-$; ○, *TAA*4$^+ \cdot$ *pic*$^-$; ■, *TAA*6$^+ \cdot ClO_4^-$; ◆, *TAA*7$^+ \cdot Br^-$; ●, *TAA*6$^+ \cdot Br^-$; ▲, *TAA*6$^+ \cdot Cl^-$; solid lines are the normalized curves (see text)

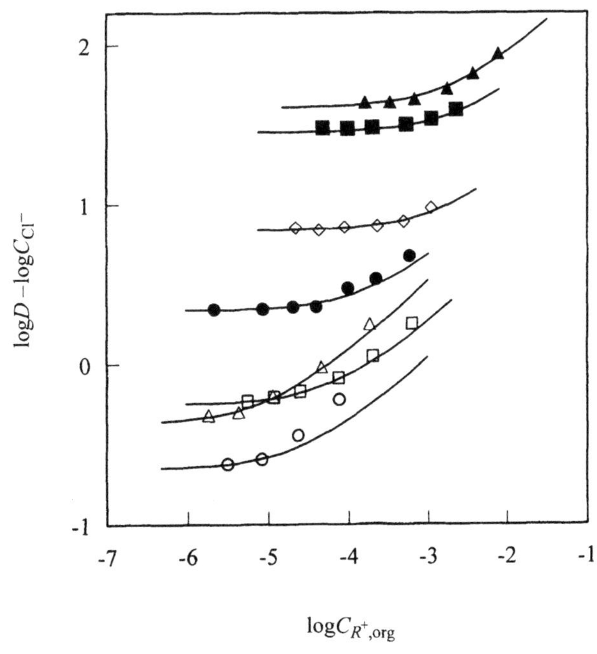

Fig. 2. Plot of $\log D - \log C_{Cl^-}$ as a function of $\log C_{R^+,org}$ for the *TAA*6$^+ \cdot Cl^-$ system; □, toluene; △, xylene; ○, carbon tetrachloride; ◇, 1,1,1-trichloroethane; ■, chlorobenzene; ●, benzene; ▲, bromobenzene; solid lines are the normalized curves

$R^+ \cdot X^-$ into benzene) and Fig. 2 (distribution of $TAA6^+ \cdot Cl^-$ into various solvents) as examples. Similar results were obtained for other $TAAn^+ \cdot X^-$ and solvent systems.

By the assumption of the presence of only a monomeric species in the organic phase, the value of $\log D - \log C_{R^+}$ ($\log K_{dist}$) must be constant, irrespective of $[R^+ \cdot X^-]$, *i.e.* $\log C_{R^+,org}$ (Eq. (3)). Nevertheless, the slope of the plot increases with increasing $\log C_{R^+,org}$, although it approaches zero at low $C_{R^+,org}$. These facts indicate the formation of polymeric species of $R^+ \cdot X^-$ ion pairs in the organic phase at higher $C_{R^+,org}$. By assuming a formation of dimeric species $(R^+ \cdot X^-)_2$ (quadruple ions) in the organic phase,

$$2(R^+ \cdot X^-)_{org} \overset{K_{dim}}{\rightleftharpoons} (R^+ \cdot X^-)_{2,org} \tag{4}$$

the dimerization constant is defined by

$$K_{dim} = \frac{[(R^+ \cdot X^-)_2]_{org}}{[R^+ \cdot X^-]^2_{org}}. \tag{5}$$

The total concentration of R^+ in the organic phase is given by

$$C_{R^+,org} = [R^+ \cdot X^-]_{org} + 2[(R^+ \cdot X^-)_2]_{org} = [R^+ \cdot X^-]_{org} + 2K_{dim}[R^+ \cdot X^-]^2_{org} \tag{6}$$

Solving the quadratic equation for $[R^+ \cdot X^-]_{org}$, we obtain

$$[R^+ \cdot X^-]_{org} = \frac{-1 + (1 + 8K_{dim}C_{R^+,org})^{1/2}}{4K_{dim}} \tag{7}$$

The distribution ratio of R^+ can be rewritten as

$$D = \frac{[R^+ \cdot X^-]_{org} + 2K_{dim}[R^+ \cdot X^-]^2_{org}}{[R^+]}. \tag{8}$$

Substitution of Eq. (2) into Eq. (8) leads to

$$D = K_{dist}[X^-](1 + 2K_{dim}[R^+ \cdot X^-]_{org}). \tag{9}$$

Substitution of Eq. (7) into Eq. (9) finally affords

$$\log D - \log C_{X^-} = \log K_{dist} + \log F \tag{10}$$

where $F = 1/2 + 1/2(1 + y)^{1/2}$ and $y = 8K_{dim}C_{R^+,org}$.

The normalized curve $\log F$ *vs.* $\log y$ was compared with the experimental results, *i.e.* the plot of $\log D - \log C_{X^-}$ *vs.* $\log C_{R^+,org}$. As shown in Figs. 1 and 2, the experimental results comply well with the normalized curve (solid lines); a similar fits was obtained for other systems. Thus, it was confirmed that the ion pairs of tetraalkylammonium salts in the organic phase are generally in equilibrium between the monomeric and dimeric species. The logarithmic distribution constant, $\log K_{dist}$, and the logarithmic dimerization constant, $\log K_{dim}$, evaluated by this curve fitting are listed in Tables 1 and 2, respectively. The values of distribution constant listed in Table 1 were converted to the mole fraction constants, K_{dist}(mf), because the mole fraction unit is the most suitable concentration unit to discuss thermodynamic constants among different solvents. These constants are the

Table 1. Logarithmic ion-pair distribution constants of $R^+ \cdot X^-$ ($\log K_{dist}(mf)$)

Solvent	$TAA4^+ \cdot pic^-$	$TAA5^+ \cdot pic^-$	$TAA6^+ \cdot pic^-$	$TAA6^+ \cdot ClO_4^-$	$TAA7^+ \cdot Br^-$	$TAA6^+ \cdot Br^-$	$TAA6^+ \cdot Cl^-$
Chloroform	8.24	10.24	11.03				
Bromobenzene						5.53	3.97
Chlorobenzene	7.39	9.68	10.90	9.50		5.75	4.11
1,1,1-Trichloroethane						4.87	3.32
Benzene	6.20	8.53	9.89	7.98	6.76	4.06	2.77
Toluene				7.67		3.53	2.27
Xylene	5.69	8.06	9.88	7.27	5.83	3.42	1.93
Carbon tetrachloride	4.65	7.14	9.62	6.22	5.69	3.32	1.81
n-Hexane	2.31	4.91	7.63				

Table 2. Logarithmic ion-pair dimerization constants for $(R^+ \cdot X^-)_2$ in the organic phase ($\log K_{dim}$)

Solvent	$TAA4^+ \cdot pic^-$	$TAA5^+ \cdot pic^-$	$TAA6^+ \cdot pic^-$	$TAA6^+ \cdot ClO_4^-$	$TAA7^+ \cdot Br^-$	$TAA6^+ \cdot Br^-$	$TAA6^+ \cdot Cl^-$
Chloroform	~2	3.1	3.9				
Bromobenzene						2.1	2.0
Chlorobenzene	~2	3.1	3.6	2.2		2.1	2.1
1,1,1-Trichloroethane						2.2	2.3
Benzene	~2	2.9	3.8	3.3	3.2	3.7	3.3
Toluene				3.5		4.2	3.6
Xylene	~2	2.7	3.8	4.2	4.2	4.4	4.4
Carbon tetrachloride	~2	2.8	3.6	4.9	5.1	4.7	3.9
n-Hexane	~2		3.7				

thermodynamic constants, *i.e.* the activity coefficients of ions in the aqueous phase were corrected with the help of the *Davies* equation [12]. The results for systems in which the distribution of $TAAn^+ \cdot X^-$ is too high or too low are omitted from the Tables because of very large experimental errors. For some systems, no reliable results could be obtained because of the formation of an emulsion at the interface of the aqueous and organic phases.

Some plots of $\log D - \log C_{X^-}$ *vs.* $\log C_{R^+,\mathrm{org}}$ show a deviation from the normalized curve at high $C_{R^+,\mathrm{org}}$, *i.e.* the slope of the experimental plot is higher than predicted from the formation of dimeric species. These results suggest a formation of trimeric or higher polymeric species, $(R^+ \cdot X^-)_{m,\mathrm{org}}$ $(m > 2)$, in the higher $TAAn^+ \cdot X^-$ concentration region. On the other hand, in the case of $TAA4^+ \cdot pic^-$ the value of $\log D - \log C_{X^-}$ hardly changes with increasing $C_{R^+,\mathrm{org}}$ in any solvent system as can be seen from Fig. 1. This fact indicates that $TAA4^+ \cdot pic^-$ does not form the dimeric species under the present experimental conditions, *i.e.* the dimerization constant is very small. The dimerization constant of $TAA4^+ \cdot pic^-$ was tentatively determined as $\log K_{\mathrm{dim}} \approx 2$ from the small increase in $\log D - \log C_{X^-}$ at high $C_{R^+,\mathrm{org}}$. In the case of the solvents of relatively high relative permittivity such as 1,2-dichloroethane ($\varepsilon = 10.4$), the value $\log D - \log C_{X^-}$ decreases as $C_{R^+,\mathrm{org}}$ increases. This result can be reasonably interpreted by the dissociation of ion pair in the organic phase. In the present systems, as can be seen from Figs. 1 and 2 no dissociation of ion pairs was observed because of the low relative permittivity ($\varepsilon < 6$) of the solvents.

Distribution constant

In order to elucidate the solvent effect on the distribution of $TAAn^+ \cdot X^-$, the correlation of $\log K_{\mathrm{dist}}$ with typical solvent parameters, such as the relative permittivity (ε), the *Dimroth-Reichardt* parameter of polarity (E_{T}), and the *Hildebrand* solubility parameter (δ), were examined. The values of these solvent parameters of water-saturated solvents are listed in Table 3. The values of ε and δ are hardly changed by the saturation of solvent with water because of the relatively low solubility of water. The E_{T} values, however, are significantly affected by the presence of water as shown in Table 3 (E_{T}, dry solvent; E_{T}(WS), water-saturated solvent). The donating ability of the solvent used in the present study is very low (Donor Number = 0) [13]. The logarithmic distribution constants, $\log K_{\mathrm{dist}}$, are plotted as a function of E_{T}(WS) in Fig. 3. Those values show a good correlation with E_{T}(WS), *i.e.* it smoothly increases as E_{T} increases. The values of $\log K_{\mathrm{dist}}$ of chloride and bromide systems linearly increase with increasing E_{T}, whereas the slopes of picrate systems decrease with increasing E_{T}(WS). These results indicate that the solvent effect on the distribution of the ion pairs is mainly controlled by the polarity or acceptor nature of the solvent [13], *i.e.* the stability of the ion pair increases with increasing in the interaction of the solvent with the anion moiety of the ion pair. As can be seen from the plots of $\log K_{\mathrm{dist}}$ *vs.* ε (Fig. 4a) and δ (Fig. 4b), $\log K_{\mathrm{dist}}$ shows no reasonable correlation with these parameters; thus, the electrostatic interaction (dipole–dipole interaction) and cohesive energy between the ion pair and solvent may not be a dominant factor of the difference in the stability of the ion pairs among the solvents.

Table 3. Solvent parameters

Solvent	E_T[a] kcal · mol^{-1}	E_T(WS)[b] kcal · mol^{-1}	ε[c]	δ[d] cal$^{1/2}$ · cm$^{-3/2}$
Chloroform	39.1	40.2	4.81	9.94
Bromobenzene	36.6	39.0	5.45	9.87
Chlorobenzene	36.8	38.9	5.69	9.97
1,1,1-Trichloroethane	36.2	36.9	7.24	8.50
Benzene	34.3	35.5	2.28	9.51
Toluene	33.9	35.2	2.38	8.90
Xylene	33.1	34.3	2.40	9.13
Carbon tetrachloride	32.4	33.4	2.24	8.93
n-Hexane	31.0	32.2	1.89	7.57

[a] *Dimroth-Reichardt* parameter of pure solvent; [b] E_T value of water-saturated solvent; [c] relative permittivity; [d] solubility parameter

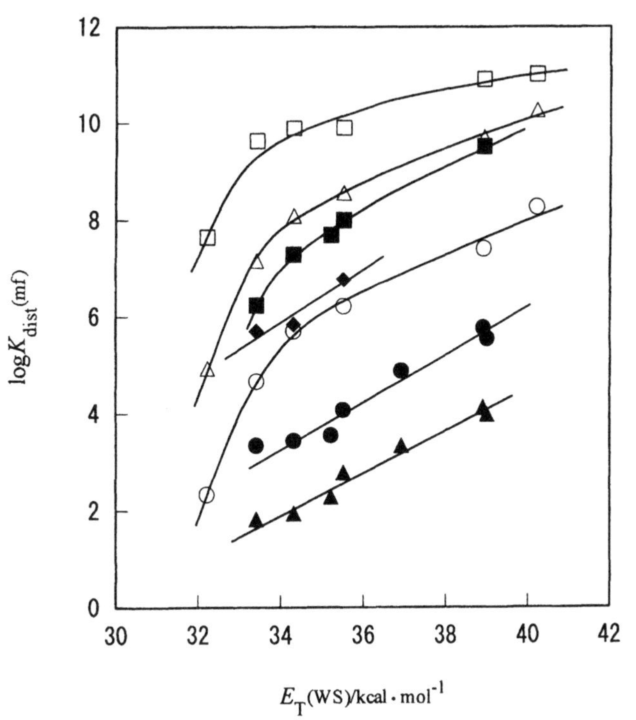

Fig. 3. Plot of the logarithmic ion-pair distribution constant ($\log K_{dist}$(mf)) as a function of E_T; for symbols, see Fig. 1

As can be seen from Table 1, the values of $\log K_{dist}$ of the $TAA6^+$ systems decrease as the ionic radius of the anion decreases. Similar results have been obtained for the $TAA4^+ \cdot X^- - TAA7^+ \cdot X^-$ systems in xylene [10]. A linear correlation of $\log K_{dist}$ with the reciprocal crystal radius of X^- has been reported, and these results were reasonably explained by the hydration energy of the anion in the aqueous phase. The value of K_{dist} increases with increasing in the chain length (n) of the alkyl group of $TAAn^+$ and a linear correlation of $\log K_{dist}$ with n has been observed.

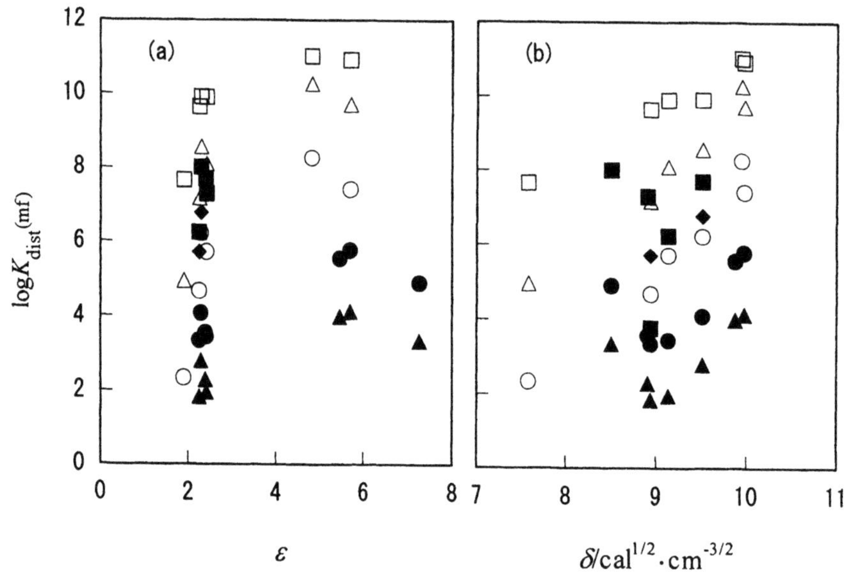

Fig. 4. Plot of the logarithmic ion-pair distribution constant ($\log K_{dist}$(mf)) as a function of ε (a) and δ (b); for symbols, see Fig. 1

The increment of $\log K_{dist}$ ($\Delta\log K_{dist}$) by increasing n by 1 (*i.e.* the total number of methyl groups of $TAAn^+$ is increased by 4) has been reported as 2.5–3.2, irrespective of X^- [10] and solvent [3].

In order to discuss the effect of the solvent on this increment, the difference of $\log K_{dist}$ between two systems of $TAAn^+ \cdot X^-$, $\Delta\log K_{dist}$, was plotted as a function of E_T in Fig. 5. The values of $\Delta\log K_{dist}$ between the inorganic anions are almost the same, irrespective of solvent, although the value of K_{dist} of ClO_4^- in carbon tetrachloride is very small compared with those of other systems. On the other hand, the values of $\Delta\log K_{dist}$ between different $TAAn^+$ of picrates (*i.e.* between $TAAn^+ \cdot pic^-$ and $TAA(n-1)^+ \cdot pic^-$) decrease with decreasing E_T; in particular, $\Delta\log K_{dist}$ between $TAA6^+ \cdot pic^-$ and $TAA5^+ \cdot pic^-$ shows a significant decrease. The values of $\Delta\log K_{dist}$ between $TAAn^+ \cdot X^-$ and $TAA(n-1)^+ \cdot X^-$ are almost the same among the low E_T solvents, and these results agree with those determined for various kinds of inorganic anions in xylene ($\Delta\log K_{dist} = 2.6$). Consequently, it is concluded that the effect of chain length on the distribution is diminished for high E_T solvents, although it hardly changes for low E_T solvents.

Dimerization constant

The logarithmic dimerization constants, $\log K_{dim}$, of $TAAn^+ \cdot X^-$ are plotted as a function of E_T of the solvent in Fig. 6. In the same manner as the extraction constants, the dimerization constants show a good correlation with E_T value, whereas they do not show a reasonable correlation with the parameters δ and ε. The changes of $\log K_{dim}$ by changing the cation, anion, and solvent are significantly different between the picrate ion and inorganic anion systems. In the case of picrate, the value of K_{dim} hardly differs when changing the solvent, and it deceases

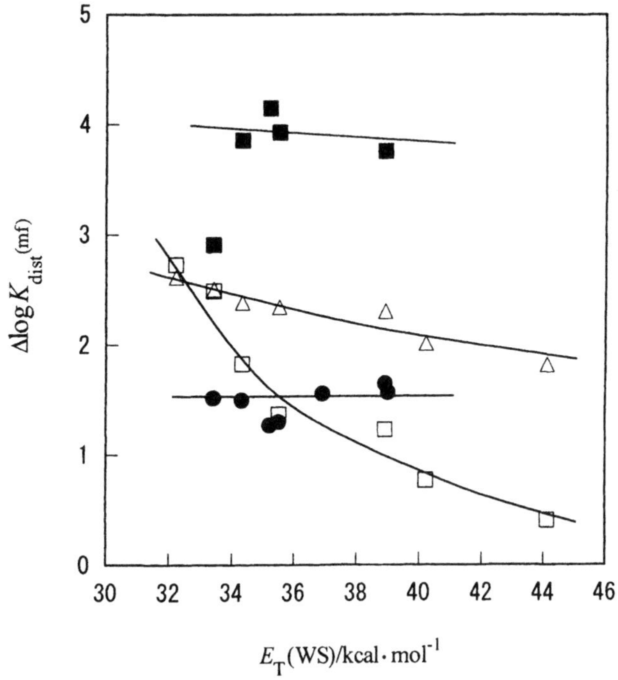

Fig. 5. Plot of the difference of the logarithmic ion-pair distribution constant between two $R^+ \cdot X^-$ species ($\Delta\log K_{dist}$(mf)) as a function of E_T; □, *TAA*6$^+ \cdot pic^-$ – *TAA*5$^+ \cdot pic^-$; △, *TAA*5$^+ \cdot pic^-$ – *TAA*4$^+ \cdot pic^-$; ■, *TAA*6$^+ \cdot ClO_4^-$ – *TAA*6$^+ \cdot Br^-$; ●, *TAA*6$^+ \cdot Br^-$ – *TAA*6$^+ \cdot Cl^-$

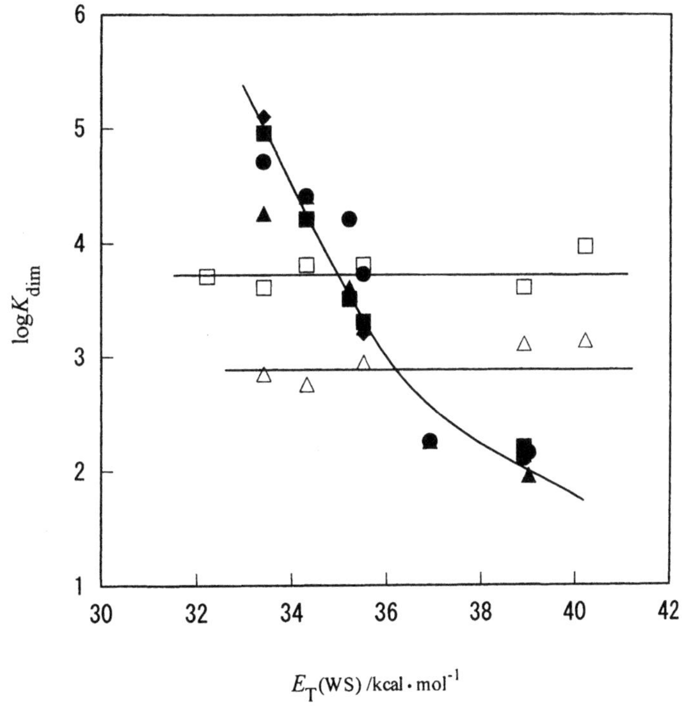

Fig. 6. Plot of the logarithmic dimerization constant ($\log K_{dim}$) as a function of E_T; for symbols, see Fig. 1

by decreasing chain length. On the other hand, logK_{dim} values of $TAAn^+ \cdot X^-$ significantly decrease as the E_T value of the solvent decreases, and the changes are almost the same for a given solvent. As can be seen from the results of $TAA7^+ \cdot X^-$, the dimerization constant of inorganic X^- hardly changes by changing chain length of $TAAn^+$. It also has been confirmed for the xylene system [10] that K_{dim} scarcely changes in the series $TAA4^+$–$TAA7^+$ for any inorganic anion.

The markedly different behaviour of the dimerization constant between the picrate and inorganic ions might be explained by the structure of the ion pairs. The structures of the monomeric ion pairs in the organic solvents have been estimated by means of electroconductivity and distribution equilibria as shown in Figs. 7a (inorganic anion) [8] and 7b (picrate ion) [14]. That is, the interionic distance of $TAAn^+ \cdot X^-$ indicates that the counter anion directly contacts with the positive charge of the central nitrogen atom of $TAAn^+$ in the non-solvating solvents. MO calculation and electronic spectra of the picrate ion suggest that the phenolate oxygen site of the picrate ion approaches to the cation because the highest negative electrostatic potential exists in the vicinity of the phenolate O^- [14], although the negative charge is delocalized to the nitro groups. Molecular models of the dimeric species estimated by molecular mechanics calculation with the help of CS Chem 3D® [15] are shown in Fig. 7c (($TAA6^+ \cdot Cl^-)_2$) and 7d (($TAA6^+ \cdot pic^-)_2$). In the case of inorganic anions of relatively small ion size, the anions of the dimeric form are mostly surrounded by the alkyl groups of the $TAAn^+$ ion. The solvent molecules solvated to the anion of the monomeric $TAAn^+ \cdot X^-$ have to be desolvated for dimer formation. Thus, the formation of the dimer of $TAAn^+ \cdot X^-$ is

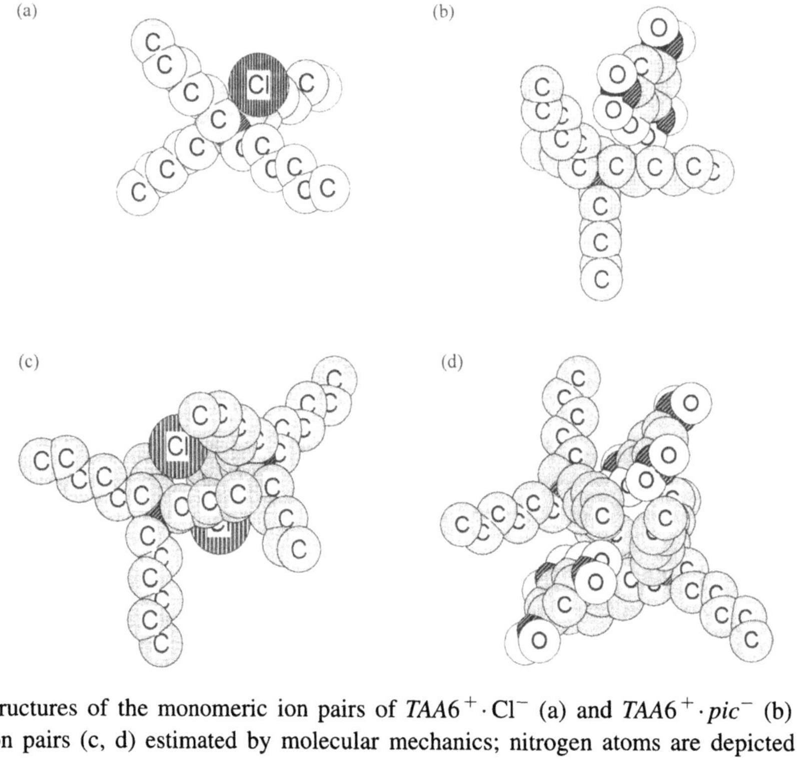

Fig. 7. Structures of the monomeric ion pairs of $TAA6^+ \cdot Cl^-$ (a) and $TAA6^+ \cdot pic^-$ (b) and their dimeric ion pairs (c, d) estimated by molecular mechanics; nitrogen atoms are depicted by black circles, hydrogen atoms are omitted

unfavourable in solvents having a higher solvating ability to the anion, *i.e.* of higher E_T value. On the other hand, the picrate ion is largely exposed to the solvent even in the dimeric form. Consequently, the number of the solvent molecules released by dimerization is small, and the dimerization constant is hardly affected by changing the solvent. The increase in the dimerization constant of inorganic ion systems by increasing the chain length n of the tetraalkylammonium ion may indicate that a peculiar intermolecular interaction operates between the alkyl groups of another $TAAn^+$.

Experimental

Reagents

Reagent grade $TAAn^+ \cdot Br^-$ and $TAA6^+ \cdot Cl^-$ were purchased from Nakarai Tesque and Aldrich Chemicals, respectively. Picric acid and organic solvents were purchased from Wako Pure Chemicals. Sodium picrate was prepared from picric acid by neutralization with sodium hydroxide. Other tetraalkylammonium salts ($TAAn^+ \cdot X^-$) were prepared from $TAAn^+ \cdot Br^-$. An aliquot of a $0.1\,mol \cdot dm^{-3}$ $TAAn^+ \cdot Br^-$ solution in C_6H_6 ($100\,cm^3$) was shaken with $100\,cm^3$ of $1.0\,mol \cdot dm^{-3}$ aqueous $Na^+ \cdot X^-$ solution ($X^- = ClO_4^-$ or pic^-). After separation by centrifugation, the aqueous solution was discarded. By three repetitions of this treatment, the bromide ion in the organic phase was completely replaced by X^-. The organic solvent was washed five times with an equivalent volume of deionized water (Puric Z, Organo).

In order to determine the *Dimroth-Reichardt* E_T values of the water-saturated solvents, the UV/Vis absorption spectra of 2,6-diphenyl-4-(2,4,6-triphenyl-1-pyridinio)-phenoxide (Aldrich Chemicals) in these solvents were measured using a Shimadzu UV-1600 spectrophotometer. E_T(WS) values of organic solvents were calculated from the wavelengths of the absorption maximum [16].

Procedure

A quantity of organic solution containing 10^{-2}–$10^{-5}\,mol \cdot dm^{-3}$ $TAAn^+ \cdot X^-$ ($20\,cm^3$) was shaken for 30 min with $20\,cm^3$ of distilled H_2O or a 0.01 M aqueous $Na^+ \cdot X^-$ solution in a thermostatted water bath ($25.0 \pm 0.05°C$). This shaking time is sufficient for equilibrium to be reached. After the separation by centrifugation, the organic and aqueous phases were evaporated to dryness at 70°C under reduced pressure. After dissolving the residue in a proper amount of benzene, it was shaken with an aqueous $0.1\,mol \cdot dm^{-3}$ sodium picrate solution. It was confirmed that the $TAAn^+$ ion was quantitatively extracted in the benzene phase as $TAAn^+ \cdot pic^-$. The concentration of the picrate ion, *i.e.* the concentration of $TAAn^+$, was determined spectrophotometrically (absorption maximum: 362.0 nm, molar extinction coefficient: $1.77 \times 10^4\,mol \cdot dm^{-3}$). It was confirmed that the sodium ion is not extracted in any of the systems and that any anions are not extracted in the absence of $TAAn^+$.

References

[1] Sawada K, Kikuchi Y (1995) Bunseki 989

[2] a) Czapkiewicz J, Czapkiewicz-Tutaj B (1980) J Chem Soc Faraday Trans 1, **76**: 1663; b) Taft RW, Abbound JLM, Kamlet MJ, Abraham MH (1985) J Solution Chem **14**: 153; c) Taft RW, Abbound JLM, Abraham MH, Doherty RM, Kamlet MJ (1985) J Am Chem Soc **107**: 3105; d) Iwamoto E, Ito K, Yamamoto Y (1981) J Phys Chem **85**: 894; e) Rangana S, Kundu KK (1998) Indian J Chem Sect A **37A**: 678; f) Talukdar H, Kundu KK (1992) J Phys Chem **96**: 970; g) Ueno M, Tsuchihashi N, Imai H, Yoshida K, Ibuki K (1998) High Press Sci Tech 1162; h) Ohtsuka C, Mori Y, Hayashi S, Tsuda T, Wada H (2000) J Liq Chromatogr Relat Technol **23**: 669

[3] Motomizu S (1989) Bunseki Kagaku **38**: 147
[4] Yamamoto K, Adachi K (1998) Talanta **47**: 1065
[5] Miyabe K, Taguchi S, Kasahara I, Goto K (2000) J Phys Chem B **104**: 8481
[6] Charlot G, Tremillon B (1963) Chemical Reactions in Solutions and Melts. Gauthier-Villars, Paris
[7] a) Fuoss RM, Kraus CA (1933) J Am Chem Soc **55**: 2387, 3614; b) Batson FM, Kraus CA (1934) J Am Chem Soc **56**: 2017; c) Robinson RA, Stokes RH (1959) Electrolyte Solutions, Butterworths, London
[8] Sawada K, Chigira F (1995) J Mol Liq **65/66**: 265
[9] a) Kikichi Y, Takahashi N, Suzuki T, Sawada K (1992) Anal Chim Acta **256**: 311; b) Kikuchi Y, Sakamoto Y, Sawada K (1998) J Chem Soc Faraday Trans **94**: 105
[10] Sawada K, Sohara T, Kikuchi Y (1995) J Chem Soc Faraday Trans **91**: 643
[11] Mishustin AI (1997) Zh Fiz Khim **71**: 1959
[12] Robinson RA, Stokes RH (1959) Electrolyte Solutions. Butterworths, London
[13] Burger K (1983) Solvation, Ionic and Complex Formation Reaction in Non-Aqueous Solvents. Academy of Chemistry, Budapest
[14] Sawada K, Chigira F, Satoh K, Komatsuzaki T (1997) J Chem Soc Faraday Trans **93**: 1903
[15] Cambridgesoft (1999) Cambridge, USA
[16] Reichardt C (1988) Solvents and Solvent Effects in Organic Chemistry. VCH, Weinheim

Received May 15, 2001. Accepted (revised) July 18, 2001

SpringerChemistry

Werner J. Blau, Panagiotis Lianos, Ulrich Schubert (eds.)

Molecular Materials and Functional Polymers

2001. VI, 192 pages. 97 figures, partly in colour.
Hardcover DM 228,–, öS 1596,–, as of Jan. 2002 EUR 115,90
(Recommended retail prices)
All prices are net-prices subject to local VAT.
(Special Edition of "Monatshefte für Chemie/
Chemical Monthly", Vol. 132, No. 1, 2001)
ISBN 3-211-83597-0

The articles in this book summarize the work presented at the final workshop of the COST (European Cooperation in the Field of Scientific and Technical Research) Action on Molecular Materials and Functional Polymers for Advanced Devices, which was held in June 2000 in Patras, Greece.

The collection gives an excellent overview of the state of the art in this field and the progress made by the coordinated research projects. The results range over the synthesis, physical properties, and applications of molecular materials (nanotubes, fullerenes, phthalocyanines), inorganic and inorganic-organic hybrid materials, and functional polymers (electronic conduction, photoluminiscence, optical storage, photovoltaic devices).

Please visit our new website: **www.springer.at**

SpringerWienNewYork

A-1201 Wien, Sachsenplatz 4–6, P.O.Box 89, Fax +43.1.330 24 26, e-mail: books@springer.at, Internet: **www.springer.at**
D-69126 Heidelberg, Haberstraße 7, Fax +49.6221.345-229, e-mail: orders@springer.de
USA, Secaucus, NJ 07096-2485, P.O. Box 2485, Fax +1.201.348-4505, e-mail: orders@springer-ny.com
Eastern Book Service, Japan, Tokyo 113, 3–13, Hongo 3-chome, Bunkyo-ku, Fax +81.3.38 18 08 64, e-mail: orders@svt-ebs.co.jp

GPSR Compliance

The European Union's (EU) General Product Safety Regulation (GPSR) is a set of rules that requires consumer products to be safe and our obligations to ensure this.

If you have any concerns about our products, you can contact us on ProductSafety@springernature.com

In case Publisher is established outside the EU, the EU authorized representative is:

Springer Nature Customer Service Center GmbH
Europaplatz 3
69115 Heidelberg, Germany

Batch number: 10578463

Printed by Printforce, the Netherlands